Lincoln Christian College

P9-DFW-103

A HISTORY
OF THE
LIFE SCIENCES

A HISTORY
OF THE
LIFE SCIENCES

LOIS N. MAGNER
History Department
Purdue University
West Lafayette, Indiana

MARCEL DEKKER, INC.   New York and Basel

Library of Congress Cataloging in Publication Data

Magner, Lois N   [Date]
    A history of the life sciences.

    Includes bibliographies and index.
    1. Life sciences--history. 2. Biology--History.
I. Title.
QH305.M22   574'.09   79-13545
ISBN 0-8247-6824-8

COPYRIGHT © 1979 by Marcel Dekker, Inc. All Rights Reserved

Neither this book nor any part may be reproduced nor transmitted in any form
or by any means, electronic or mechanical, including photocopying, microfilming,
and recording, or by any information storage and retrieval system, without permission in writing from the publisher.

Marcel Dekker, Inc.
270 Madison Avenue, New York, New York 10016

Current printing (last digit):
10 9 8 7 6 5 4 3 2 1

Printed in the United States of America

574.09
M19

B2J

22 33/

8 April 80

59567

To my husband

CONTENTS

PREFACE

In the past the term "scientific revolution" conjured up advances generally limited to the physical sciences. Threats to man's mental and physical world generated by science were also associated with the physical sciences. Today we are living in the midst of a biological revolution that some find profoundly disturbing. Biology has been less a subject of historical analysis than the physical sciences because, paradoxically, it is the oldest and the youngest of the sciences. Although biological problems are among the oldest to interest man, scientifically valid answers to such questions are of relatively recent date. Yet the pace of progress in the life sciences has accelerated so rapidly that events prior to 1953 are seen as ancient history by most scientists and students.

Were biologists really totally ignorant of basic scientific principles before the present century? Did they struggle with questions they lacked the tools to solve? Or are we so complacent with our new-found answers that we fail to appreciate the past? These are some of the questions that a general history of the life sciences must address.

This book is not intended as a comprehensive or exhaustive history of the life sciences, but rather as an introduction to a vast, complicated, and fascinating story. My purpose has been to call attention to the main themes of biology—the continuity and conflicts of theories and techniques, the interactions between various disciplines, the diverse attitudes with which scientists have approached the phenomena of life—and to trace the interplay of theories that have arisen out of these factors. Thus, this book emphasizes major questions, obstructions, and events which seem to have acted as the "hinges of history." Therefore, some problems that absorbed the major energies of natural scientists at particular periods—such as taxonomy or the discovery of particular organs, glands, secretions, compounds, enzymes, and metabolic pathways—may seem somewhat slighted. The history of the life sciences is exciting; indeed, it is often highly dramatic. However, as any scientist or creative worker knows, such periods of reward and exhilaration punctuate long periods of rather dull and plodding work. This pattern applies to the history of science as well as to the career of the individual scientist or artist. The author believes that a complete exposition of such minutiae is to the understanding of the development and workings of science, as the reading of a telephone directory is to the understanding of the life of a city.

Although science and technology are of fundamental importance in the complex modern world, there is a serious imbalance in our approach to science education. Our curriculum and our culture are divided by a vast gap between scientists on the one hand and nonscientists on the other. Although scientists understand the concepts and methods of their own science, they are apt to be as ignorant as laymen in their knowledge of other sciences and of the effects of science on society. Can a society effectively use the ideas and inventions of twentieth century science and technology when it remains essentially scientifically illiterate?

Attitudes toward science are in a tense transition phase at this time. After a boom period, when science and technology were accorded great respect and vast sums of money, a reaction has set in. Many modern problems (pollution, overcrowding, chronic diseases, etc.) are regarded as the result of excessive developments in science. There are demands that we either stop further progress in science or limit the practice of science to such fields as can provide an immediate pay-off in practical terms, such as a cure for cancer or a blight-resistant strain of corn. There are critics of science who focus on presumptive psychological damage done by science in terms of loss of religious or artistic sensitivity in a "clockwork" mechanical universe which science is supposed to give us as the only rationally acceptable model for modern man. Often one finds the criticism directed against the outmoded mechanical concepts of the previous century and quite unrelated to the thrust of contemporary science.

Studies of the history of science could do much to bridge this gap. Yet this opportunity has largely been neglected by both historians and scientists. Although science, as a human invention and mode of thought, deserves a prominent place in the drama of history, because of the basic split in our culture it is generally absent from history courses. Thus, science students tend to find history courses boring and "irrelevant," merely a chronicle of wars, treaties made and treaties broken, kings, queens, politicians, and more wars. Scientists whose work saved more lives from the ravages of disease than even all our wars have been able to eliminate are accorded little or no place in general history books.

The history of science can be enlightening for other reasons. Once beyond the elementary school level of history, we realize that history is not so much the record of what happened, but rather the story of what survived. This is as true for the history of science as for any other branch of history. This distortion is not simply dependent on the passage of time but is also quite apparent upon critical examination of recent textbook accounts of science. When the actual doubts and loose ends, ambiguities and implicit assumptions are glossed over, science falsely emerges as a static thing, with knowledge seemingly handed down from some great celestial source. More properly, we should see science as a dynamic concept, encompassing both a body of knowledge and a means of obtaining and using more knowledge.

Science today is a vast and productive human endeavor. Its scope is such that no one individual can properly understand its content and the implications of its development on society in general. Scientists, like physicians today tend to be specialists. While science is dynamic and ever-changing, some of its goals and methods have a permanence and spiritual continuity worthy of study on their own merits. Here, a study of the history of science can give us some understanding of the fundamentals of the ways of science and scientists. Although some scientists have claimed that all the interesting questions have been answered, a broader, historical perspective seems to support the opposite view—that science is an "endless frontier."

I should like to express my deep appreciation to Dr. Vernard Foley and Dr. John Parascondola for their invaluable advice and criticism.

Lois N. Magner

1

"IN THE BEGINNING"

Origins and Definitions

Biology, like all the sciences, has roots reaching back into the eons of time preceding the first civilizations. Myths, superstition, religion, craft techniques, and the empiricism that predates purposeful experimentation and theorizing were the means of dealing with nature. As the magical and empirical aspects of knowledge slowly separated, science as we know it began its gradual evolution.

Biology is most simply defined as the science of living things. The most important lessons of all to human survival are those that fall within the purview of biology—the medical arts, agriculture, and animal husbandry. Progress in these areas among primitive populations would involve the most ephemeral products of human endeavors. Pottery and weapons, glass and metallurgical techniques will leave a more permanent record than a new understanding of the cycles of nature. Yet it is ultimately on such biological wisdom that human survival and evolution depended. Thus, biology is a mixture of both the oldest and the newest of the sciences. The word "biology" itself was not coined until the beginning of the nineteenth century to denote a departure from the old ways of the "natural philosopher." Today, biology encompasses the most intriguing areas of science: genetics, molecular biology, development and differentiation, cytology, evolution, and ecology.

Although the ancient Greeks are often credited with the invention of science, primitive man was also a kind of naturalist with keen interest and insight into his environment. The anthropologist, Claude Levi-Strauss,[1] found that so-called primitives establish a fairly sophisticated classification and ordering scheme for their environment. Things are arranged into categories such as the raw and the cooked, the wet and the dry. Certainly it is important to know the difference between spinach and poison ivy. Even such a seemingly simple division as edible and inedible requires a certain degree of sophistication and considerable experimentation. Primitive people also learned through experience that a series of operations could transform things from one class to another. For example, in the preparation of manioc both a food and a poison are produced.

Biology as the study of living things probably began with the transition from the original "animal man" or "naked ape" to "psychosocial man" some 40,000 years ago. This is the stage in human evolution marked by the invention of fire, speech, abstract thought, religion, and magic. At this stage, humans must have become conscious of themselves as special and set out to deal with problems of human existence—death and disease, hunger and discomfort—in a new way. Since this transition, cultural evolution, which is unique to man, took precedence over biological evolution, which is shared with the rest of the organic world. Another great transition took place about 10,000 years ago when agriculture was invented. In the process of domesticating plants and animals, man, too, became domesticated and enmeshed in a life style and mode of thinking that revolved around the natural forces controlling his new enterprises.

Anatomy, both human and animal, was one of the earliest components of biological knowledge. For early man, fear and respect for the dead led to elaborate death rites involving manipulation of the corpse. Sometimes certain organs, such as the brain, would be removed and eaten. The body might be cremated or embalmed, there might be attempts to preserve the whole body or just the skeleton or skull. There might even be more elaborate first and second burial rites. Burial customs may show compassion and respect—as in very ancient rites in which flowers were buried along with the individual—in addition to attempts to placate the spirits of the dead and make it impossible for them to return (as they sometimes seemed to do in dreams) and harm the living.

Death need not have been accepted without magical and rational attempts to ward it off. Primitive attempts at surgery, even such dramatic operations as trepanation, would give further knowledge of human anatomy and of the location of vital points on the body. Anticipation of death may have allowed gathering of other physiological information during the death watch, such as information about the importance of respiration and heartbeat.

Knowledge of animal anatomy and behavior is important to the herdsman as well as the hunter. The herdsman who follows the natural migrations of animals knows much about them, but the domestication of animals can lead to more definite studies of their properties, behavior, and ultimately to actual attempts to select and breed the best of the herd. In addition to purely practical knowledge of animals for hunting, herding, butchering, and cooking, such knowledge would be useful to priests, diviners, and magicians. Animals could serve as totems of the tribe, animal organs or behavior could be used as omens. Since divination and prophecy depended on structure and behavior, any peculiarity of action, shape, or color could decide the fortunes of the tribe. For the diviner, as for the modern physician, knowledge of normality must precede knowledge of pathology.

Where society was dominated by magical beliefs, much biological knowledge became the professional secrets of a special caste of priests, sorcerers, and magicians. Such a monopoly on biological and often astronomical knowledge on the part of the priesthood was inimical to the development of pure science. The separation of information into the sacred and the secular was indispensible for a free inquiry into nature. A world view dominated by capricious demons, who could be propitiated only through the rituals of the priesthood, was not conducive to rational inquiry.

Biology and Ancient Civilizations

The invention of science is generally credited to the Greek natural philosophers who lived during the sixth century B.C. This probably reflects more of our own cultural bias and ignorance about preceding cultures than a lack of achievements in them. Great civilizations developed along the valleys of the Nile, Tigris-Euphrates, and Indus rivers as much as 5000 years ago. Although these cultures left their mark in terms of technological advances, complex social organization, and a written record, only a small portion of what they achieved was recorded and preserved. Because of the paucity of sources, we undoubtedly have a greatly distorted view of their real accomplishments.

The emphasis on the traditional Western heritage from Greek culture has resulted in the neglect of earlier civilizations and other traditions. China and India have been especially neglected, except for the notice of certain exotic medicines and the invention of printing and rockets. Many important discoveries in pharmacology and nutrition could be listed as part of China's contribution to the history of biology, but to merely pick out a few items that have found acceptance in the West is a totally inadequate treatment for a complex civilization which has existed for thousands of years. To understand biology and medicine in China and India really requires a thorough knowledge of the sciences in the context of traditional Chinese or Indian civilization. Such an exploration has only recently been attempted, largely stimulated by the work of Joseph Needham.[2] Although of great interest, unfortunately these studies are beyond the scope of this book.

Agriculture was among the earliest forms of "applied" biology and probably the cause of the first population explosion.[3] Modern methods of archaeology have led to descriptions of when and where agriculture began, the nature of the first plants and animals that man cultivated and domesticated, and the effects these steps had for further stages of cultural evolution. Settlements connected with the cultivation of crops seem to have appeared first in the region which has been called "the Fertile Crescent," the area which surrounds the Tigris and Euphrates basins, and a second region which runs from New Mexico to Guatemala and Ecuador. The crops in the Old World were first wheat and barley, and later peas, lentils, and flax. In the New World the major crops were beans, squash, and maize. Settlements that go back to 7000 B.C. have been excavated, but these were probably preceded by others that may have appeared 2000 years earlier.

A settled agriculture led to advances in agricultural implements made of wood and stone, including hoes, sickles, grinding stones, and storage vessels of pottery and wood. Most important, the agriculture practiced in these river valleys, where natural flooding deposited a fresh layer of fertile silt every year, avoided the customs of more primitive agriculture which quickly exhausted the fertility of the soil. The new agriculture produced population expansions and migrations, as communities grew from villages to towns to cities. The transfer of techniques at this early stage of human development apparently involved major migrations of peoples, rather than the transmission of information alone. Thus, the first steps in human cultural evolution are measured in thousands of years as people migrated, exchanged ideas, and settled into new areas.

Along the great river valleys, productivity, life, and death depended on cycles of the flood and retreat of the rivers. Complex activities had to be planned, directed, and supervised by a centralized system of administration. The priesthoods who directed this central government dominated the way their subject populations thought about nature. The ancient myths of the separation of chaos into dry land and water and the development of living things from the mud are in essence the life story of these first civilizations. These myths served the powerful central government in dealing with the mundane problems of agriculture—draining of marshes, land surveying, irrigation and flood control, storage of food for bad years—and in explaining the capriciousness of nature, which they coped with through divination and propitiation of dangerous spirits.

Cosmology of Mesopotamia and Egypt

The Mesopotamians and Egyptians had similar ideas about the structure of the universe. Waters were prominent in the Mesopotamian cosmology, as might be expected from the experience of the dwellers in river valleys subject to rather unpredictable flooding. Early Mesopotamian myths described the earth and heavens as two flat discs supported by water. Somewhat later the heavens were described as a hemispherical vault which rested on the waters that surrounded the earth. There were waters even above the vault of the heavens. The heavenly bodies were regarded as gods who dwelt beyond the waters and came out of their dwelling place daily to travel on definite paths over the heavens. Since the gods controlled events on earth, the motions of the heavenly bodies were closely studied to reveal what the gods had in store for man.

In Egypt the world was viewed as essentially a rectangular box. The earth was at the bottom, making that part of the box somewhat concave. The sky at the top was supported by four mountain peaks at the corners of the earth, while the Nile was a branch of a universal river that flowed around the earth. This river carried the boat of the sun-god on his journey across the sky. Seasonal changes were explained by deviations in the path taken by the boat. During the time of the Nile flood, it came nearer to the earth than it did in the winter.

Creation myths explained how the world came into existence from a primeval chaos of waters. Matings of male and female gods in the time of chaos brought forth the heavens, earth, air and the natural forces and objects that were personified by various gods. The gods organized the universe out of the original chaos, separating the land from the waters much as the pioneers of the river valley civilizations had reclaimed the earth from the waters. Egyptian and Mesopotamian civilizations, in their political organization, scientific interests, and religions reflect the differences in the patterns of the great rivers. The Nile floods were quite predictable and valued, Egyptian dynasties long-lived, even rigid and complacent. The floods of the Tigris and Euphrates were unpredictable and feared. Governments in Mesopotamia were more transient and the future always uncertain. Astrology and divination were the tools and weapons with which the Mesopotamians struggled with the gods and the chaotic forces of nature.

Despite great accomplishments during the high bronze age, these civilizations remained essentially static while great innovations were being made by cultures at the fringes of bronze-age civilization. The smelting of iron and the use of a simple alphabetic script created a new cultural revolution. Sometime in the second millennium B.C. a tribe in the Armenian mountains developed an efficient method of smelting iron. Originally quite secret, the method became more generally known after about 1100 B.C. Iron was used in weapons and in agricultural implements, allowing the exploitation of the more difficult lands to the north. With iron weapons, barbarians like the Greek tribes, were able to conquer rich bronze-age cultures.

Notes

1. See C. Levi-Strauss (1973). *From Honey to Ashes. Introduction to a Science of Mythology.* Vol. 1. Translated by J. Weightman and D. Weightman. New York: Harper and Row.
2. See J. Needham (1954–76). *Science and Civilization in China.* 5 vol. Cambridge, Mass.: Cambridge University Press.
3. See M. N. Cohen (1977). *The Food Crisis in Prehistory: Overpopulation and the Origins of Agriculture.* New Haven: Yale University Press.

References

Adams, R. M. (1960). The Origin of Cities. *Sci. Am. 203:* 153–168.

Braidwood, R. J. (1960). The Agricultural Revolution. *Sci. Am. 203:* 131–148.

Breasted, J. H. (1934). *The Dawn of Conscience.* New York: Scribners.

Buettner-Janusch, J. (1966). *Origins of Man: Physical Anthropology.* New York: Wiley.

Campbell, B. (1966). *Human Evolution.* Chicago, Illinois: Aldine.

Ceram, C. W. (1967). *Gods, Graves, and Scholars, The Story of Archaeology.* New York: Knopf.

Childe, V. G. (1934). *New Light on the Most Ancient East.* New York: Evergreen.

Childe, V. G. (1951). *Man Makes Himself.* New York: New American Library.

Clark, G. and Piggott, S. (1965). *Prehistoric Societies.* New York: Knopf.

Cohen, M. N. (1977). *The Food Crisis in Prehistory: Overpopulation and the Origins of Agriculture.* New Haven: Yale University Press.

Darby, W. J., and Grivetti, L. (1977). *Food: The Gift of Osiris.* New York: Academic.

Dubos, R. (1968). *So Human an Animal.* New York: Scribner.

Graziosi, P. (1960). *Paleolithic Art.* New York: McGraw-Hill.

Hawkes, J. (1973). *The First Great Civilizations, Life in Mesopotamia, the Indus Valley, and Egypt.* New York: Knopf.

Hawkes, J. (1976). *The Atlas of Early Man.* New York: St. Martin.

Klein, R. G. (1969). *Man and Culture in the Late Pleistacene: A Case Study.* San Francisco: Chandler.

Kramer, S. N. (1959). *History Begins at Sumer.* Garden City, N. Y.: Doubleday Anchor.

Leakey, L. S. B. (1960). *Adam's Ancestors, The Evolution of Man and His Culture.* 4th Ed. New York: Harper and Row.

Levi-Strauss, C. (1973). *From Honey to Ashes. Introduction to a Science of Mythology.* Vol. 1. Trans. by J. Weightman and D. Weightman. New York: Harper and Row.

Ley, W. (1968). *Dawn of Zoology.* Englewood Cliffs, N.J.: Prentice-Hall.

Muller, H. J. (1952). *The Uses of the Past. Profiles of Former Societies.* New York: Mentor.

Nakayama, S., and Siven, N., eds. (1973). *Chinese Science: Explorations of an Ancient Tradition.* Cambridge, Mass.: MIT Press.

Needham, J. (1954–76). *Science and Civilization in China.* 5 vols. Cambridge, Mass.: Cambridge University Press.

Piggott, S., ed. (1961). *The Dawn of Civilization.* New York: McGraw-Hill.

Starr, C. G. (1973). *Early Man: Prehistory and the Civilizations of the Ancient Near East.* New York: Oxford.

Tannahill, R. (1973). *Food in History.* New York: Stein and Day.

Washburn, S. L., ed. (1961). *Social Life of Early Man.* Chicago: Aldine.

Wilson, J. A. (1959). *The Culture of Egypt.* Chicago: University of Chicago Press.

2

THE GREEKS: NATURAL PHILOSOPHERS AND SCIENTISTS

The Pre-Socratic Philosophers

Although the first civilizations could claim great achievements in agriculture, technology, medicine, and astronomy, a strong secular tradition of free inquiry did not develop because religious and superstitious attitudes toward the pattern and purpose of life remained dominant. Creativity and innovation eventually faded, but the richness of the river valley settlements drew invasions of less civilized peoples who exploited the new opportunities of the iron age.

The Greeks, a mixed collection of tribes, invaded the eastern part of the Mediterranean and settled there about 2000 B.C. The city-state which became the unit of organization of these peoples was quite different from the systems of the older civilizations so entrapped in the unity of priesthood and state. Unlike the previous high cultures, the Greeks were not organized only around agriculture. Trade and shipping were important, and since Greece was relatively overpopulated in relation to cultivatable land, colonization and industry were encouraged.

The earliest Greek "scientists" were the natural philosophers of the sixth and fifth centuries B.C. Their interest was in the natural world and speculations about how things arose and why the world and its creatures were formed and organized as they were. The first natural philosophers saw man as a part and product of natural evolution, separated from other animals by the unique powers of his mind, which were in turn the product of social evolution. Science was regarded as an integral part of man's attempts to understand and control his environment.

Natural science developed first, not in the Athens of Socrates, Plato, and Aristotle, but on the Aegean fringe of the mainland of Asia Minor known as Ionia. Political power was in the hands of a mercantile aristocracy which was interested in promoting the rapid development of new techniques to improve

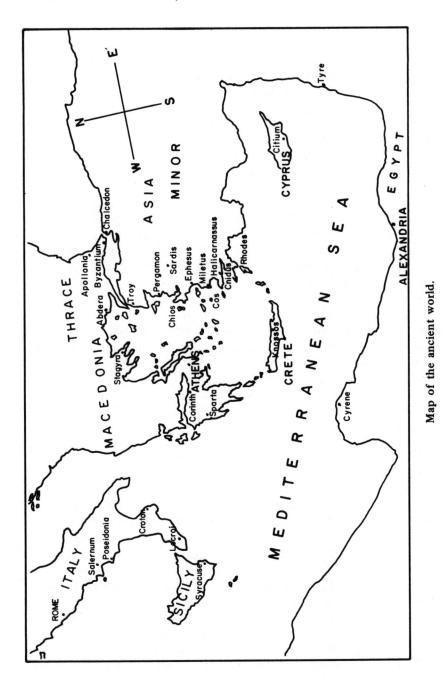

Map of the ancient world.

their commercially based prosperity. Miletus, home of the first of the natural philosophers, was the mother city of numerous colonies on the Black Sea and had commercial interests all over the Mediterranean as well as contacts with the old civilizations of Mesopotamia and Egypt.

The first philosophers seem to have been active, politically aware people, with an interest in trade, travel, and politics. Their speculations on the nature of things were novel in that they explained the workings of the universe in terms of everyday experience and analogies with craft processes rather than religious dogma and superstition. Their achievements were remarkable, considering the very limited opportunities for such studies at the time. Scholars in the ancient world were generally highly respected, as shown by the fact that various city-states would ask them to serve as lawgivers, rulers, or tutors to the sons of rulers. On the other hand, in times of political turmoil scholars might find themselves persecuted by hostile political factions. Thus, while the natural philosopher had more potential freedom of inquiry than the intelligensia—the priesthood—of earlier cultures, he lacked the protection that the interpreters of the gods had assumed.[1]

The Ionian philosophers were preoccupied with finding the regularities they believed existed beneath the apparent changes in the world. They reasoned that there must be some chain of cause and effect which must stem from some common primary cause of all the apparent variations. Thus, there must be a common element from which everything originated. Despite the different answers which satisfied various inquirers, the basic questions they asked were the same: What is the primary element? How have things originated from it? How do things attain their proper place?

Thales of Miletus (ca. 640–546 B.C.)

Thales, the first of the Greek natural philosophers, flourished at about the same time as Buddha, Confucius, Zoroaster, and Hippocrates. He was apparently a man of business, a merchant, statesman, engineer, mathematician, astronomer, and lover of knowledge. Yet with all this to his credit, he left no writings. Although the Phoenician alphabet was probably adapted to the Greek language in Miletus about 800 B.C., it is possible that Thales was illiterate. He apparently learned in the manner of his time, by travels to the older civilizations. He visited Egypt, probably on business, and brought back knowledge of geometry which he put to a new and practical use. By means of the doctrine of similar triangles, he devised a method to determine the distance of ships at sea. In travels to Mesopotamia, he learned Phoenician astronomy. Using Babylonian astronomical tables, he is said to have predicted an eclipse in 585 B.C., transforming what had been a very terrifying and mysterious phenomenon into a natural, predictable event.

According to a story told by Aristotle, Thales was able to put his observations of nature to practical use when it pleased him. Thales' friends had reproached him for his poverty, suggesting that philosophy was of no use. But through his study of the heavenly bodies the philosopher had observed that there would be a large olive crop. In the winter he raised some money and rented all the olive presses in Miletus and Chios very cheaply. At harvest time the demand for the presses was great and he was able to hire them all out on very favorable terms and make a great profit. He proved "it is easy for philosophers to be rich, if they wish, but that it is not in this that they are interested."[2]

Probably influenced by the creation myths of the older civilizations, Thales reshaped these ideas into a more or less self-consistent and wholly natural view of the universe. It is remarkable for the way it explains much of what was known at the time without recourse to supernatural agencies. As in the ancient creation myths, all things in Thales' cosmology originally came from water, which he viewed as the primary element of the universe. The earth was a disk which floated on water, whereas the sky was a cover of rarefied water. Experience tells us that this is reasonable, since there are oceans and rivers; wells bring us water from below the earth and the sky lets down great quantities of water in the form of rain.

In addition to the primary material substance, water, Thales' cosmology required two processes or forces. These were consolidation and expansion. Water was expanded by heat until it turned into air. Water could also be consolidated until it turned into earth. Anyone who has observed the appearance of the inside of a kettle from which hard water has boiled away will know that everyday experience also validates this idea.

Despite his business successes, Thales seems to have become the prototype of the absent-minded professor. Plato recounted a story of how Thales was mocked for falling into a well while he was observing the stars and gazing upward. He was so "eager to know things in the sky, that what was behind him and just by his feet escaped his notice."[3]

Anaximander (ca. 611–547 B.C.)

Anaximander seems to have been a pupil and perhaps the nephew of Thales. His cosmology is more sophisticated in that he argued that the primary material basis of the universe could not have any individual characteristics of the things that it becomes. Thus the "universal stuff" cannot even be water. According to the Egyptians and Babylonians, the primary constituents of the world were water, earth, and air. As befitting an iron age philosopher who respected the crafts of his culture, Anaximander added a fourth element, fire.

To avoid the trap of the primary element being the same as any one of the

four known elements and a part of the other three, Anaximander postulated
some rather ill-defined primary stuff or "aperion." When acted on by heat
and cold this primary material was transformed into the four elements which
made up the world. Anaximander also provided a dynamic but naturalistic
explanation for the formation of the universe out of the original chaos. Again
this involved some indeterminant primary force which caused a vortex to be
formed. The whirling of this vortex led to the separation of the elements
through their stratification according to density. Since earth is the heaviest
element it tends to remain at the center of the universe. Water covers the
earth and is in turn enveloped in a layer of mist. Fire, which is the lightest
element, has its natural place at the outside of the universe. It was the action
of the vortex which had caused the world and its creatures to become formed
and organized.

While the stratification was in progress, fire acted on the water to produce
dry land, along with more mist. Increasing pressures broke up the orderly
shells of the elements and created wheels of fire enclosed by tubes of mist.
According to Anaximander, the heavenly bodies were actually holes in the
tubes which allow us to see the fire at the outside of the universe. Because
the tubes are of various sizes, we see the sun as larger than the moon, and
the moon as larger than the stars.

Anaximander's scheme could even account for the development of living
things. As the sun warmed the mud, it foamed and bubbled and brought
forth various animals. The first creatures were fish formed during the period
when water predominated. As separation proceeded and dry land appeared,
some of these fishy creatures came onto the land and changed as conditions
on earth changed. Even human beings had been produced from lower forms
of life. Man developed from a fish-like creature which had adapted to life on
land and discarded its fish skin. Because man is originally so helpless, he
could not emerge as an infant, but had to be nurtured inside the fish skin
until the land was ready to receive him. Although Anaximander is sometimes
listed as a precursor of evolutionary theory, he did not believe that change
was directed along one path. On the contrary, he believed that there were
cycles of change. Thus, the forces that led to separation of the elements and
the development of life as we know it now could be reversed periodically and
chaos would once more predominate.

Anaximander's views were presented in a poem entitled "On Nature,"
which was apparently very well known and respected in antiquity. It was
much cited and criticized by Aristotle, but the original composition seems to
have been lost by the time of the later classical period. It is unfortunate that
so little is known about Anaximander. His theories are certainly ingenious,
if somewhat naive, and his conception of nature was among the most
imaginative and sweeping accounts of the ancients.

Anaximenes of Miletus (ca. 550–475? B.C.)

Anaximenes was the last of the Milesian philosophers. In 494 B.C., Miletus was destroyed by the Persians. Almost nothing is known of his life and work; even his lifespan and dates are in question. He seems to have been a pupil of Anaximander, and perhaps of Parmenides. His ideas certainly show a debt to Anaximander, and even Thales. He apparently wrote a book, at least a part of which was known to Theophrastus, who is said to have written a monograph on Anaximenes. Unlike Anaximander, who said that the primary material was undefined, Anaximenes chose air (or mist) as the principle from which all other things are produced. He was able to explain how other substances were formed from air by adding two processes to the picture. Air could be rarefied or condensed, that is, the amount of air packed into a particular place determined its form without altering its original nature. His inspiration may have been the industrial process of felting, in which fibers are packed down.

Anaximenes believed that air was always in motion, "for things that change do not change unless there be movement." When air is in its most uniform state it is invisible to our sight. Air is revealed to us by cold, by hot and damp, and by movement. When made finer, air becomes fire. Winds are air that is becoming condensed and cloud is air produced by felting. If air is further condensed it becomes water. When thickened still more it becomes earth and then, if condensed as far as possible, stone. Exactly what Anaximenes meant by "air" is uncertain, for he offered no proof of its substantiality. This was left for Empedocles.

The conception of the earth as broad, flat, and shallow seems to have been formalized by Anaximenes. His picture of a flat earth riding on the air as a leaf will float on air was followed up by Anaxagoras and the atomists. The heavenly bodies, which were created by the rarefaction of mist into fire, also rode on air.

Anaximenes seems to have believed that air is the cosmic equivalent of the life-soul in man. Implicit in Anaximenes' philosophy is the notion that man and the rest of the world are fashioned from the same primary principle, and thus similar rules must be applicable to both. His fundamental assumption—that a change in the quantity of the primary material determines its form— was probably influential in the thinking of Heraclitus and other philosophers. But later philosophers were more interested in theological problems and in the arrangement of things, rather than the search for the material basis from which the rest of the world differentiated.

Anaximenes was succeeded by Xenophanes and Heraclitus, who were also Ionians, and on the mainland by Anaxagoras, Leucippus, and Democritus.

Xenophanes of Colophon (576–490 B.C.)

Having been expelled from his native land, Xenophanes spent the remainder of his life in Sicily. The fragments of his writings which survived suggest that he was primarily interested in theology. Although he had some keen insights into physical problems, the extent of his influence is a matter of dispute. Despite his interest in poetry and the study of the gods, he was able to make interesting deductions from observations of fossils.

Xenophanes believed that everything, including man, originated from water and earth. He noted that the earth and sea had changed places in the past and would do so again. Fossil evidence supported this idea because shells are found inland, in the mountains, and in the quarries of Syracuse and other places. These fossils proved to him that the earth had been covered with mud, so that impressions of fish and other marine objects had been left when the mud dried out. Although the Milesian philosophers had also described the world (including living creatures) as coming from mud, they had not used the physical fact of fossil forms as evidence.

Xenophanes recognized the tendency of man to make gods in his own image and the cultural relativity of human judgments on supposedly universal problems. "The Ethiopians say that their gods are snub-nosed and black, the Thracians that theirs have light blue eyes and red hair," he wrote, "but if cattle and horses or lions had hands, horses would draw the forms of the gods like horses, cattle like cattle, and they would make their bodies such as they each had themselves."[4] Although Xenophanes emphasized the limitations of human knowledge, he believed that through actively seeking truth one could reach a higher state of wisdom. Kirk and Raven say that Xenophanes' appeal to caution about the extent of human knowledge was valuable but did not noticeably inhibit the overdogmatic tendency of Greek philosophy.[5]

Heraclitus of Ephesus (556–469 B.C.)

Although Hericlitus took fire as his primary element, he won the title of the "Gloomy" or "Obscure" from his fellow Greeks. Peculiar stories about his life and death were fabricated out of fragments of his sayings and the enmity of his contemporaries. Thus Diogenes Laertius says that Heraclitus, born into an old aristocratic family, grew up to be exceptionally haughty and supercilious, as evidenced by his statement: "A great deal of learning does not bring wisdom; if it did it would have instructed Hesiod and Pythagoras, Xenophanes and Hecateaus."[6]

The "weeping philosopher" withdrew from the world to live in the mountains, where he ate grasses and plants. This caused him to develop a dropsy. Seeking aid from doctors, he tested them with a riddle. Because they could not answer him, Heraclitus rejected their medicines and buried himself in a cow stall. He expected that the heat of the manure would evaporate off the dropsy, but it did not. And so he died. This bore out his statement that "it is death for souls to become water" as well as his comment that corpses are more worthless than dung.[7]

Heraclitus' system seems more unified than his predecessors', but perhaps that is merely because it is better preserved. Much of his thought is indeed obscure, as even his contemporaries complained, and much interpretation is uncertain because of the fragmentary nature of what has survived and later additions and alterations misconstrued by both his admirers and enemies. His system does seem to explain all aspects of the world systematically, as based on his major premise—that there is a regularity and balance that governs all changes in nature. This balance existed in the mundane world of business transactions as it did in cosmic transactions. "All things are an equal exchange for fire and fire for all things," wrote Heraclitus, "as goods are for gold and gold for goods."[8]

Fire, the common element of all things, is the cause of this balance. Fire is both an element which changes and the process that causes change, whether in cooking, pottery-making, or metallurgy. Heraclitus is responsible for the idea that everything is in flux and change. According to Plato, Heraclitus said that "all things are in process and nothing stays still, and likening existing things to the stream of a river he says that you would not step twice into the same river."[9]

The philosophy of Heraclitus could explain everything from the changes in the external world to the complexities of human behavior, thought, life, and death. He taught that human beings must strive to understand the true constitution of things or their souls, which are made of fire, would become excessively moistened. For Heraclitus, a "dry soul" was wisest and best. Perhaps his fear that wine would moisten the soul helped to make him so melancholy, for wine in particular moistened the soul. To Heraclitus, understanding nature was not separate from understanding man, because the same materials and laws governed both. Thus, human behavior and changes in the outer world are of the same order.

The Milesians represent the first stage in the history of pre-Socratic Greek philosophy. Heraclitus may be regarded as an intermediate stage. The second stage of development was reached by the Pythagorean and the Eleatic philosophers.

Pythagoras of Samos (ca. 582–500 B.C.)

Bertrand Russell called Pythagoras a combination of Einstein and Mrs. Eddy. Pythagoras is said to have left Samos during the reign of the tyrant Polycrates and to have settled in southern Italy at Croton. Here he founded a brotherhood concerned with mathematical speculation and religious contemplation. It was remarkable that both men and women were admitted to the association on equal terms. All property, and even mathematical discoveries, were considered as common property within the brotherhood. He and his pupils, numbering perhaps 300, apparently wrote laws for the Italians and advanced to fame and authority. But during a rebellion, Pythagoras fled to Metapontium where he died. The name of Pythagoras had become enmeshed in mystery and legend by the time of Plato and Aristotle. The rules of secrecy which the brotherhood practiced, probably contributed to the legend building. Eventually the Pythagoreans split up into scientific and religious factions.

Certain fragments show that Pythagoras believed that human souls could be reincarnated in the form of other animals. This has been taken to mean that the Pythagoreans believed in the kinship of all living things. Pythagoras seems to have believed that religion and science were inseparable aspects of the complete life. "Life," he said, "is like a festival; just as some come to the festival to compete, some to ply their trade, but the best people come as spectators, so in life the slavish men go hunting for fame or gain, the philosophers for the truth." [10] But there remains no direct reliable evidence about the scientific teaching of Pythagoras himself and no satisfactory way to separate his ideas from those of his followers.

Alcmaeon of Croton (fl. ca. 500 B.C.)

Although very few of the contemporaries of Pythagoras are known by name, we do know that they included biologists and anatomists such as Alcmaeon of Croton. Although he may have had some link with the Pythagoreans, there is no real evidence that he was a member of the school. Most of his theories were related to medical and physiological matters, but he also had an interest in natural philosophy. Alcmaeon is said to have been the first to carry out anatomical research with the real spirit of the scientist. That is, he studied and dissected animals for the sake of learning rather than for purposes of divination by animal omens. He used the system that was to be the standard embryological model for thousands of years—the developing chick embryo. Among his reputed discoveries were the optic nerves which connect the eye with the brain and the eustachian tubes. This discovery was forgotten until repeated by Eustachius (1520-74) in the sixteenth century.

The purpose behind these physiological researches was to understand the nature of sense perception. Theophrastus wrote that among those who thought perception is of unlike by unlike, Alcmaeon was the first to define the difference between man and animals. Man is the only creature that has understanding, while other animals "perceive but do not understand." He believed that the senses were somehow connected with the functioning of the

brain. If the brain moved or changed its position, the passages through which sensations came were blocked and the senses were incapacitated. Alcmaeon's theories concerned the nature of health and the composition of the soul. He believed that the bond of health is the "equal balance" of the contrary powers: moist and dry, cold and hot, bitter and sweet, etc. Health was the "proportionate admixture of the qualities." The cause of disease was an excess or "supremacy" of one member of a pair. The seat of illness was either in the blood, the marrow, or the brain. Illness could arise in these centers from external causes, such as moisture, changes in the environment, exhaustion, excess or lack of nourishment, hardship, etc. According to Aristotle, Alcmaeon believed that the soul was immortal and always in motion. Men die because "they cannot join the beginning to the end," since, unlike the heavenly bodies, the soul of man cannot sustain circular motion."[11]

Empedocles of Agrigentum in Sicily (ca. 500–430 B.C.)

Empedocles was regarded by the ancients as an admirer and associate of Parmenides, but very much under the influence of the Pythagoreans. Like Pythagoras and Heraclitus, he inspired many legends. He has been accused of miracle mongering to exploit the common people and his disciples. Apparently he deliberately contributed to his mystical aura by boasting about his supernatural gifts, claiming to have the power to heal the sick, cure the infirmities of old age, raise the dead, change the direction of the wind and rivers, and bring the rain and the sun. Numerous references to him in later medical writings do suggest great fame and success as a doctor. Empedocles may, indeed, have changed the balance of health and disease in his native city by contributions to city planning and preventive medicine. Making a breach in the mountain wall to let in the cool north wind is said to have improved the hot and unhealthy climate; drainage of swampy areas helped eliminate malaria and improved water supplies.

Empedocles was famous for his gifts as an orator, since in the lost dialogue of Aristotle, *The Sophist,* he was apparently named as the founder of rhetoric. Also active in politics, he was said to have been an ardent democrat, and part of a revolutionary group that overthrew an organization called the Thousand. He was offered the kingship of his city but refused it, according to another legend.

Fragments of two poems by Empedocles are all that survive. Of the poems "On Nature" and "Purifications," which originally totaled 5000 lines, less than a fifth of the first and even less of the second remain. Another work on medicine was lost completely. Still this is more than remains of the work of many of the first natural philosophers. The poem "On Nature" deals mainly with a physical explanation of the universe, while "Purifications" deals with the Pythagorean belief in transmigration of souls. These interests seem contradictory, but may represent different stages in the life of the philosopher.

Empedocles was influenced by the principle that reality cannot come from unreality and that some original unity cannot produce a subsequent plurality. In accord with these strictures, Empedocles taught that there never was an original unity such as the Milesians had assumed. For Empedocles there existed four eternal and distinct substances: earth, air, fire, and water. Mixing these four eternal principles created new things, just as artists mix pigments to produce new colors. Empedocles' cosmology rested on a kind of conservation of matter. What appear to be creations and destructions are merely changes that are due to the mixture and separation of the fundamental components.

This leads to the question, what produces the motions of the four elements that lead to changes in their local combinations? According to Empedocles these changes are due to two forces, Love and Strife. Love causes attraction and the creation of new worlds, but when strife predominates, worlds are torn apart. Thus, in the Cosmic Cycle of Empedocles the universe oscillates between stages of creation and destruction.[12] Although Empedocles is sometimes cited as a precursor of Darwin, his evolutionary scheme was not a unilinear pathway, but an oscillating system.

Unlike Parmenides, Empedocles believed that the senses were a valid guide to the philosopher seeking the truth, if he carefully and critically employed each sense for its appropriate purpose. "Eyes," he said, "are more accurate witnesses than ears." According to Empedocles, perception is due to the recognition of an element outside the body by virtue of that same element inside the person: "For with earth do we see earth, with water water, with air bright air, with fire consuming fire; with Love do we see Love, Strife with dread Strife." All things—animals, plants, earth and sea, even stones, brass, and iron—give off effluences. If the effluences were of the appropriate size they would fit into the pores of the sense organ, allowing the meeting of elements requisite to perception.

The passages for each mode of perception differed so that "one cannot judge the objects of another, since the passages of some are too wide, of others too narrow . . . so that some things pass straight through without making contact while others cannot enter at all." This theory was also used to account for consciousness and thought. According to Empedocles, thought processes take place mainly in the blood around the heart. He speaks of the heart as "dwelling in the sea of blood which surges back and forth, where especially is what is called thought by men; for the blood around men's hearts is their thought." Blood was regarded as another temporary combination of the four elements, but "in the blood above all other parts the elements are blended." Because of this the blood was most fitted to perceive all the four

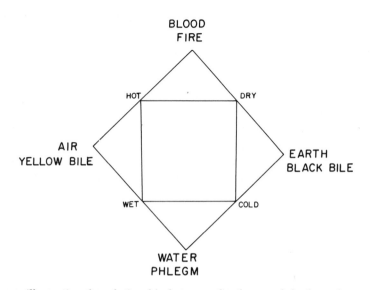

Diagram illustrating the relationship between the theory of the four elements and the four humors.

elements outside and thus serve as the chief seat of perception. Men were not unique in having the power of thought, because, Empedocles said, "all things possess thought." However, not all things or all men possesses the same degree of intelligent thought.[13]

Man and all other creatures, made up of the same four elements, were the product of a complex and peculiar evolution that characterized the cosmic cycle of Empedocles. His cycle consisted of two polar and two transitional stages, and also of two distinct stages in the evolution of living things during the transitions. Basically, Empedocles taught that the Earth at an earlier stage had powers she now lacks. At this time a great variety of living things was created, but the first products were incomplete, crudely and randomly formed. At this early stage, there were "faces without necks, arms wandered without shoulders, unattached, and eyes strayed alone, in need of foreheads." In the second stage, these solitary limbs which had "wandered about seeking for union" began to combine. But since the combinations were made at random, many monstrosities were created. Strange creatures were formed "with faces and breasts on both sides, man-faced ox-progeny," or even "ox-headed offspring of man, creatures compounded partly of male, partly of the nature of female, and fitted with sterile parts." Thus, among this great variety of living forms there were many creatures that could not reproduce their own kind through deformity; or because they could not survive competition with other, better kinds of creatures; or because the parts needed for generation were missing.

Even though the joinings in the second stage were merely chance events, sometimes "whole-natured forms" were produced. Creatures that were well combined survived, but the others perished and "are perishing still." "Whole-natured forms" which were still neither male nor female were typical of the third stage. As separation continued, the sexes eventually become distinct and separate, leading to the next stage. In the fourth stage new forms no longer arose directly from the earth or water, but now arose from generation—sexual reproduction and all the other methods of generation that the term originally included. All surviving species have some special talent or courage or speed to protect and preserve them.[14] This scheme obviously lends itself to simplification as "survival of the fittest" in the Darwinian sense.

The world that we are living in is the fourth stage in this evolutionary process. In describing this stage, Empedocles made many comments on sciences that today are each complex specialities such as botany, physiology, and embryology. He seems to have been the first of the Greek natural philosophers to deal with botany in a serious manner. He believed that plants were the first living things to appear, but like the "whole-natured forms" of the earlier stage, in plants the sexes have not separated. Neither are fire and earth as separated in the plants as in the animals, because plants still have their roots in the ground.

All living things in this stage are a mixture of elements, but are compounded of different proportions. Thus each kind of being seeks its natural place according to the mixture of elements in it. Trees are still attached to the earth because they have more earth in their nature, fish obviously have more water, while birds have more of air and fire in their composition. The two sexes are combined in one individual in the plants, which reproduce through "eggs," but man and the higher animals reproduce through sexual generation. Unlike the Pythagoreans, who believed the child to be derived from the father, Empedocles believed that the embryo received some parts from the father and others from the mother. The child would resemble the parent that contributed most to that particular case. He explained the sex of the offspring on the basis that male children were conceived in the warmer part of the womb and contain a greater proportion of warmth than the female. The growth of the child was due to the increasing warmth of the womb, and old age was due to a decrease in warmth.

Empedocles' views on respiration, which lead to a very interesting demonstration of the nature of air, are particularly important in the history of science. While parts of the theory, as reported by Aristotle, are naïve and easily dismissed, it reveals some close observation of everyday phenomena and keen insight. Empedocles said that all things inhale and exhale by means of bloodless channels in the flesh throughout the body. The surface of the skin has many pores which allow an easy path for the air to pass through, but keep the blood inside the body. When the blood rushes away from the skin, "the bubbling air rushes in with violent surge; and when the blood leaps up, the air is breathed out again."[15] Ignorant of the circulation of the blood, he thought it oscillated in the body like the tides.

Empedocles compared the movement of the air into and out of the body to the use of the Greek *clepsydra.* This device was used to transfer small quantities of liquid from one vessel to another. It consisted of a hollow cylinder with a strainer at the bottom and was used like a pipette. He described a girl playing with a clepsydra of gleaming brass. When she covered the mouth of the cylinder and dipped it into the fluid, no liquid entered the vessel because the air inside prevented liquid from entering . Uncovering the device allowed air to escape as liquid entered. Through these simple observations, Empedocles proved that air is not simply empty space, but that air has the nature of a substantial although invisible body. His proof of the substantiality of the air showed that matter could exert considerable power even though it existed in a form too fine to be visible. Once aware that nature works by unseen bodies, we can overcome the limitations of the senses and learn about things that are not directly perceptible.

Exactly what this insight meant in terms of the history of science is controversial. Some historians of science have claimed a place for Empedocles as the originator of the experimental method.[16] Other have argued that Empedocles merely applied a common observation to a preconceived theory. If Empedocles really used the experimental method, he could have tested his theory by sitting in a bath up to his neck to see whether any air bubbles passed through the water into or out of his chest as he breathed.

Anaxagoras of Clazomenae (ca. 500–428 B.C.)

Anaxagoras, like Empedocles, helped rescue Greek natural philosophy for the senses from the excess of reason and logic represented by Parmenides. Although a pupil of Anaximenes, Anaxagoras represents the new Athenian phase of Greek development. After the conquest of the Ionian towns by Persia, Athens had grown in wealth, commerce, and political power. In 490 B.C. the Persians were defeated at Marathon, under the leadership of the Athenians, and 10 years later the Greeks won a naval engagement which finally eclipsed Persian power. Athenians at this time still respected the active life, the arts and the crafts, and the inventor, as much or more than the philosopher.

Like the early Ionian natural philosophers, Anaxagoras combined an interest in practical matters with a passion for understanding how the universe worked. He believed that the earth was flat rather than a sphere and that the heavenly bodies were not divine but made up of the same stuff that exists on earth. The sun was a mass of red-hot metal, larger than the Peloponnesus. The moon, he said, was not a perfect sphere but had hills and valleys where creatures dwelt.

Pericles brought the philosopher to Athens to add to the cultural life of the city. In the *Phaedrus,* Plato says that the statesman was attracted by the high-mindedness of Anaxagoras and steeped himself in natural speculation which served as the source of his pre-eminence in the art of debate. Despite his friendship with Pericles, Anaxagoras found himself persecuted for impiety and perhaps other charges. One account says that his pupil Pericles made a speech in his defense, and the judgment was a fine of five talents and exile. Another account says he was condemned to death *in absentia.* In any case, he left Athens for Lampsacus, where he was honored and eventually buried. It was said that when asked what privilege he wished to be granted, he requested a holiday for children every year in the month in which he died.

Anaxagoras had broad interests which included astronomy, meteorology, and what is most important for our purposes, physiology and sense perception. Unfortunately, his surviving writings are fragmentary and subject to various interpretations.

His first principles were homogeneous molecules or "seeds." As gold consists of fine particles called gold dust, the universe is compounded of minute bodies. The first principles in his system, the "seeds," were infinite in both number and variety, and yet every one contained a little of all the qualities to which our senses respond.

Anaxagoras believed that: "All other things have a portion of everything, but Mind is infinite and self-ruled, and is mixed with nothing but is all alone by itself." Mind was the "finest of all things and the purest, it has all knowledge about everything and the greatest power; and mind controls all things, both the greater and the smaller, that have life."[17] In the beginning it was Mind that initiated motion, as a rotation that began in a small area, but is always increasing, and is mingling, separating, and dividing off all things. The rotation caused the separation of the dense from the fine, the hot from the cold, the bright from the dark, and the dry from the moist. Thus, Mind in Anaxagoras' scheme takes the role Empedocles ascribed to Love and Strife. Heavy bodies, such as earth, settled in the lower regions whereas fire, which was lightest, occupied the upper region—but water and air were intermediate.

To answer his own question, "How could hair come from what is not hair or flesh from what is not flesh?," Anaxagoras argued that there must be seeds of everything that is to emerge present from the beginning. Starting with the idea that it is impossible for anything to come into existence from what is nonexistent, he added the simple observation that we consume foods that are simple and homogeneous, such as bread or water. From these simple nutrients we build up "hair, veins, arteries, flesh, sinews, bones, and all the other parts of the body." Thus he reasoned there must be in the foods we eat some parts that "are productive of blood, some of sinews, some of bones, and so on—parts which reason alone can apprehend."[18] These things must exist in foods in some hidden form. During the process of digestion, the seeds or elements are sorted out. Growth occurs by assimilating the seeds into their proper places by the natural attraction of like to like.

Anaxagoras repeated the observations of Empedocles on the clepsydra, but he seems to have gone further in the direction of deliberate experimentation to show the nature of air. He is said to have demonstrated the power of air by blowing up bladders and showing their resistance to compression. In his description of the universe he used the strength of the air to keep the earth afloat. Although he seems to have been interested in many scientific questions, only bits survive as testimony. For example, he was interested in the composition of the bird's egg. Like Anaximander, he said that animals originally arose from moisture, heat, and earth. At a later stage they were propagated by generation from each other.

His notion of perception was opposite to and more subtle than that of Empedocles. He believed that perception occurs through opposites rather than by like to like. That is, we do not notice things that are the same—for example, something that is as warm or as cold as we are does not warm or cool us. He suggested that every perception is accompanied by pain, according to a fragment preserved in Theophrastus. Since everything unlike was supposed to excite perception by its contrast, too long a duration or an excess of sensation would become clearly painful.

In describing the difference between man and the beasts, Anaxagoras said that it was "possession of hands that makes man the wisest of living things." Although Anaxagoras may have contributed to observational and experimental science, he was cautious about the evidence presented by the senses. "From the weakness of our senses we cannot judge the truth," he said.[19] This warning accompanied a demonstration on the imperceptible gradation of color. Imagine two vessels: one contains a white liquid and the other a black one. If either liquid is added drop by drop to the other the eye cannot perceive any change in color until several drops have been added. Thus, although our senses tell us no change has occurred, reason tells us our senses have deluded us.

The Atomists: Democritus (ca. 470–370? B.C.) and Leucippus (fl. ca. 440 B.C.)

The theories of the atomists have been glorified as an anticipation of modern atomic theory. Actually, Greek atomism was a brilliant speculation that

Benjamin Farrington calls the culmination of the tradition of inquiry initiated by Thales. The theory satisfied some of the ancient Greek philosophers but had little impact on chemistry. Quite a different line of work led John Dalton (1766–1844) to atomic theory.[20]

Much is uncertain about the lives of the fifth century atomists. Even in ancient times, some denied that there ever was a philosopher Leucippus while others said he was a pupil of Zeno and the teacher of Democritus. He is said to have been one year older than Socrates and to have met Anaxagoras, who was some 40 years older than Deomocritus. Leucippus was said to have been the first philosopher to identify atoms as first principles. The atoms were the innumerable elements, infinitely varied in shape, which perpetually moved through the void.

Like Leucippus, Democritus also postulated a world composed of atoms and the void. Whereas Aristotle normally talks of Leucippus and Democritus together (something like today's Watson-and-Crick), Cicero says that Democritus resembled Leucippus in basic philosophy but was more productive in other respects. It seems likely that Leucippus invented the fundamentals of atomism and that Democritus refined the theory and elucidated its implications.[21] In any event, Democritus is regarded as the elaborator of atomism and its chief exponent. Democritus seems to have been a prolific author, writing on subjects as diverse as ethics, physics, nature, mathematics, music, astronomy, medicine, agriculture, and other technical subjects. Unfortunately, little remains of all this work, and the few fragments preserved are primarily from the works on ethics. Some works that were probably not genuine, such as accounts of foreign travel, became associated with Democritus, a fact which coincides with the idea that he traveled widely—to Egypt, Persia, India, and Ethiopia.

Leucippus made the bold assertion that the void, which is non-being, does exist and provides the place where atoms move. The atoms were uncreated and eternal, and did not themselves suffer change or destruction. But since they were in perpetual motion throughout the void, things appear to come into being through their combination, whereas their separation is interpreted as destruction. The idea that variety and change depend on the combination and separation of some primary material that was itself unchanging goes back to Empedocles and Anaxagoras. However, these philosophers, while accepting motion, had not postulated the existence of the void. The atomists accepted atoms and the void as the material causes of all things. The differences in shape, arrangement, and position of the atoms accounted for all apparent modifications in the universe.

Democritus believed that the atoms are too small to impinge directly on our senses. Atoms were indivisible, impenetrable, and compact because they had no void in them. Compound bodies, on the other hand, were divisible because of the void between their constituent atoms. The number of shapes atoms could assume was probably infinite because there was no reason that any atom should be of one shape rather than another.

Certainly the ancient theory of atomism was novel and today seems closer to "truth" than competing theories. But we must remember, even as we admire the theory, that it was based on logic and speculation rather than fact. The idea that rearrangements of atoms could produce different materials just as different arrangements of letters produce different words is a very ancient one. Yet the argument remained formal and logical rather than experimental, until the nineteenth century. The clarity and simplicity with which Democritus must have arranged his arguments are more impressive than the details of his theories.

From the void and the atoms, Democritus believed that innumerable worlds came into existence through mechanical means. These worlds were not all the same. Some had no sun and moon; in some the suns and moons were larger or more numerous than in our own. The intervals between worlds were unequal and the stage of becoming or dissolution varied. Sometimes worlds were destroyed by collisions with each other. Some worlds had living creatures, whereas others were "devoid of living creatures or plants or any moisture." Exactly how things happened is unclear—no supernatural agencies were called in—there is nothing but "chance and necessity." (Democritus called the vortex necessity.) "Nothing occurs at random," according to Leucippus, "but everything for a reason and by necessity." Although events seem random to us, they are merely the result of a chain of collisions between atoms. The shape and motion of the atoms involved determine the observable outcome of the obscure individual events. The atomists believed that the attraction of like for like was important, as shown in the passage: "Creatures, flock together with their kind, doves with doves, cranes with cranes and so on. And the same happens even with inanimate things, as can be seen with seeds in a sieve and pebbles on the sea-shore." [22]

There is some dispute over whether Democritus believed that the atoms had weight. This of course makes his theory quite different from Dalton's. Theophrastus said that Democritus distinguished atoms by size and shape and that Epicurus added weight to the other properties. But Aristotle says that Democritus believed that atoms had weight, and that this depended on their size. [23]

Later commentaries ridiculed the views of Democritus on sensation. Aristotle called the views of Democritus and other such natural philosophers "a great absurdity" because "they represent all perception as being by touch." According to Democritus, all perceptible things are arrangements of atoms which differ only in size and shape. [24] We assign certain qualities to these arrangements—which we call color, taste, odor, noise, touch. But, he said, these qualities are not in the bodies themselves—they are the effects of the bodies on our sense organs. Tradition has it that Democritus tested his own sense impressions by withdrawing into solitary places such as tombs. Democritus accounted for sight in what seemed to Theophrastus a "peculiar way." All things continuously give off some effluences, causing the air between the eye and the object of sight to become contracted and marked. The visual image, therefore, does not originate in the pupil, but in the air which is admitted to the eye.

Thus, according to the dictates of atomist concepts, all sensation and perception must be explained through some kind of contact or touch. The soul, composed of very mobile spherical atoms, is diffused throughout the body, but the soul-atoms were more concentrated in the mind. When mind-atoms were set in motion by collisions with appropriate atoms, thought was the result. Thus, for the atomists, images from outside were the cause of perception and thought. The other senses were explained in terms of the effects of different sizes and shapes of atoms on the appropriate organs. For example, a bitter taste was the result of contact with "small, smooth, rounded atoms," while large atoms with jagged edges caused a salty taste. But qualities such as sweet and bitter, hot and cold, and even colors exist only by convention. Only atoms and the void truly exist in nature.

The idea that man was a world in miniature—a microcosm, reflecting the whole universe, the macrocosm—is clearly part of the system of Democritus. Man was explained in material terms as much as the world around him. Man was a microcosm that had to contain atoms of every kind in the universe. Even consciousness, sleep, sickness, and death could be explained in terms of the quality of atoms or loss of atoms.

Democritus seems to have been interested in biology and to have practiced dissection of higher and lower animals. The first principle used later by Aristotle as the major division among animals—sanguiferous (vertebrates) and bloodless (invertebrates) was noted originally by Democritus. Unfortunately, most of his biological writings were lost and his views are known mainly through Aristotle's polemics against them. Although this surely must distort much of his actual achievement, often Democritus proves more correct than Aristotle. He called the brain the organ of thought, whereas Aristotle believed it served only to cool the blood. Democritus apparently described the heart as the organ of courage and ascribed sensuality to the liver. He also commented on a problem that was later to be important in genetics—especially in determining the contributions of the two sexes to the offspring—that of the mule and its peculiar sterility. We are perhaps fortunate to have even the few fragments from Democritus, for one tradition holds that Plato tried to burn all of his writings.

Democritus was a contemporary of Hippocrates and according to ancient tradition the two did meet. It is pleasant to imagine such a meeting, for atomism along with the rational medical theories of the Hippocratic school can be counted the highest achievements of Greek thought until the time of Plato and Aristotle. In many ways the works of the Hippocratics and atomists are more satisfying to the modern mind than the philosophy of the later Golden Age of Greece.

Greek Medicine

Greek medicine, particularly in the work of Hippocrates, also contributed to the development of biology.[25] For the most part, medicine is outside the scope of this book. We can only mention the main aspects of the Greek medical heritage. The first, which the Greeks shared with the other ancient cultures, derived from the religious tradition of healing. In the great temples dedicated to Asclepius, the god of healing, the sick sought medical diagnosis and treatment. No doubt the temple physicians owed their success as much to their shared experience and keen observation as to the gods.

The concern of the Greeks with the balanced life—the healthy mind in the healthy body—required much athletic training. This in turn led to practical knowledge of the muscles and tendons, the parts of the body most important to and most likely to suffer damage in athletic events. The Pythagorean school also contributed a more mystical philosophical view of the proper regimen in sickness and health. Most important was the practical school of rational medicine associated with Hippocrates (ca. 460–377 B.C.). For Hippocrates and the philosophers who were his contemporaries, man was a part of the natural world. Medicine was above all a craft in which success in helping nature to cure the patient was the ultimate measure of the physician. Philosophy was less important than technique, but the Hippocratic school did develop a theory of health and disease. This was mainly derived from the four elements of Empedocles. The human being, like the rest of the universe, was composed of the four elements, earth, air, fire, and water. Each of these elements was related to one of the four humors: black bile, yellow bile, blood, and phlegm. Individual variations in composition determined personality and states of health and disease. When the four humors were well balanced, a state of health existed. Deviations from perfect balance produced disease. This all-encompassing but vague theory of the body juices was to influence medicine up to the nineteenth century.

Greek philosophy entered into a new phase with the end of the pre-Socratic period. The changing climate of Greek thought is best appreciated from the vantage point of Athens, scene of the next phase in the development of Greek philosophy. The Peloponnesian Wars of 431–404 B.C. resulted in the conquest of Athens by Sparta. The impulse that had fired the Ionian philosophers—to find a purely naturalistic explanation of the universe and man's place within it—no longer suited the times. The kind of practical and naturalistic ideas which had appealed to the old school were no longer welcome. The crafts and techniques that had once inspired philosophers came to carry a social stigma and, as Xenophon, a pupil of Socrates, tells us, were "rightly dishonored" in Athens.

The Age of Socrates, Plato, and Aristotle

Although Socrates, Plato, and Aristotle are of the greatest interest in the history of philosophy, in many ways at least the first two provide less satisfaction than the pre-Socratics for the historian of science. The earlier philosophers lived in an age that respected not only abstract thought, but also technical progress, manufacturers, merchants, the crafts, trades, and practical inventions. There was what Benjamin Farrington called the "propaganda of enlightenment,"[26] the definite and conscious idea that through rational thought and practical techniques man could banish mystery and ignorance from the world and understand man, life, and all of nature.

Although Plato and later philosophers often speak of Socrates as rescuing Greek thought from the abstract and materialistic concerns of his Ionian predecessors, for science the new intellectual milieu of Greece was a regression. The older philosophers were not preoccupied with the heavens to the exclusion of earthly matters. Because man was made of the same materials and was subject to the same laws as the great universe, understanding the heavens enriched inquiry into the nature of man.

Socrates (470-399 B.C.)

According to Plato, philosophy treated only physics before Socrates added a second subject: ethics. While it is possible that Socrates may have been interested in natural science, his major interests were ethics and politics. His teacher, Archelaus of Athens, certainly had been interested in natural philosophy and had been himself a pupil of Anaxagoras. Socrates is said to have warned his disciples that the study of science could drive men mad—as in the case of Anaxagoras. Because Anaxagoras made no mention of the gods in his lectures, Socrates resolved to reject science. It is possible that Socrates realized how dangerous the study of nature could be when he saw Anaxagoras banished from Athens for impiety. Anaxagoras had said that the sun was not a god but merely a hot rock. If this is true, it is even more ironic that Socrates was condemned to death by the Athenians for "godlessness"—refusing to acknowledge the gods recognized by the state. Socrates was accused of introducing new and different gods and corrupting the youth. For these "crimes" Socrates was sentenced to death. Although his friends urged him to escape, Socrates ended his life in prison with hemlock, the potion of the condemned.

Socrates taught that there was only one good, which is knowledge, and one evil, which is ignorance. He claimed that he had been given a divine mission to question and test all statements by any man. Yet Socrates said his role was that of midwife who did not teach others but, by making them think, gave life to their thoughts. Spending his time in discussion and teaching, Socrates never wrote down his ideas. Plato, his favorite pupil, later developed a system of philosophy ostensibly based on the teachings of Socrates.

Plato (429–347 B.C.)

Although Plato included a theory of the nature of the universe in his system of philosophy, ethics, politics, and theology were his primary concerns. His comments on nature were designed to support his vision of the ideal society. After the death of Socrates, Plato is said to have traveled widely to learn from other philosophers, mathematicians, and even the priests of Egypt. Returning to Athens, he founded the Academy in an olive grove on the outskirts of Athens. Here, surrounded by his pupils, he lectured and wrote out his memories of the teachings of Socrates and his own philosophical system. Not all of his students were totally devoted to philosophy, for when Plato read his dialogue *On the Soul* only the faithful Aristotle is said to have stayed to hear the end. The Academy survived in one form or another for almost 900 years. Plato is the first of all the Greek philosophers whose major writings survived in full.

Philosophy was viewed as a hunger for divine wisdom. Plato believed the task of the philosopher was to elucidate the eternal values. Knowledge of these was reached not through direct observation of nature, but only through abstract thought and mathematics. Plato added the third component of philosophy—dialectics—to physics and ethics. Unlike the materialistic theories of his predecessors, Plato's system of nature was heavily imbued with the supernatural or "divine providence."

As revealed in the *Timaeus,* Plato's vision of the creation is highly religious.[27] The world was created out of an original chaos by an eternal and perfect god in accordance with a rational, intelligent design. In Plato's version of the creation, man was not derived from simpler creatures, but was the first being to appear. The first part of man to be formed was the head because it was most nearly spherical and served as the organ of the immortal soul. The soul was tripartite, with the rational part located in the head, the heart the locus of the passionate part, and the liver the site of the appetitive part. In a classic reversal of evolution, the other species were created by degeneration from man. Even woman was a secondary production, being created from the souls of men who were "cowardly and spent their life in wrongdoing." With more wit than compassion for his predecessors, Plato wrote that "harmless but light-witted men, who studied the heavens but in their simplicity supposed that the surest evidence in these matters is that of the eye" were turned into birds. Men who had no use for philosophy were transformed into four-legged beasts. Poor Anaximander was given an even worse fate for his speculation about men evolving from fish-like creatures. According to Plato, creatures that lived in the water were derived from the most foolish and stupid men. These souls were so polluted by their transgressions that the gods found them unfit to breathe pure air. Thus Plato dealt with men who had tried to develop a thoroughly naturalistic explanation of the universe and those who believed that the data obtained by the senses were at least of equal value to abstract thought. In the world of Plato, true reality is found only in the sphere of abstract thought. What is perceived by the senses is but an imperfect image of the eternal ideal, like shadows cast on a wall.

Although Plato did little for biology directly, his ideas were to shape and direct the inquiries of later scientists. The concept of species and genera, the foundations of systematic taxonomy, come from Plato's search for the eternal ideal forms.

Aristotle (384–322 B.C.)

Aristotle, disciple of Plato and disappointed heir-apparent, seems to have diverged more and more from the views of his mentor during a long and rather turbulent life. More than 2000 years after his death, Aristotle's ideas still intrigue philosophers and some of our great modern scientists. Max Delbrück, for example, recently spoke of his admiration for Aristotle. In a burst of hyperbole, Delbrück credited Aristotle with having anticipated by 23 centuries the basic concept of heredity as transmitted by DNA.[28]

Just as Plato is the first of the philosophers whose writings have survived in bulk, Aristotle is the first great scientist to have this distinction, and his writings are voluminous.[29] Aristotle's life is of special interest to us for several reasons. First, the intellectual history of Aristotle may throw some light on his work; second, some aspects of life in the Greek world of the fourth century B.C.—its science, philosophy, politics, and customs—are revealed in the biography of the man Aristotle. Aristotle was born in Stagira, a town in Macedonia regarded as semibarbaric by the more sophisticated, older Greek states. Macedonia, largest of the Greek states at that time, was governed by Philip, father of Alexander the Great. Nicomachus, Aristotle's father, was a successful and wealthy physician at the royal court. As a youth, Aristotle must have learned the secrets of the brotherhood of the Asclepiades, the tradition of Hippocrates, and the advantages of royal patronage. Although he could be instructed in medicine as technique, science, and religion at Macedon, a properly educated physician also needed training in philosophy. Therefore, Aristotle was sent to Plato's Academy when he was 17. Here he was exposed to a very different kind of thinking—one that placed the abstract over the practical. Plato's philosophy was very different from that of the practicing physician. Apparently this new world suited Aristotle, for he remained at the Academy until the death of Plato, 20 years later.

During this time, Aristotle had already established himself as a philosopher and author. He even became critical of some aspects of the philosophy of his mentor. Probably Plato had noticed this with displeasure, since he appointed his nephew, the otherwise undistinguished Speusippus, as his successor instead of Aristotle. Embittered and disappointed, Aristotle abandoned the Academy, Athens, and mathematics. Biological researches, carried out mainly on the island of Lesbos, were his main interest during this period of his life. When a revolution deposed his patron, Hermias, tyrant of Atarneus, he had to flee to Macedonia to stay at Philip's court. For three years, Aristotle served as tutor to Alexander (338–335). It was not the most successful student-teacher relationship in history, although it is one of the most famous. Certainly, Alexander was more interested in war and conquest than in philosophy. It seems he did remember his old tutor well enough to send back reports and specimens of the peculiar animals he found in exotic places during his conquest of the known world. There is also the story that he kept a copy of Aristotle's edition of the *Iliad*, along with a dagger, under his pillow.

In 334 B.C., at the age of 50, Aristotle returned to Athens. Here he set up his own school, called the Lyceum, and assumed a rigorous schedule of lecturing and writing. Unfortunately this productive and relatively serene period soon came to an end when political considerations again made him a refugee. After the death of Alexander, the Athenians rebelled against Macedonian rule. Faced with an indictment for impiety, Aristotle fled the city to "prevent a fresh crime against philosophy." After a year spent in despondence and exile at Chalcis, the philosopher died.

Aristotle left a remarkable body of writings on many diverse subjects. Although the bulk of his works have survived, there are some gaps and some uncertainty as to the order in which they were written. Some of the works were apparently edited by his collaborators and some may be student notes, which might present a garbled version of what he really said and thought. A gift for organizing, arranging, and even appropriating other people's material and recognizing affinities enabled Aristotle to discuss almost every field of human knowledge. Very often all that is known of his predecessors comes from

Aristotle's comments on them. Where he disagreed he could be quite sarcastic and perhaps misleading.

His work falls into four major divisions: (1) Physics, especially astronomy: This is the Earth-centered, closed, finite universe that dominated Western thought until the time of Nicolaus Copernicus (1473–1543). (2) Ethics and politics: Although these works are outside the scope of this book, it should be noted that even his most philosophical works contain many links between his views on nature and society; biological examples and analogies are often used to explicate difficult points. (3) Logic and metaphysics: Much of this work is a criticism of the work of his predecessors, particularly of Plato's theory of the ideal forms. (4) Biology: Here Aristotle reveals himself as a remarkably acute observer of nature who, despite his prejudices against manual labor, was willing to dirty his hands in dissections of even the lowliest creatures: his major writings on biology are *Natural History of Animals, On the Parts of Animals, On the Generation of Animals* and *On Psyche.*

From his writings, two Aristotles emerge. One was the theorist and logician, who believed that the logical and the real were likely to be identical. The second Aristotle was the observer, who realized he was working in unchartered territory and advised his followers to rely on their own observations rather than his words. Aristotle took Plato's highly abstract theory of the Ideas or Forms which exist only in some world of reason, and turned it into a guide to the study of nature. For Aristotle, the Ideas or Forms exist in nature and not apart from it.

The true natural philosopher, according to Aristotle, would find every realm of nature to be marvelous and would not recoil with "childish aversion from the examination of the humbler animals." As in his remarkable studies of marine creatures, the scientist must "study every kind of animal without distaste; for each and all will reveal to us something natural and something beautiful." Contemplation of nature's works would reveal "her generations and combinations" as always "a form of the beautiful."[30]

The theory of Forms was important in stimulating the study of species of plants and animals. Even though Aristotle knew some 520 different species and dissected some 50 kinds of animals, he realized that this was only the first step into a whole new domain. Although Aristotle is one of the great pioneers of systematic taxonomy, there is no formal taxonomy in his work. Aristotle used *genus* and *species* as categories but admitted that the task of sorting out and describing animals was a monumental one. Still, the parts and habits of animals had to be studied before any reasonable classification scheme could be constructed.

Aristotle's biological investigations must have been very enjoyable and satisfying to him, but as theories they were only a link in his general world system. Thus, although his descriptions of animals, even those as small and delicate as the eggs and embryos of fish and mollusks, are remarkably accurate, sometimes his great powers of observation were distorted by his preconceived notions about the way things should be. As Einstein said, "It is the theory that decides what we can observe." Thus, Aristotle had some mistaken notions of the functions of important organs of the body. He believed that the heart was the seat of the soul and intelligence. Dissection is not invariably an instrument of understanding because function is not obvious in structure. The brain, according to Aristotle, merely served to secrete mucous and cool the blood.

Although Aristotle did not explicitly create a classification system, he did approach the problem in sufficient detail to enable one to draw up a crude scheme from his discussions of the animals and the requirements of a classification scheme. Unlike Plato, who used simple bifurcations to separate out groups of animals, Aristotle recognized that many characteristics must be studied to determine natural affinities among animals. One must study structure, habits, environment, means of locomotion, and reproduction to begin to know the animals.

A natural scheme of classification is one that gives attention to many characteristics and weighs them to determine the natural affinities beneath the variety of structures and functions. For example, it would be absurd to use habitat alone to order animals, although this very criterion served for classification schemes used by some Chinese botanists and medieval natural philosophers. Such a division would put together bats, birds, and flying squirrels. To lump together those that lived in water would link fish with marine mammals such as porpoises and whales. In addition to the ridiculous groups such a scheme would engender, it had the the more important disadvantage for Aristotle's purposes that it would not allow one to say which animals were higher or lower on the great "scale of being."[31]

In searching through all the characteristics of animals, Aristotle found one that suited him. Using means of reproduction as the major determinant of groupings generated a satisfactory continuous ladder of creation. Like Democritus, Aristotle began by dividing animals into groups with and without red blood. This corresponds to vertebrates and invertebrates. In his writings we find only the terms "genus," which stands for family or kinship, and "species," from the Greek for "form" used to deal with the common names of things.

In Aristotle's hierarchical scheme, animals were arranged according to the level of development at birth. The group that we call mammals was at the top because the members of this group are warm and moist, not earthy. Their young are born alive and complete, and are nursed by the females of the species. Next were creatures that were less warm but were still moist. These animals were born alive, but developed from eggs that grew inside the body of the mother. Lower down the scale were creatures that were warm and dry and laid complete eggs such as the birds and reptiles. Lower still were the cold and earthy animals which produced incomplete eggs, such as the frog and the ink fish. The lowest of all the sexually reproducing forms were worms, which produced eggs. On a rung below these were the vermin that came into being by spontaneous generation. Thus a ladder of creation from plants at the bottom to man at the top was created. The quality of the soul or souls determined the habits and functions, as well as the structure and perfection, of each species. Plants have only a vegetative soul, for growth and reproduction; animals have also a sensitive soul for movement and sensation; and man has, in addition, a rational soul, whose seat is in the heart, not the brain.

There were three major means of generation known to Aristotle. Among these were spontaneous generation, which routinely produced fleas, mosquitoes, and various kinds of vermin. Animals such as the star fish, worms, and shellfish arose through asexual reproduction. Regeneration, common in marine organisms, was regarded as part of generation. The higher animals and even some of the insects, such as the grasshopper, were produced by sexual reproduction. Unfortunately for the history of biology, Aristotle gave his approval to the theory of spontaneous generation, and rejected the evolutionary theory of the Milesian philosophers. Belief in spontaneous generation was common up to the seventeenth century and again during the nineteenth century in a different form. According to Aristotle, even spontaneous generation obeys rules—certain kinds of mud or slime give rise to specific kinds of insects and vermin.

Not surprisingly, sexual reproduction is due to the occurrence of the two sexes. In Aristotle's scheme the male is the more complete and "warmer" entity, while the female is less complete and "colder." As to the old question of the contribution of the two parents to the offspring, Aristotle taught that the masculine contribution is form, motion, and activity. The male sex product or "seed" is produced in the blood through a complete "cooking" process and received the purest and most form-creating qualities.

The female contribution, according to Aristotle, is matter which is entirely passive. The menstrual blood, a kind of undeveloped semen—perhaps "half-cooked"—is the female sex product. Just as a block of marble is the matter out of which the sculptor creates the statue or form, the matter contributed by the female serves as the potentiality through which the form principle contributed by the male achieves reality. During pregnancy the blood that was normally lost during menstruation went to build up and increase the size of the embryo. Aristotle's theory of generation had the major virtues of accounting for the known facts, accommodating the prejudices of his time, and the minor defect of being completely wrong. Not until William Harvey (1578–1657) investigated the subject was Aristotle's theory of generation seriously challenged.

Nevertheless, Delbrück sees Aristotle's theory as an anticipation of the discovery of the role of DNA. Like DNA, the "form factor" furnished by the male semen contains the blueprint for every part of the embryo and controls growth without itself becoming altered in the process. Delbrück would award a Nobel Prize for this insight despite the obvious criticism that it entirely neglects the contribution of the female parent—a contribution that predecessors of Aristotle, like Hippocrates, had already acknowledged.

The vague metaphor of cooking or "coction" was used by Aristotle to explain nutrition as well as embryonic development. Some foods were cooked in the stomach and produced food vapors which ascended to the heart. This organ transformed them into blood and, through the movement of the blood, nutriments were transferred to the body and directly assimilated.

Although Aristotle had many disciples, he completely overshadows them. His ideas dominated biology as well as other aspects of human thought for generations. Indeed the biology of antiquity and the Middle Ages never really got beyond Aristotle's conceptions of the phenomena of life. The great tragedy is that while Aristotle taught the basics of the scientific method— close and accurate observation, collection of many facts, deductions from facts—this part of his teachings was almost completely forgotten. Later workers were slavish followers of the words, rather than the spirit, of Aristotle.

Aristotle's Followers

The disciples of Aristotle tended to take more modest portions of all possible knowledge for their investigations. The old tradition of formulating a complete cosmology essentially died with Aristotle. Theophrastus and Strato are the only followers of Aristotle at the Lyceum to really deserve a place alongside their master. Yet the Lyceum continued to function until it was succeeded as a center of intellectual activity by the Museum of Alexandria.

Theophrastus (ca. 373-ca. 285 B.C.)

Theophrastus was born at Eresus on the island of Lesbos. The fact that his father was a fuller probably gave Theophrastus a more practical outlook on life and science than many other philosophers. He began his instruction in philosophy at the Academy when Aristotle was still associated with Plato. When Plato died in 347 B.C., Theophrastus followed Aristotle and became his favorite pupil, friend, and chief assistant. When Aristotle went into exile for the last year of his life, Theophrastus became head of the Lyceum where his lectures were very well attended. He was said to be so respected in Athens that when he was accused of impiety, his prosecutors narrowly escaped punishment.

Although Theophrastus is overshadowed by his teacher, the years in which he served as head of the Lyceum were not barren times for science. While the virtue of industry—as a prodigious worker and voluminous writer—has been readily associated with Theophrastus, only recently has he been seen as an independent and even original thinker. Perhaps he has been neglected because only a fraction—perhaps one-tenth—of his writings survived. Although he is remembered most as the father of scientific botany, he actually worked on broader problems in metaphysics, biology, medicine, and the doctrine of the four elements.

In some respects, Theophrastus returned to concepts of the pre-Socratic philosophers. It seems that he had some disagreements with Aristotle over the actual make-up of nature and the role of observation in understanding things. Theophrastus protested against the preoccupation of natural philosophers with the study of "first principles," which he called "metaphysics," and their denigration of the study of nature herself through observational means, which should be called physics. Scientific progress was being inhibited by a confusion between the two and the assumption that metaphysics was the more worthy pursuit. The other factor hindering science was the reckless use of teleological arguments, also known as arguments from design. According to the teleological principle, used explicitly and implicitly by Aristotle, all things are for the sake of an end, for nature does nothing in vain. If one chooses to observe natural phenomena, it is easy enough to show that nature does not always act with universality of purpose and intelligent design. Theophrastus pointed out the existence of droughts and floods, breasts in the male, and superfluous hair. This same argument against design in nature, based on rudimentary or vestigal organs, was to be very important to Charles Darwin and other evolutionists.

Theophrastus realized that the teleological explanation was often an oversimplification and at times a disguise for sloppy thinking. Such thinking provides instant answers to all questions, but militates against asking hard questions and answering them in a meaningful way. In conclusion he warned: "We must try to set a limit to the assigning of final causes. This is the prerequisite of all scientific inquiry into the universe, that is into the conditions of existence of real things and their relations with one another."[32]

Having this valuable insight into the preconditions for scientific progress, did Theophrastus then apply it to his own work? The answer to this question seems to be a qualified yes, based on his fundamental work on botany.[33] He seems to have used the students who came to the Lyceum from distant places to expand his botanical research by having them observe the plants in their own regions. He was able to describe some 500 species and varieties of wild and cultivated plants.

One of the most basic contributions of Theophrastus was in drawing a clear distinction between plants and animals. Aristotle had derived animals from man through a process of degeneration. He went on to derive plants from animals by further degeneration. Theophrastus explicated the fundamental differences between animal and plant structures. For example, plants readily renew their parts, whereas loss of "parts" of animals is generally accidental and regeneration is limited. He warned against drawing analogies between the morphology of plants and animals, but unfortunately later botanists and anatomists were to ignore his sound advice.

Botany did not truly advance beyond Theophrastus until quite modern times. The classification scheme he used was continued until the time of Linnaeus. Plants were divided into the categories of trees, shrubs, herbs, annuals, biennials, and perennials. His description of plants included items of practical and scientific interest, such as plant juices, products, and properties like viscosity, gumminess, odor, and color. Also recorded were various superstitions about the safest ways to gather plants. Since plant products were the principle source of medicines at the time, and gathering "simples" was the precursor to pharmacology, these beliefs about plants were quite important to scientist and layman alike. Although Theophrastus was interested in the reproduction of plants, he did not recognize plant sexuality, except in the case of the date palm and some higher forms. According to Theophrastus, the higher forms of plants reproduced through their seeds and the lower forms by spontaneous generation. Both Aristotle and Theophrastus believed that plants did not necessarily breed true through their seeds because some of their characteristics could be changed by the environment.

A fragment of a treatise, *On Fire,* reveals a very clever critique of the venerable doctrine of the four elements. Fire is peculiar because we can generate it by force, and unlike the other elements, it requires a substratum for its existence. "Fire is a form of motion," he wrote, "always in search of nutriment." In a detailed argument, Theophrastus clearly shows the difference between arguing from observation and arguing from logic alone.[34] In this fragment are many examples of observation of natural and technological or craft processes.

Theophrastus thus points the way back to argument by analogy with craft processes and forward to experimental science. These arguments did not directly lead to a modern experimental science, but did prepare the way for Strato.

Strato of Lampsacus (fl. 286–268 B.C.)

Strato was called "The Physicist," and has been credited with taking scientific inquiry beyond passive observation and into the realm of experimentation.[35] Abandoning the study of ethics, Strato turned to the study of pneumatics.

Although little is known of him or his writings, he was apparently a well-respected scholar and absorbed valuable lessons in survival and patronage from his mentor, Theophrastus. The latter had described scholars as true citizens of the world, able to survive changes of fortune and obtain employment in any realm. As the tutor of Ptolemy II, Strato was able to collect enormous fees. The first Ptolemy, Soter, had summoned the philosopher to supervise the education of his son, Philadelphus. Thus, Strato had the interesting experience of living for a time in Alexandria. In 287 B.C. when he must have been between 40 and 50 years of age, he succeeded Theophrastus as head of the Lyceum and retained that position until 269 B.C. Although he may have written about 40 books, all of Strato's writings have been lost. Some of his works seem to have been plagiarized or appropriated by later scientists. In particular, his work in pneumatics may have been used rather freely by Hero of Alexandria. Not only was Strato interested in experimentation, he also believed that such inquiry could result in practical applications.

In a passage that Hermann Diels ascribed to Strato rather than Hero of Alexandria, Strato explicitly said that he would prove the existence of a vacuum by "experimental tests." First, he discussed an experiment that proved that "empty" vessels were really full of air. If a vessel was carefully plunged into a dish of water, the air would keep the water from entering. But if a hole was made in the bottom of the vessel the water would enter the mouth while the air escaped through the hole. This experiment also showed that wind is air in motion, because one could easily feel the air being driven out by the water.

Strato went on to make an analogy between the particles of air and the grains of sand at the beach. His position was that small vacuums exist in a scattered state in air, water, and other bodies. There were spaces between the particles of air, just as there were spaces between the grains of sand. Thus, the particles of air could be forced together more compactly or could separate and increase the empty space between them. A simple experiment to demonstrate the existence of a small vacuum was done with a vessel having a narrow mouth. After the air had been sucked out, the vessel would remain suspended from the lips because the vacuum tended to pull the mouth in to fill the empty space. Another proof involved a glass vessel with a narrow mouth used by medical doctors. To fill these with liquid, the air was sucked out, the mouth of the vessel was covered with the finger, and the vessel inverted in the liquid.

Strato challenged his opponents to reply not with logic alone which he called "verbal gymnast," but with experiments. These examples show that Strato carried out deliberate experiments with objects at hand and accepted the evidence of the senses. This was a major advance over Empedocles, but Strato went even further and designed special apparatus to prove particular points about the nature of air and the vacuum. Simplicius tells us that Strato was bored by a debate over whether change of position could occur without sup-position of a continuous vacuum. He put a stone into a closed vessel full of water and showed that the stone changed its position when the vessel was inverted.

With many ingenious arguments, Strato "proved" the existence of void within sensible bodies—whether air, liquid, or solid. Thus the sun's rays get to the bottom of a bucket of water, but some of them hit particles and are reflected, so that only a few penetrate to the bottom. If there were no spaces this would not occur and the rays would make the water overflow. Another proof was the way wine poured into water visibly disperses. Book 4 of Aristotle's *Meteorology,* is said to be written by Strato. It is a unique fragment because it is the only Greek work dealing with chemical problems before the time of the Alexandrian alchemists.

Strato also carried out experiments on the effect of fire. He is said to have weighed a piece of wood before and after heating. Although the charcoal had the same volume as the wood, it weighed significantly less. This proved that some of the substance had been driven out and left larger spaces between the remaining particles. Apparently, Strato also had a fairly good idea about the nature of sound and its transmission. Strato rejected Aristotle's notion of weight as related to the "natural place" of each element and returned to the theory of Democritus that weight is motion towards the center. He further discarded Aristotelian concepts of gravity and levity and said that all the elements have gravity and none have levity.

Some of Strato's views on psychology and the nature of man have been rescued from obscurity. Unlike Plato, he believed that true knowledge was the result of experience. He rejected Plato's view of true knowledge as something independent of experience which was first and only the possession of the soul. He had critically examined the relationship between the senses and the mind and concluded that it is in the mind that a stimulus generated from the out-side is actually transformed into a sensation. According to Strato, perception and thought were the province of the same soul. Since animals have sense organs like ours, they must also have minds. Unlike the Aristotelian view that animals are degenerate forms of man, Strato regarded man as merely a better kind of an animal.

For Strato whatever exists has been made by purely natural forces and movements. That he was out of temper with his time is evident in the loss of his writings and the criticism heaped on him by later commentators.

Although the Lyceum did not totally die with Strato, after his death little work of real scientific merit was produced in Athens. The next phase of scientific investigation was carried out at the great Museum of Alexandria. Strato knew the Lyceum was in trouble during his tenure as he watched the number of students decline. In his will Strato left the school to Lyco "since the others are either too old or too busy."[36]

The new dynasty of the Ptolemies in Egypt recognized the useful aspects of science and scholarship and were eager to transfer all that could serve them to Alexandria, as we shall see in the next chapter.

The Decline of Greek Science

Although the modern student of science would tend to view the transition from the purely speculative to the experimental approach to science represented by Thales and Strato as a promising one, the actual status of science over this period suffered a serious decline. It seems paradoxical that just as Athens was at the point of producing a nearly modern form of science the judgment of history must be that Greek science suffered a failure of nerve. Although he might have fit well into our own age, Strato was a man out of step with his contemporaries.

The failure of Greek science is actually a part of the decline of Greek culture in general and thus tied to political conditions which continuously worsened as the Greek city-states lost their independence and power. Intellectuals turned away from the study of nature to ethics and politics—subjects to help them survive and even prosper in difficult conditions. In so far as philosophers abetted and participated in this decline, it can be traced to the unfortunate tendency to separate "mind" and "experience." Living in and enjoying the fruits of a slave society, philosophers generated a rationale for the leisure class view of managing the affairs of men.

Part of the work of Plato and Aristotle must be seen as a justification for a slave society. They shared a contempt for the crafts, the craftsman, and the practical. Plato did not believe that the human craftsman could even originate anything. All inventions came from the Idea or Form created by God and transmitted to the worker by the philosopher. He regarded any manual work as demeaning, even if engaged in to study abstract questions of scientific or mathematical interest. Plato even criticized colleagues who had constructed a piece of apparatus to help solve a geometrical problem, for he regarded even such "labor" as "contaminating thought."

The classical tradition favored the life of the mind. The concept of the dignity of work and the relationship between work, technology, facts, science, and idea were unknown or rejected. Thus, Seneca believed that the inventions of his period were necessarily the province of slaves, since only slaves were concerned with things instead of ideas. Archimedes, according to Plutarch, was ashamed of the machines he had built. As Lynn White put it, "in Antiquity learned men did not work, and workers were not learned."[37] Plato implied that no further improvements of the techniques were necessary or possible since the status quo must be the "divine plan." For Aristotle, "applied science" had already completed its task. Science need not be useful since all the needs for comfort and social refinement already existed. Theophrastus wrote that life "now lacks nothing in the cultural amenities which promote enjoyment of leisure."[38]

Labor-saving devices were unneeded since only slaves worked with their hands and there were plenty of slaves in the ancient world. For Plato the slave was only an animated machine. There was no need for more effective exploitation of nature, since there were sufficient creature comforts for the important people. The important people were those who led the life of the mind. Never mind that only a few could enjoy the refinements of life. Human society was a hierarchy, just as the universe was, from the lowliest unformed matter through slaves and masters to the great Prime Mover.

Notes

1. T. W. Africa (1968). *Science and the State in Greece and Rome.* New York: Wiley.

2. G. S. Kirk and J. E. Raven (1957). *The Presocratic Philosophers: A Critical History with a Selection of Texts.* Cambridge, Mass.: Cambridge University Press, p. 78. (Aristotle, Politics AII, 1259a9.)

3. Kirk and Raven, p. 78. (Plato, Theaetetus, 174 A.)

4. Kirk and Raven, p. 168–169.

5. Kirk and Raven, p. 181.

6. Diogenes Laertius. *Lives of the Philosophers.* Translated and edited by A. R. Caponigri (1969). Chicago: Regnery, p. 47.

7. Kirk and Raven, p. 183.

8. Kirk and Raven, p. 199.

9. Kirk and Raven, p. 197.

10. Kirk and Raven, p. 228.

11. Kirk and Raven, pp. 232–235.

12. D. O'Brien (1969). *Empedocles Cosmic Cycle: A Reconstruction from the Fragments and Secondary Sources.* London: Cambridge University Press.

13. Kirk and Raven, fragments concerning sense-perception and consciousness, pp. 343–344.

14. Kirk and Raven, "Four Stages of Evolution," pp. 336–340.

15. Kirk and Raven, p. 341.

16. J. Burnett (1930). *Early Greek Philosophy.* 4th ed. London: Black, p. 27.

17. Kirk and Raven, Anaxagoras, "Mind," pp. 372–375.

18. Kirk and Raven, Anaxagoras, "In Everything a Portion of Everything," pp. 375–377.

19. Kirk and Raven, p. 393–394.

20. See, for example, F. M. Cornford (1932). *Before* and *After Socrates,* Cambridge, Mass.: Cambridge University Press, p. 25; B. Farrington (1969). *Greek Science, Its Meaning for Us.* Baltimore: Penguin, pp. 62–63; S. Toulmin and J. Goodfield (1962). *The Architecture of Matter.* New York: Harper and Row.

21. C. Bailey (1928). *The Greek Atomists and Epicurus.* Oxford: Clarendon; T. Cole (1967). *Democritus and the Sources of Greek Anthropology.* Cleveland: Western Reserve University Press.

22. Kirk and Raven, Leucippus and Democritus, "The Formation of Worlds," pp. 409–414.

23. Kirk and Raven, Leucippus and Democritus, on "The Behaviour of Atoms," pp. 414–421; D. J. Farley (1967). *Two Studies in the Greek Atomists.* Princeton: Princeton University Press.

24. Kirk and Raven, Leucippus and Democritus, on "Sensation, Thought and Knowledge," pp. 421–424.

25. See for example, H. O. Taylor (1963). *Greek Biology and Medicine.* New York: Cooper Square; A. J. Brock (1929). *Greek Medicine.* New York: Dutton; E. J. Edeltstein (1945). *Asclysius.* Baltimore: Johns Hopkins Press; Hippocrates. *The Genuine Works of Hippocrates.* Translated by F. Adams (1939). Baltimore: Williams and Wilkins.

26. Farrington, p. 81.

27. F. M. Cornford, ed. (1937). *Plato's Cosmology: The Timaeus of Plato.* London: Paul.

28. M. Delbrück (1971). Aristotle–totle–totle, in *Of Microbes and Life.* Edited by J. Monod and E. Borek. New York: Columbia University Press, pp. 50–55.

29. W. Jaeger (1948). *Aristotle.* 2nd ed. London: Oxford University Press; R. McKeon, ed. (1941). *The Basic Works of Aristotle.* 4th ed. New York: Random House; W. D. Ross, ed. (1952). *Aristotle's Works.* 12 vols. Oxford: Clarendon.

30. Aristotle, *Parts of Animals.* I 5.645a, in *The Works of Aristotle.* Vol. V. Translated by William Ogle (1912). Oxford: Clarendon Press; p. 644.

31. A. O. Lovejoy (1936). *The Great Chain of Being.* Cambridge, Mass.: Harvard University Press.

32. W. D. Ross and F. H. Forbes, eds. and trans. (1929), *Theophrastus Metaphysics.* Oxford: Clarendon.

33. Sir Arthur Hort (1916). *Theophrastus Enquiry into Plants,* 2 vols. Loeb edition. Cambridge, Mass. Harvard University Press.

34. Farrington, p. 167.

35. H. B. Gottschalk (1964–66). Strato of Lampsacus: some texts, in *Proceedings of the Leeds Philosophical and Literary Society–Literary and Historical Section.* Vol. 11 (1964–66), Part 6, 1965. M. R. Cohen and I. E. Drabkin (1958). *A Source Book in Greek Science.* Cambridge: Harvard University Press.
36. Farrington, p. 191.
37. L. White (1968). *Dynamo and Virgin Reconsidered.* Cambridge: MIT Press, p. 65.
38. Theophrastus, *Athenaeus,* 511d.

References

Adcock, F. E. (1957). *The Greek and Macedonian Art of War.* Berkeley: University of California Press.

Africa, T. W. (1968). *Science and the State in Greece and Rome.* New York: Wiley.

Aristotle. *Historia Animalium.* Translated by T. Taylor (1809). London: Wilks.

Aristotle. *De Partibus Animalium.* Translated by A. L. Peck (1937). London: Heinemann.

Aristotle. *De Generatione Animalium.* Translated by A. L. Peck (1943). London: Heinemann.

Bailey, C. (1928). *The Greek Anatomists and Epicurus.* Oxford: Clarendon.

Breasted, J. H. (1935). *Ancient Times.* 2nd ed. Boston: Ginn.

Brock, A. J. (1929). *Greek Medicine.* New York: Dutton.

Burford, A. (1972). *Craftsmen in Greek and Roman Society.* Ithaca: Cornell University Press.

Burnett, J. (1930). *Early Greek Philosophy.* 4th Ed. London: Black.

Casson, L. (1974). *Travel in the Ancient World.* Toronto: Hakkert.

Cohen, M. R. and I. E. Drabkin (1948). *A Sourcebook in Greek Science.* New York: McGraw-Hill.

Cole, T. (1967). *Democritus and the Sources of Greek Anthropology.* Cleveland: Western Reserve University Press.

Coonen, L. P. (1957). Theophrastus revisited. *Centennial Review* 1: 404–418.

Cornford, F. M., ed. (1937). *Plato's Cosmology: The Timaeus of Plato.* London: Paul.

Cornford, F. M. (1932). *Before and After Socrates.* Cambridge: Cambridge University Press.

Delbrück, M. (1971). Aristotle–totle–totle, in *Of Microbes and Life.* Edited by J. Monod and Ernest Borek. New York: Columbia University Press, pp. 50–55.

Diogenes Laertius. *Lives of the Philosophers.* Translated and edited by A. R. Caponigri (1969). Chicago: Regnery.

Dodds, E. R. (1951). *The Greeks and the Irrational.* Berkeley: University of California Press.

Edelstein, E. J. (1945). *Asclepius.* Baltimore: Johns Hopkins Press.

Edelstein, L. (1967). *The Idea of Progress in Classical Antiquity.* Baltimore: Johns Hopkins Press.

Edelstein, L. (1967). *Ancient Medicine: Selected Papers of Ludwig Edelstein.* Edited by L. Temkin and C. L. Temkin. Baltimore: Johns Hopkins Press.

Eichnolz, D. E., ed. (1965). *De Lapidibus.* Oxford: Clarendon.

Farrington, B. (1965). *Science and Politics in the Ancient World.* New York: Barnes and Noble.

Farrington, B. (1969). *Greek Science, Its Meaning for Us.* Baltimore: Penguin.

Finley, M. I. (1972). *The Ancient Economy.* Berkeley: University of California Press.

Finley, M. I. (1970). *Early Greece: The Bronze and Archaic Age.* New York: Norton.

Finley, M. I., ed. (1960). *Slavery in Classical Antiquity.* Cambridge, England: Heffer.

Furley, D. J. (1967). *Two Studies in the Greek Atomists.* Princeton: Princeton University Press.

Furley, D. J., and Allen, R. E. (1975). *Studies in Presocratic Philosophy.* New York: Humanities.

Guthrie, W. K. C. (1962–69). *A History of Greek Philosophy.* 3 vols. Cambridge: Cambridge University Press.

Harvey-Gibson, R. J. (1919). *Outlines of the History of Botany.* London: Black.

Hawkes, E. (1928). *The Pioneers of Plant Study.* London: Sheldon.

Heath, T. (1921). *A History of Greek Mathematics.* 2 vols. Oxford: Clarendon.

Hippocrates. *The Genuine Works of Hippocrates.* Translated by F. Adams (1939). Baltimore: Williams and Wilkins.

Hughes, J. D. (1975). *Ecology in Ancient Civilizations.* Alberquerque: University of New Mexico Press.

Jaeger, W. (1948). *Aristotle.* 2nd Ed. London: Oxford University Press.

Kirk, G. S. and Raven, J. E. (1957). *The Presocratic Philosophers: A Critical History with a Selection of Texts.* Cambridge: Cambridge University Press.

Lewes, G. H. (1864). *Aristotle.* London: Smith, Elder.

Lloyd, G. E. R. (1970). *Early Greek Science: Thales to Aristotle.* New York: Norton.

Lloyd, G. E. R. (1973). *Greek Science After Aristotle.* New York: Norton.

Lovejoy, A. O. (1936). *The Great Chain of Being: A Study of the History of an Idea.* Cambridge, Mass.: Harvard University Press.

Majno, G. (1975). *The Healing Hand: Man and Wound in the Ancient World.* Cambridge: Harvard University Press.

McKeon, R., ed. (1941). *The Basic Works of Aristotle.* 4th Ed. New York: Random House.

Mosse, C. (1970). *The Ancient World at Work.* New York: Norton.

Needham, J. (1934). *A History of Embryology.* Cambridge: Cambridge University Press.

Neugebauer, O. (1957). *The Exact Sciences in Antiquity.* Providence: Brown University Press.

O'Brien, D. (1969). *Empedocles Cosmic Cycle: A Reconstruction from the Fragments and Secondary Sources.* London: Cambridge University Press.

Osborn, H. F. (1894). *From Greeks to Darwin.* New York: Macmillan.

Parsons, Edward A. (1952). *The Alexandrian Library.* Amsterdam: The American Elsevier.

Phillips, E. D. (1973). *Aspects of Greek Medicine.* New York: St. Martin.

Reymond, A. (1927). *History of Sciences in Greco-Roman Antiquity.* New York: Dutton.

Romanes, G. J. (1891). Aristotle as a naturalist. *Contemporary Review 59:* 275–289.

Ross, W. D., ed. (1952). *Aristotle Works.* 12 vols. Oxford: Clarendon.

Ross, W. D. (1956). *Aristotle.* 5th Ed. New York: Barnes and Noble.

Ross, W. D. and Forbes, F. H. (1929). *Metaphysics of Theophrastus.* Oxford: Clarendon.

Sambursky, S. (1956). *The Physical World of the Greeks.* Translated by Merton Dogut. London: Routledge and Paul.

Sandys, J. (1903). *History of Classical Scholarship.* Vol. 1. Cambridge: Cambridge University Press.

Singer, C. (1957). *A Short History of Anatomy from the Greeks to Harvey.* New York: Dover.

Taylor, A. E. (1955). *Aristotle.* New York: Dover.

Taylor, H. O. (1963). *Greek Biology and Medicine.* New York: Cooper Square.

Thompson, D'A. (1913). *On Aristotle as a Biologist.* Oxford: Clarendon.

Thompson, D'A. (1940). *Science and the Classics.* London: Oxford University Press.

Theophrastus. *Enquiry into Plants.* Translated by Sir Arthur Holt (1916). 2 vols. Loeb edition, Cambridge, Mass.: Harvard University Press.

Toulmin, S. and Goodfield, J. (1962). *The Architecture of Matter.* New York: Harper and Row.

White, K. D. (1970). *Roman Farming.* Ithaca: Cornell University Press.

White, L. (1968). *Dynamo and Virgin Reconsidered: Essays in the Dynamism of Western Culture.* Cambridge: MIT Press.

Withington, E. T. (1921). The Asclepiadea and the Priests of Asclepius, in *Studies in the History and Method of Science.* Edited by C. Singer. Oxford: Clarendon, Vol. 2, pp. 192–205.

3

THE GREEK LEGACY

The Alexandrian Age

As the vitality of the Lyceum declined, a new center of scholarship was called into being by the Ptolemies of Alexandria, heirs to the dynamic spirit of Alexander the Great. The rulers of Alexandria recognized that science could be made to serve government and exerted their influence to transfer all that was valuable from the Lyceum to the new Museum of Alexandria.

The ascent of Macedonia, once a poor and backward province, began in earnest during the reign of Philip (393–336 B.C.) son of Maynatas. In 359 B.C. Philip became ruler of Macedon and worked to build a professional army that could conquer the world. With his great energy, originality, and initiative it took Philip 22 years to pacify his own country, conquer his barbarian neighbors and the city-states of Greece. Skilled in the arts of war and diplomacy, he came close to fulfilling his goal—the conquest and unification of the known world. This clever and powerful man was cut down by an obscure assassin who may have been an old boyfriend of Philip. It is uncertain whether the assassin acted on his own, for Persian or Athenian agents, or for Queen Olympias and her son Alexander (356–323 B.C.).

Although Alexander was only 20 when his father died, he efficiently set about securing his place in history. He disposed of three possible rival claimants to the throne and then began his famous military career. When Alexander attacked the Persian Empire in 334 B.C. he led a force of 40,000 men and 6000 horses. He also brought along scholars and historians, botanists and geographers, engineers, and surveyors. Thus his campaigns produced a wealth of information on the conquered areas, their resources, natural history and geography. Perhaps the sheer bulk of new information may have contributed to a change in the outlook of Greek natural scientists from speculative universal concerns to more specific and practical matters.

Having liberated Egypt, Alexander was hailed as Pharaoh. The Oracle of Ammon proclaimed him a god and the son of a god. Alexander defeated Darius, Great King of the Persian Empire, conquered Babylonia, and turned to India. Shortly after his marriage to the daughter of Darius, Alexander, not yet 33 years of age, fell ill with a fever and died. Alexander retained the status of godhood in certain parts of the Empire. In his Asiatic and Egyptian provinces he was worshiped as a god and many myths grew around him. His body was stolen from Babylon by Ptolemy and enshrined in Alexandria. A thousand years later, the Moslems regarded him as another prophet. In the Roman Empire, Augustus Caesar, for political reasons, declared that Alexander was the thirteenth god of Olympus.

After the death of Alexander, his generals struggled for control of the army and conquered territories. Finally three of the successors gained control and divided the Empire. These were Antigonus Cyclops, the One-eyed (382–301 B.C.), who took the European part, Macedonia and Greece; Seleucus Nicator (356–281 B.C.) took the Asiatic part; Ptolemy Soter (367–283 B.C.) ruled over Egypt. His dynasty lasted for about 250 years. This dynasty had the most lasting effect on history, particularly the history of science.

The Ptolemies of Alexandria

Ptolemy, like Alexander, had studied under Aristotle. A literate man, he provided the only first-hand record of the life of his former commander and king. He took the richest part of the Empire for himself and assumed the name Soter, meaning saviour. Ptolemy continued the tradition of private education by calling Strato to tutor his son, Philadelphus. The son succeeded Ptolemy I two years before his death and ruled from 285 to 247 B.C. The city of Alexandria in Egypt, symbolized by the lighthouse, one of the seven wonders of the world, became the capital of the Ptolemies. Here they had the aid of the native Egyptian administration and priesthood who were able to add the achievements of Greek culture to the already considerable learning of Egypt. Alexandria became the intellectual capital of the world. Scholars, poets, and philosophers came from all parts of the Empire, especially from Athens.

The Ptolemies founded and supported development of the House of Muses (or Museum), and the great Library of Alexandria.[1] While the Museum was nominally under the direction of a high priest, its real purpose was as a research institute. As a place for teaching, it was modeled after the Lyceum, but dwarfed that institution in the size, if not the quality, of faculty and student body. Little is known for certain about the Museum because no contemporary account of its is known to have been written or survived. According to tradition and later descriptions the facilities and surroundings were magnificent, including a promenade, a place with seats for conferences, and a great hall where the scholars had their meals in common. The meals were free while the salary of the professors was provided by the king and was tax exempt. Theocritus called Ptolemy "the best paymaster a free man can have."[2] The Museum contained rooms for research, lectures, private and group study, an observatory, zoo, botanical garden, and dissecting rooms.

The magnificent Library had about half a million rolls. The librarian seems to have had a major influence on the direction of research and teaching. Many precious manuscripts were collected and recorded in this sanctuary. The library of Aristotle and perhaps that of the Hippocratic school came to rest here. The librarians bought, copied, or stole all the literature available. Books were collected by purchase or force. Travellers had to declare any books they had in their possession. If of interest, the books were seized and copied. Even the concept of the "text-book" seems to have developed in Alexandria. Experts in various fields systematized their knowledge in the format which proceeds from general principles through well-established concepts on to the latest ideas at the frontiers of research.

Illustrative of the practical emphasis at Alexandria are many engineering advances in hydraulics and pneumatics and inventions such as the valve, pump, and screw. Still, in a slave-based society there was little incentive to turn invention toward labor-saving devices. Most inventions from Alexandria seem to have stayed at the level of toys—except for development of weapons, temple magic, and, to a lesser extent, medical instruments such as the syringe. The Ptolemies did not totally devote themselves to encouraging science and engineering and scholarship, but were very much interested in using religion to enhance their power. This is a fact that should not be forgotten by overemphasis on the literary and scientific accomplishments of the Museum. Science became the servant of religion. Although Strato had declared that he did not need the help of the gods to create a world, the science he created was to help the gods. Hero of Alexandria used pneumatics to create "miracles" in the temples for the ignorant, but religious masses.

The most creative period at the Museum was its first two hundred years. During and after its most brilliant phase the Museum served as an efficient and orderly way of producing engineers, physicians, geographers, astronomers, and mathematicians. Scientists and technicians were needed by the State, which was after all a Greek enclave in the timeless Egyptian world.

Unfortunately little of the work at the Museum and Library was to survive. The exact sequence of events in the destruction of the Library is uncertain and controversial.[3] During the reign of Cleopatra, the last of the Ptolemies, the Library is said to have contained 700,000 volumes. When Julius Caesar visited Alexandria in 48 B.C. with more than 3000 soldiers, Cleopatra gave him many precious works from the Library as gifts. During the riots against the Romans, fire destroyed as many as 40,000 manuscripts. The losses that occurred in 48 B.C. were compounded by later Christian attacks on pagan institutions. Theophilus, Patriarch of Alexandria (385–412 A.D.) led a mob against the Temple which caused the destruction of many more books. According to legend, in 415 the last scholar at the Museum, a woman mathematician named Hypatia, was dragged out by Christian mobs and beaten to death in a church. Again, in the tradition of religious fervor and excess, the Arabs continued the wrecking efforts when they conquered the city in 646 A.D.

Anatomical Researches at Alexandria

Many sciences flourished at Alexandria, although research was primarily oriented toward specialized science and its practical application. Because progress in medicine was encouraged, great advances were made in anatomical researches. The salient factor here was the encouragement of dissections of human beings. For the first time in history dissections were methodically executed. Perhaps the Egyptian tradition of cutting open the body and removing certain organs as part of the embalming ritual helped overcome the Greek antipathy to human dissection. Although the practice seems to have been discontinued by the end of the second century B.C., anatomical researches did not cease since living and dead animals were still dissected.

Herophilus (fl. 300 B.C.)

Sophisticated anatomical studies were carried out by two great rivals: Herophilus, the anatomist, and Erasistratus, who called himself the physiologist. Unfortunately, the writings of both these scientists were lost and we only know of them because later authors quoted from and commented on their work. When the church fathers, Tertullian and St. Augustine, wanted evidence of the heinous acts committed by the pagans, they pointed to the notorious Herophilus and accused him of torturing to death 600 human beings. Celsus, a Roman writer of the first century B.C., says that the government awarded to Herophilus and Erasistratus condemned criminals for vivisection. Some historians accept the allegations, while others remain skeptical.[4] The extent or existence of the practice of human vivisection at Alexandria is still controversial, since the best evidence—the writings of the Alexandrian anatomists—is lost and only the diatribes of their enemies have been preserved. It is quite certain that the Ptolemaic rulers placed the necessary materials at the disposal of the anatomists—criminals condemned to death. The anatomists were radicals who despised the traditional Greek fear of dissecting human bodies. It was later said that Herophilus took advantage of his opportunities to carry out studies on living men and examine their internal organs in the living state. However, simply carrying out public anatomies must have been enough to shock his superstitious contemporaries.

Except for the possibility that Herophilus studied with Praxagoras of Cos, little is known of his life although he is said to have come from Chalcedon in Blithynia. He wrote several books including *On Anatomy, On the Eyes,* and a handbook for midwives. He included women among his pupils, most famous of whom was Agnodice, who had to practice medicine disguised as a man. According to later writers, Herophilus described the brain as the center of the nervous system, the structure of the eye, digestive tract, and particularly noted the variability in the shape of the liver. In his investigation of the circulatory system, he noted the difference between the arteries, with their strong pulsating walls, and the veins, with their relatively weak walls. Contrary to the popular belief that veins carry blood and arteries carry air, he stated that both kinds of vessels carried blood. Although he noticed that the pulse differed in health and disease, he did not relate this to the heart beat. He also tried to quantitate the pulse with the water clock which had been invented at Alexandria. Tendons and nerves, which had not been differentiated by Aristotle, were characterized.

Herophilus disputed Aristotle's claim that the heart was the most important organ of the body and the seat of intelligence. Although Aristotle gave at least ten reasons for his position, he was quite wrong, as Herophilus showed through his study of the nervous system. He seems to have gotten a good overall picture of the nervous system, including the connection between brain, spinal cord, and nerves, even noting a distinction between the motor and sensory nerves.

For medical philosophy and practice, Herophilus relied on the Hippocratic tradition of humoral pathology. After all, it was at Alexandria that the *Corpus Hippocraticum* was assembled and edited. He seems to have preferred a theory of four faculties that moved the human body to the ancient concept of body juices. His four life-guiding faculties were a nourishing faculty in the liver and digestive organs, a warming power in the heart, a sensitive or perceptive faculty in the nerves, and a thinking force in the brain. In clinical practice he seems to have favored more active intervention by the physician than recommended in the oldest Hippocratic works. He favored more bloodletting than Hippocrates and a system of complex pharmaceuticals. The clinical practitioner should be familiar with dietetics, medicines, surgery, and obstetrics.

Erasistratus (ca. 310–250 B.C.)

Erasistratus extended the work of Herophilus, and seems to have applied Alexandrian mechanics to his medical theories. He may have authored 62 books, but all are lost. Although Herophilus seems to have been a masterful anatomist, Erasistratus apparently went beyond anatomy to physiology and experimentation—as to be expected of a student of Strato. Little is known of his life, except the tradition that having diagnosed his own illness as an incurable cancer he committed suicide rather than suffer inexorable decline. Galen (ca. 130–200 A.D.) wrote two books against Erasistratus and criticized him wherever possible.

Erasistratus rejected the Hippocratic philosophy of medicine, and in most things was a follower of Aristotle. According to one tradition, he may have been his nephew. He is said to have been a gifted practitioner who rejected the idea that a general knowledge of the body and its functioning in health was a necessary part of the practicing physician's equipment. But he believed in studying pathological anatomy as a key to the local anatomical causes of diseases. He tried to replace humoral pathology with a pathology of solids. This gave him a new key to a more specialized and problem-oriented program of anatomical research. Nevertheless, humoral pathology remained the dominant theory well into the seventeenth century.

Some of the practical concepts of his group were very valuable, particularly the invention of the ligature for torn blood vessels. In some cases he wisely recommended simple remedies and hygienic living, but another theory of Erasistratus was not so beneficial to his patients. He ascribed the cause of all diseases to "plethora," which meant an excess of blood from undigested food material. This stopped the circulation of pneuma (vital spirit) in the large arteries. The local excess of blood caused damage to surrounding tissue because blood from overloaded veins (which were designed to carry blood) spilled over into the arteries (which were designed to carry pneuma). Erasistratus believed it was therapeutic to interfere with the supply of blood at its point of origin by starvation. In addition to general starvation, he tried "local" starvation by trapping blood in other parts of the body, until the sick part had used up its plethora, by the simple expedient of throwing ligatures around the roots of the limbs.

Erasistratus gave a description of the liver and its gall ducts that was better than any previous work, despite the long history of liver inspection for divination. His work on the lacteals was an extension of the efforts of Herophilus, which was not improved upon until the work of Gasparo Aselli (1581–1626). He gave the first really accurate description of the heart, including the semilunar, tricuspid, and bicuspid valves. In his anatomical investigations, he used what Guido Majno calls the "Greek microscope," a combination of genius and wishful thinking.[5] Erasistratus traced the veins, arteries, and nerves to the finest subdivision visible to the naked eye, but was sure that they were further subdivided beyond the limits of vision.

Perhaps because of his theory that disease was caused by local excess of blood, Erasistratus paid particular attention to the heart, veins, and arteries. Unfortunately, his major conclusions were wrong. Theories as well as his own observations mislead him as to the role of veins and arteries. Ironically, dissections seemed to provide proof for the prevailing theory that veins contain blood and arteries contain air. Outstanding in Erasistratus' concept of the heart is his vision of that organ as a pump and the "membranes" (i.e., valves) as the flap valves of a one way pump. We might see this insight as the fruitful outcome of interaction between different disciplines if we imagine Erasistratus and an inventor talking over their research at the great dining room in the Museum and putting their specialties together as a new mechanical analogy for human physiology. But this idea did not accord with prevailing notions of the action of the heart, veins, and arteries. The veins and arteries were seen as completely independent systems of dead end canals. Through the proper vessels, blood and air seeped slowly to the periphery of the body where they were used up.

Freeing himself partially from the assumptions of his contemporaries, Erasistratus groped towards a true understanding of the circulatory system. It is easy to suggest simple experiments Erasistratus could have done to check his ideas. After all he conceived of the heart as a pump and imagined an invisible connection between arteries and veins, although he was wrong about the direction of flow. Why not inject a syringe (also an Alexandrian invention) full of ink into the veins and see if it came back through the arteries? Apparently on the brink of a revolutionary new theory, Erasistratus was still the prisoner of stifling misconceptions. The arteries and veins were regarded as two quite separate systems. Supposedly the arteries arose from the heart and the veins from the liver. Obviously nature intended that they be separate because when blood did spill over it caused inflammation and disease. Thus the connection between arterial and venous system was invoked to explain pathological rather than normal physiological phenomena. If a living animal was injured so that an artery was cut the air in it escaped and pulled some blood after it. Nature's horror of vacuums, as taught by Strato, was incorporated into Erasistratus physiology. Erasistratus retreated from Herophilus' conviction that both veins and arteries normally carried blood. According to tradition, Erasistratus and Strato influenced each other profoundly. Thus the anatomist's acceptance of Strato's theory of the vacuum circumscribed his contribution to physiological theory.

The theory that Erasistratus wove by combining his observations with the pneumatic theory of Strato and the atomic theory of Democritus was ingenious but fundamentally wrong. Although he called himself the "rationalist" or "materialist" he could not discard the idea of animating spirits. According to Erasistratus, the body was composed of atoms and animated by warmth that came from outside. Life processes were dependent on blood and pneuma. The supply of pneuma is constantly replenished by respiration. In the body were two types of pneuma; the vital pneuma was carried in the arteries and regulated vegetative processes. Some of the vital spirit got to the brain and was changed into animal spirits or soul-pneuma. Animal spirits were responsible for movements and sensations and were carried by the nerves which were described as hollow tubes. When animal spirits rushed into muscles, they caused distension which resulted in shortening of the muscle and thus movement. These ideas presage those of seventeenth century physiologists.

Inspired by Strato's experimental methods, Erasistratus tried to solve biological problems through quantitative and experimental means. In one experiment on what might be called "metabolic problems" he put a bird into a pot and kept a record of its weight (and its excrement). He found a progressive weight loss between feedings. This led him to conclude that some invisible emanation was being lost by vital processes.

The Decline of the Alexandrian Tradition

The Museum had created an environment for science and scholarship such as had never existed before. The great Library had created the basic tools of scholarship—a storehouse of knowledge, a place for the accurate copying of texts, a tradition of textual criticism and reverence for the written word as a gift from the past and a legacy for the future. Such brilliant periods seem all too short and isolated, like oases in time. The great period of the Museum was but two hundred years. Much of the subsequent deterioration can be blamed on politics, but scientists and scholars themselves seem to have helped pull down the structural supports of the houses of learning.

Even during the great period, rivalry between the followers of Herophilus and the disciples of Erasistratus was intense. Both sides impugned the work, methods, conclusions, and even character of members of the other group. When the atmosphere in Alexandria became less favorable to science, it seemed but common sense that if scientists called each other worthless, perhaps they really were. The tension that always exists in medicine between the purely scientific spirit and the clinical and healing spirit, led to degeneration from the greatness of Hippocrates, Herophilus, and Erasistratus. Critics charged that researches into anatomy and the nature of diseases were frivolous pursuits and distractions from treating patients as people.

The later Ptolemies were not scholarly and enlightened. Nordenskiold called them "degenerate scoundrels who neglected the interests of learning as they neglected all their other duties."[6] The later Ptolemies seem to have devoted themselves to persecution of the Greek element in Alexandria. The ninth Ptolemy, who reigned from 146-117 B.C., called himself Benefactor II. The Alexandrian Greeks called him Malefactor or Fat Belly. The historian Polybius visited Alexandria during this period. He reported that the Greeks had been corrupted and mongrelized, and that Fat Belly had massacred and exiled many of them. Other contemporary reports support this gloomy

picture. Greeks and foreign scholars were unwelcome in Alexandria and
many of them created new centers of learning in Greek areas.

Finally, the worst of fates fell upon the Alexandrian scientists, they were
persecuted, their grants cut off, and they had to turn to teaching to eke out a
living in new places. Only the practical aspects of science continued to be
respected. After Caesar conquered Egypt, Alexandria was reduced to the
status of a provincial town in the great Roman Empire and the Museum and
Library were nearly destroyed.

The Roman World

One of the major accomplishments of the Alexandrian period was the diffusion
of Greek culture throughout the conquered lands. The accomplishments of
the Greeks in literature, art and science, philosophy, history, religion, government, and war were no longer the exclusive property of Athens or Alexandria.
While the later Ptolemies were purging and despoiling the great Museum of
dissident scholars and scientists, the balance of political power was inexorably
shifting to the West.

Even as the first Ptolemies were establishing Greek rule over Egypt, the ob-
scure city of Rome was initiating her conquest and unification of Italy. Like
the Greeks, the Romans had the advantage of coming to civilization in the
Iron Age. Italy had been colonized by successive waves of invaders from the
time of the paleolithic expansion up to the Iron Age. Narrowly defeated by
Pyrrhus of Epirus in 279 B.C., in 275 B.C. the Romans were able to defeat
Pyrrhus, the king who had expected to become the Alexander of the West.
During the second century B.C. the Romans overcame the successors of
Alexander in Macedon and Syria. Soon the Romans added the Greek cities of
Asia Minor and the mainland to their collection. During the reign of Augustus,
Egypt, too, was incorporated into the Roman Empire.

Although the power was to remain Roman, the culture of this new era was
to be a combination of Greek and Roman elements. The Romans resembled
Sparta more than Athens, but did appreciate higher Greek culture to the extent of assimilating what suited their purposes. All Romans who aspired to
culture had to learn Greek while Greek scholars, in order to survive in the new
world, had to learn Latin. The resultant cultural hybrid was complex and
vigorous, if less original than its Greek parent.

Although conspicuous in practical affairs, such as warfare, medicine, hygiene, agriculture, administration, and governmental organization, the Romans
made but a modest contribution to biology and medicine as scientific enterprises. In intellectual and scientific matters the Romans appropriated and disseminated Greek culture. Since the major Roman preoccupation was empire

building, practical matters were more important than pure science and scholarship. Rome honored soldiers rather than philosophers. Indeed, some Romans strenuously objected to contamination by Greek culture, but by the first century B.C. Rome had accommodated itself to such assimilation. Still, Roman nobles were more interested in the arts of persuasion and war than in scientific inquiry. They rejected the ideal life of Greek aristocrats which was basically one of "elegant indolence." If Roman upperclass members wanted to get some scientific education, they could purchase Greek tutors in the slave market for even less than the price of a good chef.

Rome did produce some men who were interested in collecting and recording all kinds of information, even if they were not engaged in original research themselves. Such men included Lucretius, Pliny, Celsus, and Dioscorides. For some Roman authors, the motive for writing was to show that Roman work could be superior to that of the Greeks in chosen fields. This was the impulse that drove Cato the Censor (234–149 B.C.) to write on medicine and agriculture. To later readers it seems obvious he failed in his mission because his "medicine" is heavily imbued with magical formulas that the Hippocratic doctors had long since rejected. Even the herbal remedies were to be used in conjunction with magic. But this did not matter to Cato who boasted that Rome was healthy without doctors.

One defect of the Graeco-Roman phase was the failure to integrate the practical and the theoretical. For example, even though the Romans came to value Greek medicine, the practice of human dissection as a basic part of medical training was rejected by Rome. Roman writers tended to be either clearly theoretical, like Lucritius, or empirical, like Pliny.

Lucretius (96?–55? B.C.)

In his great work, *The Nature of Things,* Lucretius preserved and elaborated the atomic theory of Democritus. Although the Christian Church disapproved of his materialistic views, during the Renaissance his work was highly respected. Unlike Aristotle, Lucretius did not believe the world was the finite, eternal creation of a perfect God. Rather, he thought that the earth was fated to change and eventually to perish. Like the pre-Socratics, he developed a natural explanation of the cosmos and living things. Lucretius believed that even the soul and man's dreams could be explained in terms of atoms. Other explanations were relics of superstition unworthy of intelligent beings.

To understand the importance of Lucretius' work, one has to go back to the religious and philosophical systems that were competing for primacy among Romans with pretentions to philosophical sophistication. Lucretius, like Epicurus (342?–270 B.C.), used the atomist philosophy to combat religion and superstition. Neither was primarily interested in the purely scientific aspects of natural philosophy. Although the Epicureans taught belief in the gods, their concept of nature and society was as naturalistic as possible. The gods were only involved in directing man's inner spiritual life. Supernatural agencies were not involved in nature or the unfolding of civilization and the fortunes of war.

Lucretius, who was a follower of Epicurus, realized that men were tormented by many fears, but that superstitious fears were deliberately manipulated by politicians to manage the ignorant. *On Nature* was thus a deliberate political statement as well as a work that had the virtue of preserving much of Greek science. Lucretius tells his readers that his philosophy will make them secure against the official religious mythology. A true philosophy of nature is a difficult thing to obtain, he admits, but it will banish superstitious fears.

Illustrative of his naturalistic approach is the account of the origin and progress of civilization. Lucretius tells us that the earth first produced vegetable life and then animals. The birds which hatched from eggs were first, followed by animals born from the womb of the earth. At first the earth could feed and clothe her children and make the climate pleasant for them. As the earth aged and lost her powers, things began to multiply by their own powers of generation. During this transition, monsters that she had produced before became extinct because they could not feed, protect, or propagate themselves. Man, a product of nature like the other creatures, was first a simple but strong and long-lived food-gatherer. Civilization began when man learned to control fire. If human development was essentially a natural process, why were religion and superstition so prominent in present societies? These evils flourished because men who did not have a true philosophy of nature lived with a tormenting confusion of ideas.

Fundamentally pessimistic, Lucretius advocated a return to a simpler mode of life and saw no promise in technological advances. He saw civilization inexorably declining and blamed the religion of Rome. Even Cicero (106–43 B.C.) in his work *On Divination* objected to the religious milieu of Rome. The Romans were in the habit of consulting the will of the gods for all things. The gods gave their answers by augury, haruspicy, astrology, and other ancient devices. Cicero who elsewhere defended useful established institutions, said he would be happy to serve his country by tearing all this superstition away. Certainly Lucretius as well as Cicero knew that religion served many practical purposes in the public life of Rome. Leaders of the Roman Republic exploited the "messages of the gods" for their own purposes even as they proclaimed their piety. Omens were interpreted in terms of political needs. Q. Fabius Maximus clearly stated that "whatever serves the interests of the Republic had good auspices, and whatever is against the public interest is inauspicious."[7] Even the calendar was manipulated by the use of religion to aid politics. Because political activity was prohibited on festivals or inauspicious days, the calendar was regularly juggled to prevent unwanted legislation.

The philosophy of the Stoics fit better with Roman culture than did the Epicurian philosophy. The Stoics believed that all objects in nature were alive and growing. Every entity had developed from a "seed" which contained the form or plan that determined the characteristics of the mature object. The plan was the soul or spirit that was animated and sustained by the pneuma, the universal spirit of nature.

Pliny the Elder (23–79 A.D.)

Pliny was the most famous Roman naturalist and, along with Varro and Celsus, one of the three great encyclopedists produced by Rome. Pliny's contribution was the *Natural History* which during the Middle Ages seemed to be a storehouse of all possible information about the natural world. The work, a compilation in 37 books, deals with all the natural sciences and all the human arts. Pliny took his facts and observations from some 2000 previous works, written by 146 authors of Roman origin, and 326 Greeks. Judgments of the author and his work have fluctuated from unqualified veneration, to moderate appreciation, to outright ridicule.

Pliny's life, like his work, is worth reviewing as it gives us the spirit of his time distilled through the experience of the very model of a Roman citizen. He lived during the height of Roman power and came from a family of officials. Pliny was born in Novum Comum (now Como) in northern Italy and practiced law before joining the military service. If travel provides education, then Pliny was well educated. He travelled through Germany, Gaul, Spain, and Africa. On good terms with Vespasian from his days in the military and his successor, Titus, Pliny had many official duties bestowed upon him.

Curiosity and efficiency seem to have been his personal deities. He never wasted a minute that could be employed in reading and writing. He even worked while travelling by carriage rather than waste time on horseback. Legend has it that his secretary even read to him while he took his bath. It was his curiosity that caused his death. When Mount Vesuvius erupted in 79 A.D. demolishing Pompeii and Herculaneum, Pliny insisted on studying the awesome phenomenon himself. Described by his nephew as being naturally weak in the chest, he was overcome by the poisonous fumes of the volcano and died without sustaining any external wounds.

Pliny's *Natural History* was described by his nephew as: "A work remarkable for its comprehensiveness and erudition, and not less varied than Nature herself."[8] Until the sixteenth century when new books about animals were written, Pliny remained the best source for natural history. Even in the eighteenth century he was greatly admired by the French naturalist Buffon.

True critical standards were not applied to Pliny's work until the nineteenth century. Pliny often recorded bizarre stories which he obviously could not bear to leave out of his books, but he noted that these were the opinions of other people. Certainly Pliny exercised more zeal for collecting ideas and information than critical judgment in assembling his remarkable encyclopedia. Pliny emphasized the Roman viewpoint that knowledge and science could be useful to man either directly or as a moral lesson. This anthropomorphic teleology helped to make the work acceptable to the Church during the Middle Ages.

There is some suggestion of a plan beneath the great profusion of facts and ideas. The first book, which was written last, is a dedication to the Emperor Titus, a general preface, table of contents, and a list of authors used. Book 2 discusses the heavens and the earth. In Books 3 to 6 he is concerned with geography and ethnography. The next part of the work is on what we would call natural history, the story of animals, fish, insects, and birds. Books 12 to 19 are on botany and its useful aspects. The botanical section is quite extensive since it treats forestry, agriculture, and horticulture. Pliny tells us about the manufacture of useful products from the plant world, such as wine and oil. Books 20 to 27 deal with the medicinal uses of plants. Nature is so generous, wrote Pliny, that "even the very desert was made a drug store."[9] After a description of the medicinal uses of animals, Pliny treats the mineral substances.

When Pliny speaks of the vineyard he tells of the methods of pruning, the use of wine as tribute, and the good old days when women were not allowed to drink wine in Rome. He tells us that one matron was clubbed to death by her husband for drinking wine and that Romulus acquitted the husband on the charge of murder. This is illustrative of the way Pliny is able to air his opinions on every subject by what in another author would be a digression. Nothing in Pliny's work can rightly be termed a digression—the whole book is that.

Some of Pliny's recommendations for pest control are mixtures of religion, folk magic and "chemical methods."[10] Varro and Pliny both noted that an oak stake driven into the manure pile prevents snakes from breeding in the dung. To protect trees against the ravages of caterpillars, Pliny says to touch the tops of trees with gall of a green lizard. If that does not work, a crayfish hung in the middle of the garden or the skull of a female horse or ass on a stake in the middle of the garden will chase away the pests. Mildew can be prevented from ruining the wheat by placing branches of laurel in the ground near the growing wheat. This will attract the mildew away from the grain. A toad carried around the field at night and buried in a pot in the middle of the field will prevent diseases of millet and drive away sparrows and worms. Pliny warned the farmer to dig up the pot before sowing the

field or the land would turn sour. Not all of Pliny's pest control methods
sound quite so magical and useless. For example, he recommended soaking
seeds in wine to prevent certain fungal diseases.

Pliny begins his description of animal life with man, emphasizing the point
that all other things seem to have been produced by nature for his sake. His
description of Man included what were later to be described as the "wonder
people." These stories of strange races, accounts of "marvelous" or "monstrous" births were not invented by Pliny, but were collected from previous
accounts. There are descriptions of the Arimaspi, which have only one eye,
the Abarimon whose feet are on backwards, the Androgyni who are half man
and half woman, and many others. Although Pliny never consigns even the
wildest stories to the category of fraud, he sometimes notes that certain oddities, such as claims that people have lived to 800 years, might be based on
misunderstandings or different methods of reckoning.

There is no particular order discernible in his description of the animals;
he includes the unicorn and the phoenix with as much disinterested scholarship as the lion and the elephant. In general, Pliny considers the largest and
most remarkable animals first. Thus the elephant is described as the largest
of land animals and the most intelligent, even having the virtues of honesty,
wisdom, and justice, along with a religious respect for the stars and a veneration for the sun and the moon. Pliny is also capable of detailed descriptions
of imaginary creatures such as the unicorn which in his version was a very
fierce animal with the head of the stag, the feet of the elephant, the tail of
the boar, while the rest of the body resembled the horse. The creature has "a
deep bellow and a single black horn three feet long projecting from the middle
of the forehead. They say it is impossible to capture this creature alive."[11]
(We might add that it cannot be captured dead either.) In his accounts of
common and domestic animals, Pliny could be fairly accurate. When Renaissance writers revived the study of zoology, they took up where Pliny had left
off and until the eighteenth century continued with his methods.

Pliny was concerned with the degeneration of culture that he saw around
him. Rome had established an enormous empire, with great wealth and resources. Yet science seemed to have lost all its momentum. Pliny pondered
this paradox: Science and research had flourished during a period of constant
war between the Greek states, at a time when travel had been very dangerous.
Rome had imposed order and peace, travel was safe, and access to books was
better than ever before. Yet science was unquestionably at a standstill. Pliny
concluded that his contemporaries were concerned only with security and
wealth. The decline of science was seen by Pliny as only a part of a general
malaise in the Roman world.

Botany and Medicine in the Greco-Roman World

Biology and medicine achieved some minor successes on the mainland of Asia
Minor after the decline of the Alexandrian Museum. Because practical applications of these fields were appreciated by the Romans, such efforts were encouraged and remembered. For example, Crateuas (120–63 B.C.) who served
the king of Pontus as physician, is credited with the origin of botanical drawings. He was of course primarily interested in collecting and classifying plants
according to their medical value. Illustrations enhanced the usefulness of his
descriptions of plants. Dioscorides (fl. 60–77 A.D.) continued the development
initiated by Crateuas. Dioscorides illustrates the generalization that Roman
medicine was usually carried out by Greek physicians and scholars. Indeed,
Greek doctors were allowed to settle in Rome and were even granted Roman
citizenship in 46 B.C. by Caesar. Although the details of his life are not
known, Dioscorides seems to have studied at Tarsus and Alexandria. He became a military physician attached to the Roman forces in Asia under the
emperor Nero. This gave him the opportunity to travel widely and study
plants for their therapeutic value. One of the most famous herbals ever
written is the *De materia medica* of Dioscorides.

Later editions of the manuscript revised the arrangement used by Dioscorides and substituted an alphabetical arrangement. Some 500 plants are mentioned by Dioscorides. Perhaps 130 of them were also in the Hippocratic
collection. Dioscorides gave a description of each plant, an account of its
place of origin and habitat, and most important, the method of preparation of
the drug and its uses in medicine.

Celsus (fl. 14–37 A.D.)

Almost nothing is known about the life of Celsus and his exact relation to his
famous book, *De re medicina,* is controversial.[12] Sir Clifford Allbutt called
him the creator of scientific Latin, but historians are divided as to whether he
wrote or plagiarized one of the greatest medical works of antiquity. It is not
certain whether Celsus was a physician or not. If he was, he was not likely
to have been a "practicing physician." The Roman tradition of medicine was
one of self-help and assumed the responsibility of Roman landlords to care
for the sick on their own estates. Usually the women or skilled slaves served
as private healers for each estate. It is possible that a patrician would feel it
was desirable to know enough about medicine to supervise its practice, although he would not carry out the unpleasant menial work of the healer.

Although Celsus is supposed to have written on rhetoric, philosophy, juris-
prudence, the military arts, agriculture and medicine, which would make him

one of the great Roman encyclopedists, except for the section on medicine all his work was lost. Because he wrote in Latin, at the time a language of low prestige, rather than Greek, the language of medicine and scholarship, Celsus was ignored by the medical "establishment" of his times and remained in obscurity for some thirteen centuries. Not until 1426 and 1427 were two copies of Celsus found and recopied. The timing could not have been better since by virtue of being discovered at this time, the *De medicina* became the first medical work to be turned out on the new printing press. To the Renaissance scholars, Celsus represented a source of very pure Latin. Preserving the language and medical philosophy of the first century A.D., *De medicina* served as a standard for medical scholars until about a century ago.

Opinions on the relation of Celsus to the text vary. Some say he was a mere compiler, but the quality of the work and degree of critical judgment seem to militate against this assessment. Another guess is that the *De medicina* is simply a direct translation from a lost Greek original written in Rome. While some say that Celsus used only one original work in writing his own, others say he had access to all the medical works of antiquity, including many that have been lost. Contemporaries among Roman writers referred to Celsus as a man of but moderate talent.

De medicina consists of an introduction and eight books. The introduction, which mentions about 80 writers, is valuable as a history of ancient medicine. The author takes a balanced view of the rivalry between the methodists and empiricists of his time, but rejects inflexible approaches to medical science.

The book contains information on dietetics, the effects of foods on the body, the possible dangers of certain standard medical practices such as bleeding, purging, vomiting, massage and starvation. Although the author believes these practices are useful, they must be applied in moderation and only with careful consideration of the particular patient. There are good descriptions of fevers, madness, consumption, jaundice, palsy, and other diseases, and detailed diets for rheumatism and gout sufferers. "Acute" diseases are defined as conditions which "either finish a man quickly, or finish quickly themselves."[13] Most of the drugs listed are extracts of plants, but some animal products are also recommended.

Celsus emphasized the importance of anatomy for medical practice. He knew that the arteries contain blood under pressure and not air and that an inexperienced practitioner could endanger the patient in bloodletting because the veins were so close to arteries and nerves. The descriptions of surgery are particularly valuable and suggest that considerable progress had been made since the time of Hippocrates. The instruments available at the time are clearly described as are several difficult operations, including removal of arrowheads, goiter, hernia, cataract, and amputation. The four cardinal signs of inflammation are listed by Celsus: calor, rubor, tumor, and dolor (heat, redness, swelling, and pain).

Critical of human experimentation, Celsus is the chief witness against Herophilus and Erasistratus, relating that they did vivisections on criminals. Celsus
does not totally condemn human experimentation, but argues that accidents
often enough expose the internal organs of the body. These cases should be
put to advantage by the physician so that while striving for health the physician could turn the case into an opportunity to learn "in the course of a work
of mercy, what others would come to know by means of dire cruelty."[14]

Although only the medical section of Celsus' work survived we know that
he intended to cover several other practical fields: agriculture, rhetoric, and
warfare. This made for two sections on the physical life of man and two on
the life of the citizen.

Galen of Pergamon (ca. 130-200 A.D.)

Galen is the last of the great medical writers of antiquity. With his death the
creative period of Greek medicine comes to an end. He is a strange transitional figure, too scientific for the mystical milieu of his time, and too mystical
for later scientists. He was a combination of scientist, experimentalist, philosopher, and theologian. In authority he became second only to Hippocrates.

His works are little read today, mainly dismissed for their style, which has
been characterized as consistently boring. The published legacy of this pompous genius comprised some one hundred works under separate titles. By the
time he was 13 years of age, he had written three books. Excessively verbose,
Galen composed commentaries on Hippocrates that were longer than the
originals. Writing on many subjects, Galen demonstrated great energy and interest in all scientific, medical, philosophical, and religious currents of his
time. The only complete modern edition of his work fills 20 large volumes.
It is said to have taken 12 years to complete and almost as long to read
(Kuhn's German edition, 1821-33).

Galen was born in Pergamon, in Asia Minor near the Aegean coast opposite
the island of Lesbos. The city boasted of a library and temple of Asclepius
which made Pergamon the cultural rival of Alexandria. The Alexandrian
Ptolemies tried to arrest the intellectual growth of Pergamon by forbidding
the export of papyrus. In response to this threat, parchment was developed,
and the book as we know it now replaced the papyrus roll. The town was
also known as the site of intense Christian evangelism.

Galen remembered his father, Nikon the architect, as a kind-hearted man,
but described his mother as a quarrelsome woman, so violent she sometimes
bit the servants during her outbursts of temper. Nikon was told in a dream
that his son was destined to be a physician. In his early studies with the local
Platonists Galen learned the doctrines of the Epicureans, Stoics, Aristotle, and
Theophrastus. At seventeen, he took lessons from a doctor who specialized in
anatomy. When Nikon died, Galen who was only twenty, began his years of

wandering in pursuit of medical knowledge. His travels included Greece, Palestine, Phonecia, Cyprus, Crete, and Alexandria.

To some extent the spirit of scientific inquiry that animated Herophilus and Erasistratus had left its mark on Alexandria, but under the Romans dissection of human bodies was strictly forbidden. Thus Galen had to learn his anatomy from the study of pigs, goats, monkeys, and apes. Without much hesitation he applied his findings on animals to human beings. When he was 30, Galen returned to Pergamon just in time for the summer festival. His timing was excellent because there was no doctor available for the gladiators. Galen was so successful at treating the wounds of the gladiators that he was retained in this position, in addition to work at the Temple of Asclepius and a flourishing private practice. Soon Galen became restless at Pergamon and in 162 A.D. decided to test his skills in the Imperial City. Once again Galen was fortunate in finding an outlet for his skills.

In Rome, Galen soon enjoyed the adulation that he had become accustomed to in Pergamon. A well-known philosopher, ill with malaria, had been unsuccessfully treated by many of Rome's physicians, but Galen was able to cure him. He soon gained the reputation of being a wonder worker. He also gained the patronage of one of Caesar's chief administrators after curing his wife. His patron shared with him an enthusiasm for dissection. Four years later when he had gained a flourishing practice, and was earning great fees, Galen suddenly returned to Pergamon. He claimed that it was the hostility of less successful physicians that had driven him from Rome. However, it was noticed that his departure coincided with a vicious outbreak of an epidemic which may have been smallpox or typhus.

Not long after, Galen settled permanently in Rome at the request of the Emperor, Marcus Aurelius who honored him as the first of physicians and philosophers. He served as physician to Marcus Aurelius and the next three Roman emperors. During all these adventures, Galen kept himself busy with medical practice, anatomical research, lectures, disputations with other physicians, and the writing of books on many subjects. He claimed to have written 256 treatises of which 131 were medical (83 survive). His writings included philosophy, mathematics, grammar, law, anatomy, physiology, the pulse, hygiene, dietetics, pathology, therapeutics, pharmacy, commentaries on Hippocrates, and some autobiographical material. His writings reflected his scholarship as well as his own acute observations and experiments.

We know so much about Galen's work because he was one of his own best publicists—quite aware of the stature he was entitled to. Fearing that his own works were being corrupted by careless copyists and that lesser mortals were trying to pass off their inferior work as his, Galen composed a guide to the cautious reader called *On His Own Books* which provides a list and description of his genuine works.

Even though he was prevented from carrying out regular dissections of human beings, Galen was quite secure about the value of his own work. He used many animals, but believed that apes were anatomically close enough to man for medical anatomy. Although Galen never had the chance to systematically dissect human bodies, he had some fortuitous opportunities which he put to good advantage. He reported in his work *On Bones* that he had often been able to examine human bones where tombs or monuments had become broken up. One such occasion he described, with morbid enthusiasm, as the result of a flood which had washed a corpse out of its grave. The body had been swept downstream and deposited on the bank of the river. The flesh had putrefied away, but the bones were still closely attached to each other. Thus the skeleton "lay ready for inspection, just as though prepared by a doctor for his pupil's lesson." Another such occasion occured when the skeleton of a robber was left lying a short distance from a road. None of the people would bury this miscreant and the birds ate away the flesh. After only two days the skeleton was ready "for anyone who cared to enjoy an anatomical demonstration."[15]

Galen told his readers that the only way to learn about the wonders of the body was through the study of the great physicians, such as himself and Hippocrates, and by actual experience. Physicians must first study philosophy to understand man's nature in sickness and health. He especially praised Hippocrates, whose words should be received as if from a god. But study was not enough. "The doctrines of Hippocrates may be judged as to their truth and exactitude not only from a study of his commentators and opponents," he wrote, "but by going directly to Nature and observing the functions of animals, the subjects of natural research."[16] Like Plato and Aristotle, his philosophic view of man and the universe was dominated by the idea that a divine intelligence had created all things. His view of the body and its functions was thoroughly teleological. That is, he found in the organization of the human body proof of the power and wisdom of the Creator and the revelation of the divine design in all things.

Galen was tireless in his tirades against naturalists who would deny the evidence of the great plan and perfection of the human body as the work of God. He particularly vented his spleen against supporters of the materialistic, atheistic atomic theory. In his denunciations of the atomists, Galen revealed himself as thoroughly imbued with the mystical, religious attitudes of his age—even though he was one of the greatest anatomists, observers, and experimentalists of all time. Certainly he believed in astrology and sympathetic magic and smugly reported that the god of healing, Asclepius, spoke to him in dreams. For Galen, the body was the instrument of the soul. Such ideas made this pagan philosopher acceptable to the Church.

Galen was an expert and acute dissector whose writings would serve quite well the student who absorbed both the letter and spirit of his teachings. His anatomical work is described most fully in *On Anatomical Procedures* and *On the functions of the Parts of the Body*. The importance of Galen's anatomical work can hardly be underestimated. His contribution was the peak achievement of Greek medical science. Galen's word was the ultimate authority on medical and biological questions during the Middle Ages. His anatomical studies were not challenged until the sixteenth century and his physiological concepts remained virtually unquestioned until the seventeenth century.

Galen's anatomy had the obvious defect that he based his research on animals, not on the human body itself. He could not avoid this because of the temper of his times. Certainly he did not hide this fact, but did not regard it as too serious a problem. As Celsus suggested, he could have learned much by exploiting wounds as windows into the human body. We can easily imagine the joy he might have taken in the gruesome wounds of the gladiators that were his first patients.

Not satisfied with pure anatomical description, Galen wanted to proceed from structure to function, from pure anatomy to experimental physiology. Although he had been educated in Aristotelian philosophy, he was unsympathetic to the way his contemporaries among the Aristotelians preferred disputation to dissection. In extending medical research from anatomy to physiology, Galen hoped to turn the Hippocratic *art* of medicine into the *science* of medicine. In establishing a system of physiology based on experiments Galen was a pioneer with no rivals and few predecessors. Because Galen's work is so voluminous and tedious it has been imperfectly studied. Thus, his errors, which stimulated the revolt of anatomists in the sixteenth century, tend to be overemphasized. It is important to try to balance the merits and the defects in Galen's work since both exerted such a great influence on the history of medicine, anatomy, and physiology.

As an example, Galen's work on the circulatory system is a strange mixture of physiological confusion and correct demonstrations. In some respects he was less accurate than Erasistratus in his studies of the vascular system. Galen easily disproved the idea that the arteries of living beings contain only air. This shows Galen's knack for simple, direct experiments. He exposed the artery of a living animal for a considerable length and tied off the artery at two points. When the artery was cut between the two ligatures, blood and not air spurted forth. Blood was obviously present in the artery before the incision was made because it could not pass through the ligatures.

Nevertheless, Galen made several errors in his description of the vascular system. Among his major errors was his assumption that the liver was the source of the veins and his dogmatic assertion that the septum of the heart was perforated. These errors were components of his erroneous theory of the distribution of air, spirits, and blood. Because Galen's scheme required that blood must pass from the right ventricle to the left, and he knew that the septum was heavily pitted, he assumed that these pits must serve the purpose of allowing blood to pass from the right to the left side of the heart. The heart itself was a puzzle for Galen. He denied that it was a muscle because it was too hard and it never rested as other muscles did. Experiments proved that the heart could continue to beat after being removed from the body. This is a good demonstration that the heart beat is independent of nervous stimulation.

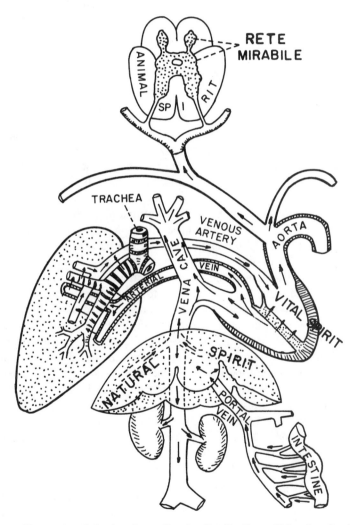

Diagram illustrating Galen's scheme for the distribution of blood and spirits.

Because he regarded the auricles as unimportant, mere dilations of the greater veins, he saw the heart as composed of only two chambers—the ventricles. The function of the heart was the production of "innate heat" by the slow combustion that occured therein. Still he gave a good description of the valves of the heart and the distribution of the blood to all parts of the body by the arteries. He even noticed that when air was added to the dark venous blood in the lungs it became light in color. Yet he found no blood in the pulmonary vein, and so he concluded in agreement with the much villified Erasistratus that it carried only air. Technical problems such as lack of injection techniques obviously contributed to Galen's difficulties with the system, but his overall frame of reference for physiological processes also misled him.

Galen's physiology rests on the doctrine of the three-fold division of life into processes governed by vegetative, animal, and rational souls. Life depended ultimately on pneuma (or air) which was the breath of the cosmos. Galen integrated his anatomical and physiological knowledge with this philosophical system to turn the human body into a place for the elaboration and distribution of three kinds of adaptations of pneuma. These were associated with the three principle organs of the body, liver, heart and brain, and the three types of vessels, veins, arteries and nerves. In essence, the pneuma was adapted by the liver to become "natural spirits," which caused growth. "Vital spirits" which governed movement were formed in the heart. The third adaptation needed for thought processes occured in the brain and resulted in "animal spirits." Thus Galen ingeniously fit what he knew of anatomy with what he believed from philosophy. To work out the details of the elaboration and distribution of the three "spirits" required some ingenious research and a wealth of speculation about the functions of the body.

Blood, according to Galen, was continuously synthesized from ingested foods. The useful part of food was transported as "chyle" from the intestines via the portal vein to the liver. The liver had the faculty of transforming chyle into dark venous blood. Since the liver was the seat of vegetative life, concerned with nourishment and growth, it imbued the blood with natural spirits. The useless part of the food was converted into black bile by the spleen. Blood moved from the liver with its natural spirits and nutritive material into the venous system for distribution to the body. The right side of the heart was regarded as a main branch of the venous system. From the right side of the heart the blood could enter different paths. Part of it discharged its impurities or "sooty vapors" into the lung by the artery-like vein (pulmonary artery) and were exhaled. Part of the blood passed through Galen's imaginary pores in the septum and entered the left heart. This blood encountered the pneuma that had been brought in by the trachea and the vein-like artery (pulmonary vein). This mixture produced the bright-red arterial blood and the vital spirits which were distributed to the body through the arteries. Some of the arteries led to the network of vessels at the base of the brain called the "rete mirable." Here the vital spirit was transformed into animal spirits which were distributed to the body via the hollow tubes called nerves.

Galen's "human" anatomy contains almost as many errors as his physiology. He described the liver as five-lobed, which it is in other animals such as the dog. The rete mirable is present in ruminants, but not in man. He argued that the heart was not a muscle and saw the active phase of heart action as dilation rather than contraction. The heart was made responsible for respiration with dilation drawing air into the body and contraction driving it out. This scheme fails to explain how the heart could drive respiration when the heart beat is much faster than the rate of respiration. Although Galen's system rests on faulty observation and incorrect inference, it was found quite satisfactory for some 1400 years. It seems to have satisfied the need for a complete and simplified synthesis of medical and philosophical traditions. It combined the Four Humors of Hippocrates with Aristotle's three-fold division of life and spirits and the great cosmic spirit of the Stoics. Beyond that it is deeply imbued with a religious spirit and admiration for the work of the Great Creator.

Galen knew that experimentation was the only path to scientific truth and often reminded his readers of this. Unfortunately, he did not follow this precept rigorously in his own work. Yet we know that Galen was a master of experimental physiology. This is particularly apparent in his contributions to the physiology of the nervous system. Here he was so advanced that no really significant improvement occurred until the nineteenth century. Galen carefully worked out the anatomy of the brain, spinal cord, and nerves. He showed that the nerves originate in the brain and cord and not, as Aristotle said, in the heart. Furthermore, he conducted a series of experiments on the spinal cord which were described in *On anatomical operations.* When the cord was injured between the first and third vertebrae, death was instantaneous. Injury between the third and fourth vertebrae arrested respiration. Section below the sixth vertebra caused paralysis of the thoracic muscles. Lesions lower in the cord produced paralysis of the lower limbs, bladder, and intestines.

It is difficult to imagine a physiological problem that Galen did not at least touch on. He went beyond his statement that the function of the kidneys was obvious to prove that urine was made by the kidney and not the bladder by tying the ureters. He attempted to study digestion by feeding different diets to pigs and then opening their stomachs to see the effect.

Although he is known to have given public lectures and demonstrations in the Temple of Peace, Galen left no "school" or disciples. Despite his brilliance and his ability to get along with the Roman emperors, his personality must have been quite disagreeable to colleagues and potential students. Galen tried to synthesize a universal system that would embrace physiology, pathology, and clinical medicine. Although it is easy to point out his errors today, his work was found satisfactory to generation after generation of scientists.

Undoubtedly Galen was one of the most important anatomists and medical experimentalists up to the time of Vesalius and then Harvey. Even these scientists began as Galenists. Perhaps their rejection of Galenism should be regarded as the triumph of the true spirit of Galen the scientist. It was from Galen that they took their instruction and inspiration. For centuries Galenism had suffocated progress in biology. It was not Galen himself who did so. The Church invested great authority in his concept of the three spirits, his anthropomorphic teleology, his religiosity, and the use of scientific methods as proof of religious dogma. For long ages, to question Galen was to risk charges of heresy. It is unfortunate for science, and for Galen's reputation, that his most flagrant errors were treated as inviolable truths by lesser minds.

While most of Galen's voluminous works were unknown during the Dark Ages in Europe, his writings were not lost or entirely neglected. His works were translated into Syriac and Arabic. Finally the Arabic versions of his works were translated into Latin and energized a new wave of study in Europe during the eleventh and twelfth centuries. His religious orientation and his medical knowledge, if not his experimental methods, became the foundation of Christian and Muslim medical studies.

The Decline of Science and Culture

Before leaving the Greco-Roman world, we must stop for a moment to consider the obvious paralysis of science—the failure to consolidate and put to use the very great accomplishments both in theory and practice that we have traced from Thales to Galen. By about 200 A.D. the old Greek culture was almost dead. The Roman era, a period of prosperity and almost 200 years of relative peace, might have been a time of great intellectual development. Many were conscious of the malaise that permeated their society and of the fact that little had been accomplished since the time of Aristotle. Pliny complained often that humanity had grown corrupt and his age was worse than all its predecessors.

After the death of the philosopher emperor Marcus Aurelius, imperial Rome experienced a sad, inexorable decline due more to internal decay than barbarian invasions. Rome suffered internal misgovernment, civil war, economic, political and moral corruption, despotism, high taxation, endemic malaria, and a decline in agriculture. Some historians see religiosity and occultism as the cause of the decline of culture and science in the second century A.D. Astrology became the "all-pervading natural law of the Greco-

Roman world and indeed of Europe before Newton."[17] A preoccupation with the occult suffocated scientific inquiry and prevented even the greatest minds from posing the questions that rationalists would ask. By the third century the Christian community was expanding. Tertullian and other Christian writers were pleased by the decline of the pagan world and the destruction of pagan learning.

A concern for ecology might be more compelling today if we remember the ecological disarray that prevailed in Roman times. The richer veins of minerals had been exhausted. The productivity of Italian lands diminished in response to deforestation, erosion, and the neglect of irrigation canals. The live birth rate in Italy and the population of the West after the time of Hadrian (76-138 A.D.) declined significantly. In the Campagna farms were offered free to anyone who would work them because so many had been abandoned. In 250 A.D. Bishop Bionysius complained that the population of Alexandria was half that of previous times.

In the third and fourth centuries, respect for science progressively declined among the educated classes. This was of no consequence to theologians who were ambivalent toward science at best or, more likely, openly hostile. Few reached the vituperativeness of Tertullian who said, "We have had no need for curiosity since Jesus Christ, nor for inquiry. . . . It was highly appropriate that Thales, while his eyes were roaming the heavens in astronomical observation, should have tumbled into a well. This mishap may well serve to illustrate the fate of all who occupy themselves with the stupidities of philosophy. . . ." Christians were warned to avoid all books of the heathen which turn the young from the faith and instead to read the Book of Kings for history, the Prophets for philosophy, and Genesis for the creation of the world. "There is no gain in knowing these things," they were told, "leave these things to God."[18] Ironically many of those who denounced pagan or secular learning, such as Tertullian and Saint Augustine, were themselves quite learned.

The world of the Christian West was to be the physical and biological world described in the Bible—a flat, unmoving earth, created as described in Genesis. Science had to be sacrificed to scriptural authority, no matter what intelligent men might say or attempt to prove. When the Emperor Constantine converted to Christianity, the power of the state could be turned against scientists and scholars. Libraries were wrecked and pagan books burned. Justinian, the Byzantine emperor, barred pagans and heretics from academic posts and confiscated their property. Thus the Academy, the living heritage of Plato, was closed in 529 A.D. and many scholars were sent into exile in Persia. Scholars and ancient writings once lost were irreplaceable. Books were produced by hand and there were few copies. The copy rate during the twelfth century was about five pages a day and it generally took about six months to copy over a Bible.

It was Petrarch (1304–1374) who labeled the period from the fourth to the fourteenth century the "Dark Ages." Today the term is generally restricted to the period from the fall of Rome to the end of the tenth century. The period has generally been portrayed as one of widespread ignorance and poverty, lacking any sort of intellectual or scientific progress. Recent scholarship into the Dark Ages has challenged this characterization in many areas. However, as far as biological science is concerned it is difficult to detect any activity worthy of the name. The era was dominated on an intellectual plane by theological concerns rather than any interest in nature.

After the collapse of Roman rule, new kingdoms were established only with great difficulty. The new masters were "barbarians" as far as the cultivated Greeks and Romans were concerned, yet their mastery could not be denied. As always it was easier to conquer the old cultures than to maintain or improve them. Disruption of lines of communication, safe travel, education, industry, and agriculture is easy compared to the task of building civilization. The Roman Empire was entirely fragmented in the West. In the East, the Byzantine Empire remained powerful and the Greek Orthodox Church became a unifying force. In Syria and Persia the Moslems were to become the new cultural power, reaching their Golden Age in the ninth to eleventh centuries. Typical of the degeneration of culture during this period were Theodoric, the Gothic king of Italy, and the Byzantine emperor Justin, in the sixth century. Neither of these rulers could read or even write their own names. Theodoric signed documents with the use of a golden stencil.

As with all learning, medicine fell into the hands of religious organizations, but priority resided in saving the soul rather than healing the body. Every cure became a religious miracle rather than a natural healing process. Hippocrates' battle to separate medicine from superstition was forgotten. Some theologians even argued that the spirit of God would not be found in healthy bodies. The mortification of the flesh was encouraged by some religious enthusiasts. Diseases that no earthly physician could cope with in any way overwhelmed the modest abilities of the physicians. A series of plagues had tormented the Roman world during her long decay, but the plague of Justinian in 540 A.D. stands out among these. Indeed, the beginning and end of the Middle Ages stand neatly bracketed by devastating visitations of the bubonic plague. The plague of the sixth century spread from Egypt via Alexandria and Palestine to the rest of the known world. Deaths were reckoned at the tens of thousands per day. According to Gibbon, some countries did not regain their previous population density for centuries. The devastation caused by the plague helped in making the Church a force in Europe. Ineffectual scientific medicine capitulated to the cult of Christ the Healer.

Biological researches were not in the spirit of the Middle Ages. Human anatomy certainly was not a worthy subject in the eyes of Church authorities.

Direct study of plants and animals was regarded as of no value. Pliny's *Natural History* remained a popular compendium of all such knowledge, and his stories, so uncritically assembled, were used to teach moral lessons and manners as well as science. Even the simple illustrated herbals became corrupted and their descriptions and pictures unrecognizable.

The Middle Ages

Technology

Although the scientific tradition inherited by the western part of the Empire was impoverished and anemic, the period generally referred to as the Middle Ages was not entirely bleak. There was a fundamental change in attitude that did prepare the way for science and society as we know it. Obscured by the "darkness" of the Middle Ages was a very real social and allied technological revolution.

The catalog of inventions either made or successfully exploited for the first time during the Middle Ages, if not brilliant, does contain items of fundamental importance. In terms of technological advances, labor-saving devices, and agricultural improvements this period was quite remarkable. At the very beginning of the period, the rotation of crops with a two- or three-field system was either invented or put to good use for the first time. Another fundamental improvement was the use of the modern system of harnessing animals from the shoulder so that riding animals and draft animals could be better exploited. The saddle horse obtained its modern harness complete with saddle, bit, and nailed iron shoes. The draft animal had its shoulder collar, shafts, and nailed shoes. A new heavy-wheeled plough with a mold board and rudder was in use by the thirteenth century. Watermills, windmills, mining, road construction, methods of drainage, as well as agriculture all improved in the Middle Ages. New crops such as rye, oats, spelt, and hops were cultivated.

A catalog of medieval inventions includes the mechanical saw, forge with tilt-hammer, bellows with stiff boards and valve, window glass and glazed windows, the domestic chimney, candle and taper, the wheel barrow, spectacles for the far-sighted, lock gates on canals, gunpowder, and the grandfather clock, and culminated with the printing press and gunpowder weapons. Even the "barbarians" contributed things that we take for granted today, such as the modern costume of trousers instead of the toga for men, improved methods of making felt, butter to replace olive oil, and the use of skis, barrels, and tubs. Textile working was also improved with the spinning wheel and the loom. The application of water power led to many improvements, especially to power the bellows of furnaces so that iron could be melted and cast. Navigation improved and larger ships were able to sustain longer voyages.

About the only technology which did not progress during the Middle Ages was weight lifting. Without the power of thousands of slaves, the builder and engineer turned to making the great monuments of their time out of small units.

By the twelfth century both food production and population were showing significant gains in Europe. As productivity and population grew so did industry, the crafts, commerce, and urban centers. The inventions and agricultural innovations of the Early Middle Ages transformed and invigorated European society. Yet these inventions are rarely accorded the respect given to copyists of the long-lost works of Aristotle. The inventors of the technologies that transformed the Middle Ages remain nameless and faceless. They may be compared to members of a great ant hill, each tiny individual a part of a vast enterprise it can hardly be expected to comprehend.

Scholarly Contributions

A revival of learning reanimated European culture between the eleventh and thirteenth centuries. Although the scholars of the Middle Ages cannot be accused of adding much to the heritage of Greece, its partial preservation and assimilation was important. Europe did not receive this legacy directly, but only with difficulty through a tortuous route from the Jews and the Arabs to the East who had been in closer contact with the ancient learning. But by the end of the thirteenth century Europeans had progressed in scholarship and science, including mathematics, astronomy, alchemy, and medicine.

Some individuals stand out for their curiosity and originality. Frederick II (1211–1250), called Wonder of the World by his admirers, studied nature, even conducting anatomical and physiological experiments. But questions of theology and political maneuvering were more likely to absorb the energies of the most talented and intelligent individuals. Questions that other eras might address through science—the origin and nature of the universe and life—were considered totally explained by religion. This religious orientation to questions undercut one basic intellectual incentive to scientific inquiry. Craft processes had inspired the pre-Socratic Greek philosophers, but the crafts of the Middle Ages were run by guilds that jealously guarded the secrets and "mysteries" of their skills, effectively blocking another path to scientific progress.

The pace of scientific and scholarly activities had definitely quickened by the thirteenth century. The interesting question is why it happened at that time and not earlier. No doubt the opportunity must have been present before, but the interest had been lacking. Church authorities had complained about Christians studying geometry, Aristotle, Theophrastus, and Galen in second century Rome. During the ninth century the same complaint was made about Christians in Spain who studied Arabic translations and

commentaries on the pagan writers. The Muslim presence in Spain presented a
clear opportunity for translation from Arabic to Latin. Direct translation
from Greek to Latin was a possibility during the occupation of Constantinople
from 1204–1261.

An important phase in the translation of ancient works dates from the
western Crusade against the Muslims in Spain. Toledo was taken in 1085 and
became a Christian archbishopric. Spain became an important locus of interaction, having bilingual Muslims and Christians and Jews, some of whom were
trilingual. The process was stimulated by Archbishop Raymond who set up a
school of translation at Toledo which attracted scholars from all over Europe.
Among them was the famous Gerard of Cremona (1114–1184), who progressed
from the astronomical works of Ptolemy to complete some eighty other works
of translation.

Sicily was another important site of interaction between scholars who knew
Latin, Greek, or Arabic. Christian forces gained control of the area in 1091
after years of Muslim domination. Although the traditions of learning had
never entirely died out here, the career of Michael Scot (d. 1235) illustrates
the new European passion for scholarly pursuits. Scot's work was encouraged
by Frederick II, Holy Roman Emperor and King of southern Italy and Sicily.
Truly an intellectual prodigy, Scot was interested in meteorology, medicine,
sociology, psychology, physiognomy, astrology, alchemy, and magic. The
Pope described Scot as "burning from boyhood with love of science."[19] Although Scot translated the biological works of Aristotle into Latin, his
"science" was primarily astrology and divination. For divining the future,
people were advised to go along the public way and bring back objects for use
as omens. Scott explained that the stars would direct their choice. Divination
was also applied to medicine through the presumed interaction between
phlebotomy and astrology. Scot warned the Emperor that he should not be
bled when the moon was in the sign of Gemini or a dangerous double punc-
ture could occur. Frederick was assured by his barber that any time was
suitable for bleeding, but the barber dropped the lancet on the Emperor's
foot—resulting in a painful tumour, and convincing him of the value of Scot's
"scientific" predictions.

Thus, Scot illustrates the strange intellectual make-up of his times. Interested in philosophy and science, he accepted astrology and alchemy. He said
that "all science is from philosophy" and advised others about the four things
that make a man wise: intelligence of reason, diligence in doing, experience of
knowledge, and a living memory. Yet he told stories about plants and animals
that were at least as bizarre as Pliny's. According to Scot's zoological wisdom,
if one of the 12 tail feathers is removed from a bird, it cannot fly well. The
number of tail feathers is twelve because that corresponds to the signs of the
zodiac. The Indian phoenix can speak to all people on all themes in any

language they use. The Hoopoe was reported to live on human dung and the panther to exude a sweet odor noticeable over a mile away. There were also animals which could kill with a single glance. Some animals could eject smoke through their nostrils, while others contained poison and antidotes, like the scorpion which had poison in the sting of its tail and theriac in its belly. Although wind is invisible to man, cows, cats, geese, cranes, and hens can see it. Scot's comments on agriculture are equally enlightening. To protect shrubbery from storms, the reader was advised to cremate three live crabs. A red frog in a new clay pot would ward off tempests from crops and vineyards. Consecrating a magpie among the vines would also protect the vineyard.

Many factors engendered the new intellectual climate in the twelfth and thirteenth centuries of the Middle Ages that provided the real stimulus to scholarly and scientific activities. Neither the Italian trade in the Levant nor the Crusades seem directly responsible, since few translations came from these sources. New religious movements were undermining the monolithic structure of faith, leaving a climate of doubt and also curiosity. Pockets of doubt and curiosity are like intellectual vacuums which have to be filled. The ancient philosophies of the Greeks promised to satisfy this need.

The work of translation was extensive but not all inclusive. Christian translators preferred the Roman and Greek authors with whom they were familiar and ignored many original contributions of the Arabs whom they saw only as intermediaries. For example, the discovery of the minor circulation by Ibn al-Nafis was unknown until the present century. Other Arabic and Persian works are still untranslated. Aristotle emerged as the Prince of Philosophers as the great corpus of his works became available. This interest in Aristotle was to help replace the Platonic view of the early Church Fathers. Thomas Aquinas (1225?-1274) encouraged this trend by editing the works of Aristotle.

The Growth of Universities

The Universities of the Middle Ages were very different from those of today, especially in the relationships among students, faculty, and administrators. These institutions were also totally different from the ancient centers of learning. Informal discussion was replaced by set lectures to large groups of students. The community of scholars was replaced by the concept that those formally licensed to teach would instruct those seeking their own credentials. General inquiry into areas of mutual interest was replaced by a set liberal arts curriculum.

The exact origin of some of the major universities is unknown, although in general it seems that ecclesiastical authorities dominated these nascent institutions. As the old monastic schools declined in prestige in the eleventh and

twelfth centuries, cathedral schools and semisecular municipal schools began
to supersede them. Some of these schools grew quite large by the twelfth
century and developed into centers of higher learning after centuries of obscurity. For example, Charlemagne (768-814) had founded schools associated
with metropolitan churches. Here young priests gave lessons in theology,
music, and other subjects that the Church believed necessary for clerics. When
the number of students grew large, a larger staff of teachers, called "magistri,"
organized to protect themselves. Such an organization was called a "universitas
magistrorum."

"University" was a very vague term in the Middle Ages which referred to
any corporate status or association of persons. The early universities were collections of colleges that sought the advantages of the corporate status. Thus
they were rather like guilds. Indeed during the eleventh century the words
university and guild were interchangeable for describing craft associations.
The term "studium generale" denoted a guild or university of scholars and
students engaged in the pursuit of higher learning. Such institutions drew
students from many countries to study with teachers known for particular
areas of excellence. Eventually the term "university" was restricted to institutions for higher education.

Large numbers of students were drawn to the medieval universities. Students, all of whom shared Latin as the language of learning, were organized
into "nations." Universities wielded considerable power and won large
measures of self-government from civic and ecclesiastical authorities, but the
relationship between "town and gown" was generally quite explosive. Some
institutions, notably those in Paris and Oxford, were governed by guilds of
instructors in the liberal arts. In contrast, Bologna, which specialized in legal
studies, was dominated by a guild of students. There was no standard class
and age distribution at the universities. Students had the reputation of being
rowdy, armed, often drunk, and always dangerous. In the absence of scholar-
ships and student loans, they often made their living by a combination of
begging and stealing. Despite their wild reputation, the universities were generally based on the eagerness of the young to acquire knowledge or at least
access to wealth and power after graduation.

Although each university had its unique origin and history, the European
universities can be divided into three main types. Those with ecclesiastical
foundations are represented by Paris, Oxford, and Cambridge. Here, the
students and teachers formed a closed corporation directed by a chancellor.
Bologna and Padua were civic universities. These were governed by a rector
elected by the students. The roots of the University of Bologna go back to
the period of the Roman Empire, before a formal organization can be documented. Here, students had a great deal of control over teachers and curriculum. The third type was the state university. These were founded by a

secular ruler with Papal recognition. The university at Naples, established by Frederick II of Sicily, and the University at Salamanca, organized under Ferdinand III of Castile, were of this type.

The university curriculum certainly did not encourage progress in science directly. But since the kinds of institutions typical of an era shape the kind of science and scientists produced, the universities are important in the history of western science. Public schools were unknown in the Middle Ages and the parish priest probably provided a boy with his first educational experience. From this beginning the boy might go on to a cathedral school. The curriculum consisted of the "seven liberal arts" which were divided in the sixth century by Boethius into the trivium: grammar, rhetoric, and logic; and the quadrivium: arithmetic, geometry, astronomy, and music.

During the thirteenth century there were only five significant universities in northwest Europe—Paris, Orleans, Angers, Oxford and Cambridge. In the south of France were three more. Italy boasted of eleven, while Spain had only three. Germany had none until the fourteenth century. Among these 22 European institutions those at Bologna and Paris were the most important. By the end of the Middle Ages some 80 universities were established in Europe, of these two-thirds were in France and Italy.

No institutions for learning had ever been as large as these medieval universities. Even though the educated, or merely literate, were still in the minority, some of the Universities had thousands of students. Students were prohibited from athletics, but in addition to their studies took up gambling, stealing, drinking, and fighting. Many "scholars" entered the universities at fourteen or fifteen. Riots between students and townspeople were common. Although the universities were powerful institutions, rarely were there any permanant university buildings. Lectures often were held in the professors' apartments or in rented rooms. Students often contracted independently to hire and pay teachers. The student might take an exam after six years of study. If he passed, he was qualified to become a teacher himself. Many rewarding and lucrative professions were open to the university graduate. He could take orders and become a church official. His scholarly abilities might be appreciated by a rich patron. He could study medicine or law or work as a copyist. Even in the thirteenth century there were students who rejected the conventional paths. They pursued the life of wandering scholars—visiting different universities and taverns all over Europe.

At all the universities, teaching was limited and regulated by religious dogma. Intellectual deviation could be punished as heresy. Particularly during the religious turmoil of the thirteenth century, secular authority bowed to the ecclesiastical powers, and orders of mendicant friars were established to combat heresy at the universities. This transition took place at about the same time that Aristotle's work became available to scholars in the original pure Greek form.

Aristotle's ideas found a welcome place in Church dogma. His concept of the earth as the center of the universe, but the site of all change and imperfection, was very congenial to the Church. The heavens, where God abided, were by contrast pure, incorruptible, and eternal. Further, his view of human life as part of a rigid hierarchy created an authoritarian and semiscientific justification of the hierarchy of feudal society. Thomas Aquinas was able to edit the works of Aristotle during the thirteenth century in a manner acceptable to the church. All points of deviation could easily be attributed to pagan ignorance. In time, Aristotle's authority became so great that the spirit of the time is well captured in an anecdote about the way learned men would argue over the number of teeth in a horse's mouth—basing argument on Aristotle, rather than counting for themselves.

Although the education of the period was *scholasticism* at its worst—a concern with learning from ancient authorities with no encouragement of original thinking—there were some individuals who made interesting observations. The *Physica* of Hildegard, a nun at Bingen on the Rhine (written ca. 1150) contains observations on animals, plants, and stones and how they can be used as remedies. Frederick II had Aristotle's works translated into Latin from the Greek. At the medical school which he founded at Salerno, human bodies were dissected. The Emperor, much interested in natural history himself, wrote a book on falconry which reflects his own observations and even criticism of Aristotle.

Other original minds of the thirteenth century included Albertus Magnus and Roger Bacon. But an interest in original observation and experimentation could be very dangerous at the time, for such intellectual aberrations could lead to charges of witchcraft and heresy. Albertus Magnus (ca. 1200–ca. 1280), a member of the Dominican order, enjoyed a great reputation for his learning. Although he had been a professor in Paris and Cologne, and Bishop of Regensburn, he left this kind of life to devote himself to science in the isolation of the monastery. He felt called to edit Scot's Latin translation of the works of Aristotle and make them compatible with Church teachings. He also had a strong interest in natural science, botany, and chemistry. Arsenic in a free form was first prepared by Albertus Magnus. But in biology he accepted the errors of Aristotle, such as air in the arteries, the brain as a cold, wet, and unimportant organ.

Roger Bacon (1214-1294) studied at Oxford and Paris before joining the Franciscans. His views were so liberal that he spent several years in prison. He was interested in both physics and chemistry, but was accused of witchcraft and necromancy. Roger Bacon has been called a "prophet of science" for the spirit with which he approached the study of nature rather than for any particular discovery. He did pursue some experiments in optics and seems to have predicted the invention of the microscope and telescope.[20] Even such unusual men could not escape the superstitions of their age. Albertus Magnus accepted many medieval superstitions about stones, for he himself had seen a sapphire that cured ulcers.

Arab Science

In 570 A.D. Mohammed was born into a poor family in the sacred city of Mecca. His father had died at Medina two months before. The infant inherited five camels, a flock of goats, a house, and a slave. When he was only six, his mother died leaving him in the care of his grandfather and uncle. Although Mohammed probably never learned to read or write, he composed the most famous and influential book in the Arab language, the *Koran*. At twenty-five he married a rich widow, Khadija, who was then forty and the mother of several children. He took no other wives until she died twenty-six years later. She left him some daughters, but two sons had died in infancy. Although he believed in a simple life without extravagances for himself, he had ten wives and two concubines. All his wives after Khadija were "barren." He had one son by a Coptic slave in the last year of his life, but the child died after 15 months. Only his daughter Fatima survived him when he died in 632.

In the year 610, while alone in a cave during a religious retreat, Mohammed had a vision in which the angel Gabriel told him he was the messenger of Allah. Thereafter, he had more visions during which the *Koran* was revealed to him in fragments. For many years Mohammed preached his new religion with little success, winning few converts outside his immediate family. Moving to Medina ("City of the Prophet") in 619, he soon established himself as ruler of the city and enlisted Bedouin support in a holy war against his enemies and the caravans of the wealthy merchants of Mecca. In 630 Mohammed returned to Mecca triumphantly, smashed the idols in the temples, and declared his birthplace the sacred city of the Islamic faith. (The word "Islam" means "to submit, or surrender absolutely to God." Those who follow the religion are called "Moslems.") Within a century of his death, Mohammed's followers had conquered half of Byzantine Asia, all of Persia, Egypt, North Africa, and Spain.

According to Moslem tradition, the prophet respected and encouraged the pursuit of knowledge although he was himself illiterate. "He who leaves his home in search of knowledge," said Mohammed, "walks in the path of God."[21] The Arabs established elementary schools, usually held in a mosque, where most boys, some girls, and some slave children were educated for little or no tuition. The curriculum was mostly religious—required prayers, reading of the Koran, and understanding theology, history, ethics and law through the Koran. Writing was a separate skill for those with higher education. Scribes were used by those who could not write themselves.

Secondary schools were eventually brought under government regulation and subsidized to serve as colleges. Here students could learn grammar, philology, rhetoric, literature, logic, mathematics, and astronomy. Tuition was free since the salaries of the teachers and the expenses of the students were paid by government and philanthropists. Students would travel great distances to hear the most famous teachers. Just as Latin served students from all over Europe, Arabic served as the language of learning for the Islamic world. Students did not get degrees, but hoped for a certificate of approval from individual teachers.

One of the major contributions of the Arabs to the history of science and scholarship was their service as agents in the transmission of important discoveries and ideas between cultures. Persia and Syria, although debilitated and easily conquered by the Arabs, were still repositories of the ancient traditions of learning. Many Greek physicians had emigrated to Persia, as had scholars displaced when Plato's Academy was closed in 529 A.D. In Syria there were still great libraries and flourishing schools of philosophy. From the ninth to thirteenth century Arab scholars could justly claim that intellectual leadership had been captured by Islam. Books were regarded as great treasures. Baghdad boasted of more than 100 booksellers in the ninth century. Libraries were attached to most mosques. Baghdad had 36 public libraries when it was destroyed by the Mongols. Private libraries were the hobby of the rich. The passion of the Arab world for books between the eighth and eleventh centuries was matched nowhere else at the time, except perhaps in China. One physician claimed that his library would require 400 camels to move it. Another bookworm left 600 boxes of books when he died. Each box required two men to lift it.

Arab leaders realized that their culture was weak in science and philosophy but that they had a wealth of Greek culture almost intact in Syria. Centers of learning at Alexandria, Beirut, Antioch, and other cities were allowed to continue their work. Moslems who knew Syriac or Greek encouraged translations into Arabic. The translators were generally Jews or Nestorian Christians. Arab princes sent to Constantinople and other Hellenistic cities for Greek books, particularly in medicine or mathematics. Just as Christians sought to reconcile Aristotle with church dogma, so did Arabs try to harmonize Greek philosophy with the Koran. Because the Koran was supposed to contain all wisdom of importance, Arabic scholars often disguised their work as commentaries on the Koran to make them more respectable.

In addition to transmitting the traditions of ancient science and philosophy, the Arabs absorbed and disseminated some of the techniques requisite to the further expansion of literacy and learning. The Arabs who captured Samarkand in the year 712 learned the art of paper-making from the Chinese who had been using paper since about 100 A.D. A paper-making plant was built in

Baghdad in 794. When the Arabs conquered Sicily and Spain, they introduced paper to Europe. While translating and assimilating the scientific work of other cultures, the Arabs produced many creative mathematicians, astronomers, physicists, physicians, and chemists. The zero and the so-called arabic numerals, obviously more useful than the cumbersome Roman system, were actually borrowed from India but adapted and transmitted to the West by the Arabs. At the House of Wisdom in Baghdad, the Old Testament and the works of Hippocrates, Euclid, Ptolemy, Plato, Aristotle, Dioscorides, and Galen were translated. Many words used in science and mathematics, such as alchemy, algebra, cipher, algorithm, alcohol, and nadir, are derived from Arabic.

Chemistry and Alchemy

Of far-reaching importance was the Arab interest in that most experimental of all sciences, chemistry. All the acids and alkalis known before the nineteenth century were known to the Arab chemists. The experimental approach had long been associated with magic and sorcery and was viewed as vulgar by the learned men of the western tradition before the Arab scientists helped to make it respectable and demonstrated its value. Arab chemists invented and named the alembic, distinguished between alkalis and acids, accomplished the chemical analysis of myriad substances, and the study and manufacture of hundreds of drugs. The roots of Arab alchemy go back to Egypt. The most famous of the alchemists was Jabir ibn Hayyan (ca. 721–ca. 815). In medieval Europe he was known as Geber or Jabir or "the Mystic." His works are known through the version set down by the tenth century mystic sect called the Brethren of Purity.

Soon after Geber's birth, his father, a Kufa druggist, was beheaded for his participation in certain political intrigues. Although Geber was a physician he was more drawn to alchemy. More than a hundred alchemical works were attributed to him in later collections. Some of these works may have been authored by his disciples, but others have been alleged to be Western forgeries. Alchemy has a great affinity for mystical religious orders and the Brethren of Purity fit this description. They produced an encyclopedia which devoted 12 out of 52 works to science, but their writings were regarded as heretical and the Sunni of Baghdad had them burnt. The Brethren were suppressed by the eleventh century.

Most medieval scientists would agree with Geber's proposition that all metals were ultimately of the same species because all metals were supposedly derived from an idealized form of mercury and sulfur. Therefore, all metals could be transmuted into each other by the proper techniques. With the "philosopher's stone," the alchemist could change the "base" metals such as iron, copper, lead, and tin into silver or gold. Minerals, along with blood, hair, excrement, and diverse other substances, were treated with various re-agents and subjected to basic chemical processes—calcination, sublimation, and distillation to drive off and purify their essence. With the long-sought elixir, one could prolong life as well as create gold.

The Golden Age of Arab Medicine

When the Caliph al-Mansur, Successor to the Prophet, and ruler of the new capital at Baghdad, became ill and his own physician could not cure him, he sent to Jundi Shapur for a Nestorian physician. A Nestorian school in that city was renowned for its teaching in theology and medicine. Having been cured, the Caliph naturally began to favor the Nestorian physicians, who were mainly Persian in origin. However Arab medicine and pharmacology drew on many sources. According to Levey's linguistic analysis of the materia medica of al-Kindi (801–866), one third of the drugs came from Mesopotamian and Semitic traditions, 23 percent from Greek, 18 percent were Persian, 13 percent Indian, 5 percent were Arabic, and 3 percent came from ancient Egyptian sources.[22] The Arabs added ambergris, camphor, cassia, cloves, mercury, senna, myrrh, sirups, juleps, and rose water to the general phamacopoeia. The first apothecary shops and dispensaries and the first medieval school of pharmacy were Moslem contributions. Even after the end of the "Golden Age" of Arab medicine, the Moslems continued to produce skilled physicians, especially for the treatment of eye diseases so common in the Near East.

However, the Arabs did not merely collect ancient theories, they also produced some very famous scholars in the top rank of medieval medicine.[23] The most famous of these were Rhazes (865–925), Avicenna (980–1037), and Averroes (1126–1198). Their works were translated into Latin and were basic to scientific studies in European universities.

Arab Biology

The contributions of the Arabs to biology were quite limited in comparison to medicine and chemistry. Abu Hanifa al-Dinawari (815–895 A.D.) based his *Book of Plants* on Dioscorides, but he did add many new plants to the stock of pharmacology. There were interesting Arab experiments in producing new fruits and in grafting interesting combinations. For example, the rose bush and the almond tree were combined. There were even speculations of an evolutionary nature by Othman Amr al-Jahiz (d. 869) who suggested that life had progressed from mineral to plant, from plant to animal, and from animal to Man.[24] Botany was revived during the twelfth to thirteenth centuries by Moslem scientists who began to stress purely botanical studies rather than the medical aspect of plants. Moslem writers also produced books on agronomy.

The Book of the Peasant gave analyses of soils and manures, and discussed the cultivation of almost 600 plants and 50 fruit trees, as well as the symptoms and cures of plant diseases.

While the Moslems were subject to attacks from Christian Europe via the Crusades and in Spain, they also were attacked from the East by the Mongols. The Mongol conquest took but 40 years. The Mongols conquered the Eastern Caliphate and the Sung dynasty, thus extending their power from one end of Asia to the other. Genghis Khan first attacked China in 1214. In 1233 Mongols captured a Chinese munitions factory at Pien Ching and took over the knowledge of gunpowder and grenades. These were used in the attacks against Europe. Thus, the Mongols brought to the rest of the world the Chinese invention of gunpowder and knowledge of printing.

Notes

1. E. A. Parsons (1952). *The Alexandrian Library.* Amsterdam: Elsevier Press.
2. Theocritus XIV 59 (quoted in Africa, p. 50.).
3. Parsons, pp. 273–412.
4. Celsus. *De medicina.* 3 vols. Translated by W. G. Spencer (1935–1938). Loeb Edition, Cambridge: Harvard University Press. See preface, especially pp. 15, 25, 41; Ludwig Edelstein (1935). The Development of Greek Anatomy. *Bulletin of the History of Medicine 3:* 235–248; Charles Singer (1957). A *Short History of Anatomy and Physiology from the Greeks to Harvey.* New York: Dover, pp. 34–35.
5. G. Majno (1975). *The Healing Hand.* Cambridge: Harvard University Press, p. 333.
6. E. Nordenskiold (1935). *The History of Biology.* New York: Tudor.
7. Africa, p. 68 (quoted from Cicero, *de senectute* I 11).
8. Pliny. *Natural History* 10 vols. Translated by H. Rackham, W. H. S. Jones, and D. E. Eichholz (1956–1966) Loeb Edition. Cambridge: Harvard University Press, Vol. 1, p. x.
9. Pliny, Vol. 7, p. 3.
10. A. E. Smith and D. M. Secoy (1975). Forerunners of Pesticides in Classical Greece and Rome. *Journal of Agricultural and Food Chemistry 23:* 1050–1055.
11. Pliny, Vol. 8, p. 57.
12. Celsus, see Introduction by W. H. S. Jones.
13. Celsus, vol. 1, p. 219.
14. Celsus, vol. 1, p. 25.

15. Galen. *On Anatomical Procedures*. Translated by W. L. H. Duckworth (1962). Cambridge: Cambridge University Press. "On Bones," Book 1, Chapter 2.

16. Galen. *On the Natural Faculties*. Translated by A. J. Brock (1963). Loeb Edition, Cambridge: Harvard University Press. "On the Kidneys," Book 1, Section 13.

17. Africa, p. 80.

18. Africa, pp. 86–87.

19. L. Thorndike (1965). *Michael Scot*. London: Nelson, p. 11.

20. C. Singer (1950). A *History of Biology*. New York: Henry Schuman, p. 72.

21. Ameer Ali, Syed (1900). *Spirit of Islam*. Calcutta: Lahiri, p. 331.

22. M. Levey (1966). *The Medical Formulary or Aqrabadhin of al-Kindi*. Madison: University of Wisconsin Press, p. 20.

23. S. H. Nasr (1968). *Science and Civilization in Islam*. Cambridge: Harvard University Press.

24. G. von Grunebaum (1946). *Medieval Islam*. Chicago: University of Chicago Press, p. 331.

References

Africa, T. W. (1968). *Science and the State in Greece and Rome*. New York: Wiley.

Allbutt, T. C. (1921). *Greek Medicine in Rome*. London: Macmillan.

Arnold, T. and Guillaum, A., Eds. (1931). *The Legacy of Islam*. Oxford: Clarendon.

Aries, P. (1962). *Centuries of Childhood*. Translated by Robert Baldick. London: Cape.

Bell, H. I. (1948). *Egypt from Alexander to the Arab Conquest*. Oxford: Clarendon.

Boissennade, P. (1964). *Life and Work in Medieval Europe*. Translated by Eileen Power. New York: Harper and Row.

Brock, A. J. (1916). *Galen on the Natural Faculties*. London: William Heinemann.

Brock, A. J. (1929). *Greek Medicine*. New York: Dent.

Browne, E. G. (1921). *Arabian Medicine*. London: Cambridge University Press.

Burn, A. R. (1962). *Alexander the Great and the Hellenistic World*. New York: Collier.

Bushnell, G. H. (1928). The Alexandrian Library. *Antiquity 2:* 196–204.

Carra de Vaux, B. (1900). *Avicenna*. Paris: Alcan.

Cary, M. (1959). *A History of the Greek World from 323 to 146 B.C.* London: Methuen.

Cato. *On Agriculture*. Translated by W. D. Hooper and H. B. Ash (1967). Loeb Classical Library. Cambridge, Mass.: Harvard University Press.

Celsus. *De medicina*. 3 vols. Translated by W. G. Spencer (1935–1938). Loeb Classical Library. Cambridge: Harvard University Press.

Clagett, M. (1955). *Greek Science in Antiquity*. New York: Abelard-Schuman.

Clendening, L. (1942). *Sourcebook of Medical History*. Reprinted 1960. New York: Dover.

Coonen, L. P. (1957). Biologists of Alexandria. *Biologist 39:* 13–18.

Crombie, A. C. (1969). "The Significance of Medieval Discussions of Scientific Method for the Scientific Revolution," in *Critical Problems in the History of Science*. Edited by Marshall Clagett. Madison: University of Wisconsin Press, pp. 79–102.

Crombie, A. C. (1959). *Medieval and Early Modern Science*. New York: Doubleday.

Dioscorides. *The Greek Herbal of Dioscorides*. Illustrated by a Byzantine A.D. 512, Englished by J. Goodyear A.D. 1655. Reprinted 1968. New York: Hafner.

Dobson, J. R. (1927). Erasistratus. *Proc. Roy. Soc. Med. 20:* 825–832.

Dobson, J. R. (1924). Herophilus of Alexandria. *Proc. Roy. Soc. Med. 18:* 19–32.

Edelstein, L. (1967). *Ancient Medicine*. Baltimore: Johns Hopkins Press.

Elgood, C. (1951). *A Medical History of Persia*. Cambridge: Cambridge University Press.

Farrington, B. (1969). *Greek Science*. Baltimore: Penguin.

Ferguson, J. (1973). *The Heritage of Hellenism: The Greek World from 323 B.C. to 31 B.C.* New York: Science History.

Finlayson, J. (1892). Bibliographical Demonstrations of Classical Medical Writers—Galen. *British Medical Journal 1:* 573, pp. 730–771.

Forster, E. M. (1961). *Alexandria, A History and a Guide*. Garden City: Anchor.

Galen. *On Medical Experience*. Translated by R. Walzer (1944). New York: Oxford University Press.

Galen. *On the Natural Faculties*. Translated by A. J. Brock (1963). Loeb Classical Library. Cambridge: Harvard University Press.

Galen. *On Anatomical Procedures*. The Later Books. Translated by W. L. H. Duckworth (1962). Cambridge: Cambridge University Press.

Galen. *On the Usefulness of the Parts of the Body*. 2 vols. Translated by M. T. May (1968). Ithaca: Cornell University Press.

Gibbon, E. (1910). *The Decline and Fall of the Roman Empire*. 6 vols. New York: Dutton.

Graubard, M. (1964). *Circulation and Respiration*. New York: Harcourt, Brace and World.

Green, R. M., trans. (1951) *Galen's Hygiene*. Springfield, Ill.: Thomas.

Greene, W. C. (1933). *The Achievement of Rome*. Cambridge: Harvard University Press.

Grunebaum, G. E. von (1961). *Medieval Islam*. 2nd Ed. New York: Phoenix.

Hadzsits, G. D. (1963). *Lucretius and His Influence*. New York: Cooper Square.

Hall, M. B. (1971). *The Pneumatics of Hero of Alexandria. A facsimile of the 1851 Woodcraft Edition*. New York: American Elsevier.

Hamilton, E. (1957). *The Echo of Greece*. New York: Norton.

Harris, C. R. S. (1973). *The Heart and the Vascular System in Ancient Greek Medicine from Alcmaeon to Galen*. Oxford: Clarendon.

Haskins, C. H. (1923). *The Rise of the Universities*. New York: Smith.

Haskins, C. H. (1924). *Studies in the History of Medical Science*. Cambridge, Mass.: Harvard University Press.

Haskins, C. H. (1927). *The Renaissance of the Twelfth Century*. Cambridge, Mass.: Harvard University Press.

Haskins, C. H. (1927). *Studies in the History of Medieval Science*. Cambridge, Mass.: Harvard University Press.

Hitti, P. K. (1946). *The Arabs, A Short History*. Princeton: Princeton University Press.

Horgarth, D. G. (1897). *Philip and Alexander of Macedon*. London: Murray.

Holmyard, E. J., ed. (1928). *The Works of Geber, Englished in the Year 1678 by Richard Russell*. London: Dent.

Holmyard, E. J. (1957). *Alchemy*. Harmondsworth: Penguin.

Jones, D. E. H. (1971). The Great Museum at Alexandria: Its Ascent to Glory. *The Smithsonian Magazine, Dec. 1971:* 53–60.

Jones, D. E. H. (1972). A Final Creative Gasp and the Museum Fell. *The Smithsonian Magazine, Jan. 1972:* 59–63.

Kibre, P. (1948). *The Nations in the Medieval Universities*. Cambridge: Harvard University Press.

Kibre, P. (1962). *Scholarly Privileges in the Middle Ages*. Cambridge: Harvard University Press.

Kilgour, F. G. (1957). Galen. *Sci. Amer. 196:* 105–114.

Latham, R. E. (1951). *Lucretius, on the Nature of the Universe.* London: Penguin.

Leach, A. F. (1915). *The Schools of Medieval England.* London: Methuen.

Levey, M. (1966). *The Medical Formulary of al-Kindi.* Madison: University of Wisconsin Press.

Lewis, B. (1960). *The Arabs in History.* New York: Harper Torchbook.

Lloyd, G. E. R. (1973). *Greek Science After Aristotle.* New York: Norton.

Lucretius. *On the Nature of Things.* Translated by H. A. J. Munro (1929). London: Bell.

Majno, G. (1975). *The Healing Hand: Man and Wound in the Ancient World.* Cambridge, Mass.: Harvard University Press.

Major, R. H. (1954). *A History of Medicine.* Vol. 1. Springfield, Ill.: Thomas.

Momigliano, A. D. (1975). *Alien Wisdom: The Limits of Hellenization.* Cambridge: Cambridge University Press.

Nasr, S. H. (1968). *Science and Civilization in Islam.* Cambridge, Mass.: Harvard University Press.

Nasr, S. H. (1964). *An Introduction to Islamic Cosmological Doctrines.* Cambridge, Mass.: Harvard University Press.

Nasr, S. H. (1964). *Three Muslim Sages: Avicenna, Suhrawardī, Ibn 'Arabī.* Cambridge, Mass.: Harvard University Press.

Nordenskiold, E. (1935). *The History of Biology.* New York: Tudor.

Onians, R. B. (1973). *The Origins of European Thought.* New York: Arno.

Oman, C. (1924). *A History of the Art of War in the Middle Ages.* New York: Houghton, Mifflin.

Parsons, E. A. (1952). *The Alexandrian Library: Glory of the Hellenic World.* Amsterdam: Elsevier.

Pirenne, H. (1936). *Economic and Social History of Medieval Europe.* Translated by I. E. Clegg. London: Paul, Trench, Trubner.

Pliny. *Natural History.* 10 vols. Translated by H. Rackham, W. H. S. Jones, and D. E. Eichholz (1956–1966). Loeb Classical Library. Cambridge, Mass.: Harvard University Press.

Postan, M. (1959). Why was Science Backward in the Middle Ages?, in *A Short History of Science: Origins and Results of the Scientific Revolution, a Symposium.* Garden City: Doubleday Anchor.

Rashdall, H. (1936). *The Universities of Europe in the Middle Ages.* Oxford: Clarendon.

Sarton, G. (1927). *Introduction to the History of Science.* Baltimore: Williams and Wilkins.

Sarton, G. (1954). *Galen of Pergamon.* Kansas: University of Kansas Press.

Sarton, G. (1959). Forgotten Men of Science. *Saturday Review 42:* 50–55.

Siegel, R. E. (1968). *Galen's System of Physiology and Medicine. An Analysis of his Doctrines and Observations on Bloodflow, Respiration, Humors and Internal Diseases.* Basel: Karger.

Siegel, R. E. (1970). *Galen on Sense Perception. His Doctrines, Observations and Experiments on Vision, Hearing, Smell, Taste, Touch, Pain, and their Historical Sources.* Basel: Karger.

Singer, C. (1956). *Galen on Anatomical Procedures.* Oxford: Oxford University Press.

Singer, C. (1957). *A Short History of Anatomy from the Greeks to Harvey.* New York: Dover.

Singer, C. (1958). *From Magic to Science.* New York: Dover.

Smith, A. E. and Secoy, D. M. (1975). Forerunners of Pesticides in Classical Greece and Rome. *Journal of Agricultural and Food Chemistry. 23:* 1050–1055.

Temkin, O. (1973). *Galenism: Rise and Decline of a Medical Philosophy.* Ithaca: Cornell University Press.

Temkin, O., and Straus, W. L. (1946). Galen's Dissection of the Liver and of the Muscles Moving the Forearm. Translated from *Anatomical Procedures. Bull. Hist. Med. 19:* 167–176.

Thompson, J. W. (1931). *Economic and Social History of Europe in the Later Middle Ages (1300-1530).* New York: Century.

Thorndike, L. (1964). *A History of Magic and Experimental Science during the First Thirteen Centuries of Our Era.* 8 vols. New York: Columbia University Press.

Thorndike, L. (1940). Elementary and Secondary Education in the Middle Ages. *Speculum 15:* 400–408.

Thorndike, L. (1965). *Michael Scot.* London: Thames Nelson.

Walsh, Jr. (1930). Galen's Second Sojourn in Italy and his Treatment of the Family of Marcus Aurelius. *Medical Life,* Vol. 37, No. 9, Series 120.

Wethered, H. N. (1937). *The Mind of the Ancient World, A Consideration of Pliny's Natural History.* New York: Longmans, Green.

Whetham, W. C. D. (1932). *A History of Science and Its Relations with Philosophy and Religion.* London: MacMillan.

White, A. D. (1896). *History of the Warfare of Science with Theology in Christendom.* New York: Appleton.

White, L. (1940). Technology and Invention in the Middle Ages. *Speculum, 15:* 141ff.

White, L. (1968). *Dynamo and Virgin Reconsidered: Essays in the Dynamism of Western Culture.* Cambridge: MIT Press.

Wickens, G. M. (1952). *Avicenna: Scientist and Philosopher.* London: Luzac.

Wilson, L. G. (1959). Erasistratus, Galen and the Pneuma. *Bull. Hist. Med. 33:* 293–314.

Winter, H. J. J. (1952). *Eastern Science.* London: Murray.

Wright, F. A. (1932). *A History of Later Greek Literature.* London: Routledge.

4

THE RENAISSANCE AND
THE SCIENTIFIC REVOLUTION

The Renaissance

The Renaissance, which literally means rebirth, has been called the age of discovery of earth and man. Some accounts simplistically date the beginning of
this period as May 29, 1453, the day that Constantinople fell to the Turks.
Actually, the term applies to the period from about 1300 to 1650. It is impossible to say just what factors caused this complex intellectual, social,
economic, and political transformation of Europe. Certainly many medieval
institutions, such as feudalism and the Holy Roman Empire, and ideals, such
as the unity of religion, the nobility of the ascetic life, and the absolute authority of the pope, had entered a period of decay by the end of the twelfth
century. Among the salient factors which encouraged cultural renewal and
scientific advances can be listed the revival of commerce, the growth of the
cities, the influence of wealthy patrons of learning, and new inventions such
as gunpowder and printing. Progress in biology and medicine was stimulated
by the return to the Greek idea that the human body was truly beautiful and
worthy of study and that nature was complex but governed by natural laws.
The discoveries of new lands eventually overwhelmed biologists with new
plants and animals absolutely unknown to Hippocrates and Galen.

In many ways the Renaissance was a natural bridge between the Middle
Ages and modern times and the natural culmination of the Middle Ages. Although new physical and intellectual discoveries abounded in this period, in
intellectual interests there was no sudden break with the past. The intellectual
preoccupation of this period became predominantly "humanist" as opposed to
the theological obsession of the Middle Ages. Like the much maligned
scholastics of the Middle Ages, the humanists too were enamored of scholarship—words and books. But their interests were more of earth than of heaven.
Humanism implied a concern for the classics of antiquity, a new delight in

nature, and the realization that man and his relationship to human society were as worthy of study as his relationship to God. Leaders of the great scientific revolution such as Vesalius, Copernicus, and Harvey were more likely to see themselves as scholars returning to the original spirit of Aristotle and Galen than as revolutionaries making a complete break with the past.

Some scholars of the Middle Ages had admired the literature of Greece and Rome. Indeed most of the Latin and Greek classics now extant were known to medieval scholars in limited areas. Yet many precious texts languished in forgotten corners of monasteries or cathedral libraries. Boccaccio found precious manuscripts either rotting away with neglect or mutilated to make psalters or amulets. Before the fall of Constantinople many manuscripts were rescued and taken to Italy, which still has a fine collection of ancient Greek manuscripts. Texts by Herodotus, Thucydides, Polybius, Demosthenes, Aristotle, Euripides, and others were regarded as treasures by Italian nobles and clergy. True scholars were not content with Latin translations but labored to recapture Greek classics in their pristine state. Italian scholars also took great pride in the literature and art of Rome in her past glory.

The influence of the humanists was the dominant factor in the intellectual life of Western Europe for a century. They brought back to secular learning a prestige undreamt of for centuries. Serving as the secretaries and advisors of popes and princes, humanist scholars influenced the clerical and secular realms. While the humanists were more concerned with art, literature, and politics than pure science, their new perspective served the needs of science as well.

But purely intellectual exercises were not the only stimulus for change in Europe. The ravages of the Black Death (1348), which may have killed one quarter to one half of the population in various areas, played a part in the broader transformation of society. Certainly the general confusion that followed this disaster had some role in breaking down the medieval system of feudalism. The plague killed peasants and nobles alike, and resulted in labor shortages and changes of landholders. Peasant uprisings broke out all over Europe. Yet trade and exploration increased. Cities were growing as was a money economy and the concept of the modern state. Banking and mining were carried out on a scale never seen before. Expansion of trade led to the discovery of a sea route to India by the Portugese and the discovery of America.

Printing and Gunpowder

The availability of printing and firearms at the end of the Middle Ages had an effect similar to that of the introduction of iron and the alphabet at the end of the bronze age. Not long after the Mongol invasions, gunpowder was mentioned in a letter written by Roger Bacon (1249). Gunpowder and probably firearms came to Europe through the Mongols. Cannons were first mentioned in 1325 and first pictured in 1327 as vase-shaped devices which shot a bolt with an arrow head. Gunpowder and firearms have been called the "great equalizers." The mounted knight and fortified castle of the very rich were no longer invincible. Military power became concentrated in the hands of those who could control the production of gunpowder and cannons. Thus the development of firearms gave some impetus to the rise of absolute monarchies in the sixteenth and seventeenth centuries.

The Mongols brought paper-making and printing from China to Europe. The Mongols brought some samples of Chinese printing with them, since playing cards appeared in Europe not long after the invasions. Block printing in Europe seems to have begun at Ravenna in 1289. Imprinting was actually a very old technique. The Babylonians made their symbols on bricks, and many cultures put theirs on coins, pottery, cloth, book covers, and seals for documents. Block printing with wood or metal blocks was known in China and Japan in the eighth century. Typography was used in China with wooden type characters by the eleventh century, but this method of printing with separate and movable type was not really useful for the Chinese system of writing. Movable metal type was first used in Korea in 1403.

The earliest known document to be printed with movable type in Europe was a letter of indulgence printed by Johann Gutenberg in 1454. Although his name is today almost synonymous with the early modern method of printing, it is likely that what he really did was to improve the technique that others had developed. Johann Gutenberg was born to a prosperous family in Mainz about 1400. Since his father's name was Gensfleisch (Gooseflesh), he preferred to use his mother's family name. He seems to have made experiments in cutting and casting metal type while he lived in Strasbourg. Clear records of his printing business go back to August 22, 1450, when he signed a contract with Johann Fust, a wealthy goldsmith. Gutenberg had mortgaged his printing press to Fust for a loan of 800 guilders. In 1455, Gutenberg had to forfeit his press to Fust as he could not repay his loan. In 1456, having borrowed money again, Gutenberg set up another press. This second venture is the one that produced the famous Gutenberg Bible. After Adolf of Nassau attacked the town of Mainz, printers fled all over Europe. By 1463 printers were working in Strasbourg, Cologne, Basel, Augsburg, Nuremberg, and Ulm. Gutenberg himself fled to Eltville because of more financial problems. Shortly thereafter, death released him from all debts.

Despite some grumbling about the vulgarity of printing as compared to hand copying and the rather snobbish fear that an excess of literacy might be subversive, books were soon being delivered to eager buyers by the wagonload. One scholar in Basel wrote to a friend asking if he wanted some of the new books to be arriving. "If you do, tell me at once," he wrote, "and send the

money, for no sooner is such a freight landed than thirty buyers rise up for
each volume, merely asking the price, and tearing one another's eyes out to
get hold of them."[1]

It cannot be said that printing created the Renaissance, which had begun in
Italy about 150 years before, but it probably did accelerate the Renaissance
and its progress in northern Europe. Printing made books more readily available and increased literacy. Not only scholarly works were important, for
more popular works written in the vernacular were also in great demand. Sir
Thomas More estimated that at the beginning of the sixteenth century 40 percent of the English could read. It has been suggested that printing paved the
way for the Enlightenment, the American and French revolutions, and demo-
cracy, but another observer said of printing, "after speech, it provided a
readier instrument for the dissemination of nonsense than the world has ever
known until our time."[2]

Art and Anatomy

During the Renaissance, progress in biology was mainly confined to botany
and anatomy. Both medicine and art required accurate anatomical knowledge.
In art there was a new emphasis on accurately representing nature, scientific
use of perspective, and above all the idea that the human body was beautiful
and worth studying.

The great artists of the Renaissance looked at the human body as the most
perfect work in creation. Knowledge of the exterior was insufficient to them.
They wanted to study the muscles and bones, how they worked and how they
were related to the interior of the body. To attain realism in their work,
many artists turned to dissection, convinced that study of the cadaver would
make their art more true to life. None exceeded Leonardo da Vinci—painter,
architect, anatomist, and many-sided inventor—in the execution of artistic and
scientific work, but other Renaissance artists were equally keen to study
nature for themselves. Giotto was one of the first painters to master the
principles of faithfully reproducing nature in art. Certainly Raphael (1483-
1520), Titian (1477-1576), Botticelli (1444-1510), Dürer (1471-1528), and
the Holbeins (1465?-1524 and 1498?-1543) reflect a precise and intense new
way of looking at nature. Michelangelo (1475-1564), whom Leonardo dis-
liked, was also a multifaceted artist—sculptor, painter, architect, and poet. Imbued with the spirit of humanism, like Leonardo he considered it necessary to
conduct dissections of musculature and internal organs to represent the human body as accurately as possible. Verrochio (1435-88)—painter, sculptor,
and the foremost teacher of art in Florence—insisted that all his pupils learn
anatomy. This included the study of surface anatomy and flayed bodies to
learn about the underlying musculature. To render nature more perfectly,

artists also did animal dissections, attended public anatomies at the universities, and took private lessons in anatomy.

Leonardo da Vinci (1452–1519): The Complete Renaissance Man

Studies of Leonardo's notebooks suggest that he was hundreds of years ahead of his time. It is tempting to speculate that if he had published extensively or completed the ambitious projects he set up for himself he might have revolutionized several scientific disciplines. To regard Leonardo as *typical* of his period is unrealistic, even though he had many brilliant contemporaries. Rather, Leonardo's work indicates what was *possible* for a man of incomparable genius with the materials available in the fifteenth century.

Leonardo was the illigitimate son of a Florentine lawyer, Ser Piero, and a peasant woman named Catarina. Yet Leonardo was brought up in his father's house by his grandparents. Although his father was to marry four times, Leonardo remained an only child until he was about 20. Ser Piero's fourth wife gave him some 12 children. At 14 years of age, Leonardo was apprenticed to Andrea del Verrochio. Within the next 10 years Leonardo was recognized as a distinguished artist and won the patronage of the wealthy and powerful Medicis. Thereafter, Leonardo led a restless and adventurous life, serving various patrons, prosecuted on charges of homosexuality, beginning and discarding numerous ambitious projects. Although his genius is unquestionable, many of his projects were left unfinished. The actual legacy of Leonardo is quite modest compared to his potential. He did not leave a single complete statue, machine, or book despite sketches and plans for many exciting and innovative constructions. Some 5000 pages of notes and sketches did not reach his contemporaries at all but went unread for centuries. Indeed, the left-handed Leonardo kept his secret notebooks in code, in a kind of mirror writing.

Although not truly a scientist, Leonardo's idea that the scientific study of nature was fundamental to art brought him closer to the modern concept of the scientific method than many of his contemporary "scientists." Observing the details and workings of nature, his mind restlessly turned from astronomy to anatomy, from music to machines. His idea that the earth might move around the sun may have come to the attention of Copernicus. His studies of music and painting led to speculations about the idea of sound and light waves. In more down-to-earth moments he described the use of rings in trees to determine their age and commented on the nature of fossil shells.

The needs of artistic expression which first led Leonardo to the study of the superficial anatomy of the human body inexorably drew him to an exploration of the general anatomy of man and to experimentation and comparative anatomy. Through dissection and experimentation Leonardo believed he would uncover the mechanisms which governed movement and even life itself. Leonardo began his dissections as an artist, but continued as a scientist concerned with anatomy not only as a means to an end, but as a study worth doing of itself. He attended public anatomies whenever possible, and received permission to study cadavers at a hospital in Florence. Leonardo may have supplemented these studies with some clandestine grave robbing, but he certainly had the opportunity to dissect at least 30 human bodies. Most interesting was his comparison of a seven-month fetus with a man who died when about 100. Such studies were in keeping with his preference for the unique and grotesque.

Leonardo carried out many studies of the muscles and even constructed models to study muscle action. He also developed models of the heart valves and did vivisection experiments to study the heartbeat. For example, he drilled through the thoracic wall and, keeping the incision open with pins, observed the heartbeat of a pig. Although he realized that the heart was actually a very powerful muscle, he generally accepted Galen's views on the circulation of the blood, including the imaginary pores in the septum.

Not all of Leonardo's methods are known, but he was the first to make wax casts of the ventricles of the brain by injecting hot wax into the hollow ventricles. When the wax solidified it was possible to do careful dissections of the delicate tissue. Leonardo used serial sections and developed other techniques for studies of soft tissues such as the eye. To facilitate dissection, the eye was placed in the white of an egg and heated until the albumen coagulated. Leonardo studied animal anatomy for both art and science. Among the animals he dissected were moths, flies, fish, frogs, crocodiles, birds, horses, oxen, sheep, bears, lions, dogs, cats, monkeys, developing chicks and calves. Such studies were the first since the time of Aristotle and Galen to seriously exploit the advantages of comparative anatomy. Although the project was never completed, his plans for an equestrian statue of Lodovico Sforza of Milan led to an intense study of horses and comparisons of the limbs and motions of man and horse.

Like so many of his projects, Leonardo's great book on human anatomy was left unfinished. When he died, his manuscripts were scattered among various libraries, some perhaps lost. Many of his anatomical studies were unknown for centuries after his death. Some were known although unpublished. William Hunter knew of the Leonardo manuscripts in the King's library at Windsor and obtained permission to publish them, as he announced in 1782, but Hunter died before he could do so. Although Leonardo may have indirectly influenced anatomical science, for the most part his great works remained secret. Fully endowed with the curiosity and skill necessary for a great scientist, Leonardo somehow lacked the perseverence to harness his energies and produce an impact commensurate with his genius.

Renaissance Anatomy

Medicine and art operated within a framework very different from that of today. In the Renaissance world these fields were so closely linked that artists, doctors, and apothecaries were members of the same guild in Florence. True anatomical illustrations appeared in the sixteenth century. Certainly older books had had pictures, but these were generally formal depictions of autopsies, "wound men," and astrological figures. The first illustrations in anatomy texts were more artistic or bizarre than accurate or illuminating. While cheerful-looking people held up flaps of skin to display their internal organs, the text rambled on about other matters. Even where scientist and artist were united in one searching genius, such as Leonardo, faulty observation and distortion still occurred. With the ideal of firsthand observation growing, some complained that illustrations would draw students away from actual dissections.

Despite some progress during the High Middle Ages and early Renaissance, the opportunity to conduct or even observe human dissections was quite rare. When public anatomies were performed, the stylized ceremony was hardly likely to produce any new insights or inspire the typical student to original research. The professor, trained in the scholastic pattern, would not have conducted the dissection himself. A technician did the actual autopsy while the professor read from Galen. If the reader and technician were well synchronized, the students would see the organs as the professor described them. But in the crowded arena-like room, with an illiterate technician, even this was unlikely. The typical student preferred scholastic disputations to the actual demonstration.

Renaissance anatomists were not the first Europeans to revive the Alexandrian practice of human dissection, since these had been done as early as the mid-thirteenth century quite routinely at Italian universities. Emperor Frederick II had made it compulsory for medical students to attend at least two anatomies, but these could not have been particularly enlightening experiences. The tools used by surgeons and anatomists were very primitive. The technician at an anatomy opened the abdomen and chest cavity to show the internal organs with only a knife. After a brief look at the mashed, mangled, and putrefying organs, the muscles, nerves, and blood vessels were supposed to be studied. But this was generally too difficult for the technician and too boring for the students. Even Renaissance students preferred the arguments between the Galenists of the medical faculty and the Aristotelians, professors of philosophy. Although religious dogma did not ban anatomical research outright, it did have a depressing effect on anatomy.

Renaissance anatomists were plagued by many problems. Probably the greatest was the shortage of bodies, but anatomists also lacked reliable guides to direct their efforts and tools with which to exploit the available materials. Furthermore, any observations they might make would be most difficult to describe to other researchers because there was no standard nomenclature for the parts of the body. Descriptions were issued in a jumble of Greek, Latin, Arabic, and the vernacular. The same structure might be referred to by different names or the same name applied to different structures.

Rejecting the medieval and Arabic versions of scholarship, the humanists of the sixteenth century stimulated progress in anatomy. The authority of Mondino de' Luzzi (ca. 1275-1325), who had at least done some dissections himself, was rejected along with his widely used textbook. This did not immediately lead to original research since the humanists acclaimed Galen as the greatest source of information on human anatomy. Mondino and others were rejected as a corruption of the superior achievements of Galen. By 1500, Galen's lost work *On the Use of the Parts* was available in editions suitable for use by students. The text set the precedent of discussing the function of each organ along with the anatomical dissection. Johannes Guinther (1487-1574), a professor at Paris, published a Latin translation of Galen's *On Anatomical Procedures*. This rediscovery and Guinther's commentary on Galen ushered in the period in which Galen was held in the highest esteem. The great Renaissance anatomist Andreas Vesalius assisted Guinther in the writing of his text, *Anatomical Institutions According to the Opinion of Galen for Students of Medicine.*

Andreas Vesalius (1514-1564): Reformer of Anatomy

Andreas Vesalius regarded his own work, *On the Fabric of the Human Body* (1543) as the first advance in anatomy since the time of Galen. Indeed Vesalius took up the study of human anatomy practically at the point where Galen had left it. The major innovation of Vesalius was to look directly at the human body, rather than at the pages of Galen. The authority of the ancient anatomist had grown so weighty that discrepancies between his descriptions and the actual state of the cadaver were explained away as abnormalities or changes in the human body since his time. Thus, the human femur must have been curved when Galen described it and the use of trousers must have straightened it out. Jacobus Sylvius (1478-1555) defended Galen's description of the human sternum as made up of seven bones since it was possible that in a more heroic age the chest might have contained more bones than later degenerate specimens.

The sixteenth century was a true inflection point for progress in anatomy and medical science. Although Western medicine is now heavily based on anatomy, this is not the only approach possible. The Chinese had little interest in anatomy although their physicians were highly skilled in internal medicine. The Moslems similarly believed that medicine should deal with

Woodcut portrait of Andreas Vesalius. [From J. B. de C. M. Saunders and C. D. O'Malley (1950). *The Illustrations from the Works of Andreas Vesalius.* **Cleveland: World Publishing Co.]**

diseases and their causes—not with the structure of the body. Although the Christian Church had not totally forbidden autopsies, the Moslems did have religious prohibitions against human dissection, as did the Chinese.

Before describing the life and work of Vesalius, it should be remembered that anatomy itself could not solve the problems of function. For this, progress in chemistry was an absolute prerequisite. Vesalius' peculiar contemporary Paracelsus as a chemist and physician disdained the practice of anatomy. Thus, although other pathways were possible, the one taken by Vesalius proved to be the more travelled by for biology until the end of the eighteenth century.

Vesalius was a rare combination of scholar and scientist—skilled as artist, humanist, and naturalist. His great work *On the Fabric of the Human Body* appeared in 1543—the same year as the revolutionary astronomical work of Copernicus. Both Copernicus and Vesalius rejected the corruption of classical learning and sought to raise modern science at least to the level attained by the ancient Greeks. But despite his rejection of Galen's errors, Vesalius shared with his predecessor a desire to explain the workings of the body in terms of structure and function as the workmanship of the Great Craftsman.

Vesalius was born in Brussels, Belgium, into a family of physicians. His father was pharmacist to Charles V and often away on travels with the Emperor. When quite young, Vesalius began to teach himself comparative anatomy as he dissected mice, moles, cats, dogs, and weasels. Although he studied at both Paris and Louvain, institutions noted for their extreme conservatism, his innate curiosity overcame the stifling effect of education. Jacobus Sylvius (1478-1555) was the dominant influence at Paris in anatomy, but his emphasis was on Galenic studies rather than original work. His method of teaching was the scholastic one and his great anatomical textbook was actually a commentary on Galen. Although Vesalius assisted Sylvius, he soon quarreled with the master. Working on his own, Vesalius won so great a reputation that he was called on by physicians and students to perform public dissections in which he demonstrated not only the internal organs, but also the muscles, nerves, blood vessels, and bones.

Becoming very critical of the medical professors who merely studied Galen and lectured at anatomies in the scholastic tradition, Vesalius quarreled with the faculty. His precarious position as an enemy alien in wartime forced him to leave Paris without graduating. During a brief visit to Louvain, he participated in the autopsy of an eighteen year old girl. Her uncle suspected that she had been poisoned, but Vesalius with his keen knowledge of human anatomy concluded her girdle had killed her. Perhaps Vesalius changed female fashions for a time, since he reported that after the public anatomy many court ladies ran home to divest themselves of their girdles.

The autopsy in Louvain was a rare and fortunate opportunity because obtaining human bodies was the greatest obstacle to anatomists. Female cadavers for autopsy were especially scarce. Vesalius himself only had the opportunity to work on six female bodies between 1537 and 1542. Since medical students were required to see human anatomies, teachers sometimes encouraged students to "extralegal" ways of obtaining the necessary material. As early as 1319, four medical students were prosecuted for grave robbing. In the entire course of studies, medical students might be able to see two males and one female dissected for their edification. Osteology was generally a required course. Getting a set of bones was left to the ingenuity of the student. Brave young men often obtained at least part of their education in graveyards while snatching bones out of the teeth of savage dogs. Vesalius himself described the risks he had to take to obtain bodies. On one occasion he stole the bones of a criminal left on the gallows, sneaking into town with some of the bones under his coat over a period of many days, and hiding the skeleton under his bed. When Vesalius became famous and travelled about giving public lecture-anatomies, waves of grave robbing incidents were reported.

Learning that the Inquisition was taking a warm interest in his unorthodox activities, Vesalius made his way to Padua. Graduating in medicine in

59567

December 1537, Vesalius quickly advanced to the Chair of Anatomy and Surgery. He had been unsuccessful in reforming the curriculum in Paris, but Padua was a relatively enlightened university. Although Vesalius encouraged many reforms, his major innovation was to dissect and lecture simultaneously, doing away with the technician. Hundreds of listeners attended the radically new anatomies conducted by Vesalius. Never one to spare the feelings of his more reactionary colleagues, Vesalius told his students they could learn more at a butcher shop than from some of their blockhead professors. He emphasized the need to re-examine the human body rather than relying on Galen, who had never had the opportunity to dissect human beings. In 1540, Vesalius conducted a dramatic demonstration of this new idea when he assembled the bones of an ape and a man. He was able to document over 200 differences in skeleton alone, in which Galen was in error with respect to human anatomy.

The thorough anatomies conducted by Vesalius occupied him from morning to night for three weeks at a time. This was a great contrast to the three-day anatomies conducted by Mondino. To minimize the problem of putrefaction, anatomies were done in the winter. Several bodies were used at a time so that different parts of the body could be studied and their relationship examined. Large diagrams were used for the guidance of his students. The anatomies began with a study of the skeleton and then proceeded to the muscles, blood vessels, and nerves. Another cadaver would be used for the demonstration of the organs of the abdomen, chest, and brain. This was essentially the organization used in his book, *The Fabric of the Human Body.* To improve the techniques of dissection, Vesalius introduced many new tools—some of his own design and some borrowed from craftsmen he consulted for that purpose.

Just as Copernicus revolutionized ideas about the structure of the universe—the macrocosm—so did Vesalius affect ideas about the structure of the microcosm—man. Although the *Fabrica* retained some errors of the Galenic tradition, Vesalius came close to his goal of describing the human body as it really is, without deference to ancient authorities. The text and illustrations combined to produce a masterpiece which must be counted among the greatest contributions to medical literature. Of course it was not received in this spirit by all of his contemporaries. The work infuriated Galen's still loyal followers, who denounced the *Fabrica* and its author. Enemies even resorted to absurd accusations of human vivisection against Vesalius. His old professor renamed him "Vesanus," meaning "madman."

Even publishing a work of this revolutionary nature took a brave and independent man. Johannes Oporinus (1507–1568) of Basle, the printer-publisher of the *Fabrica,* had participated in other dangerous enterprises. He had been secretary to the controversial Paracelsus, and defender and friend of the martyred Servetus. Oporinus even served a term in jail for publishing a Latin translation of the Koran.

Lincoln Christian College

Opposition to his work was so intense that Vesalius decided to abandon research and the academic world. Following in the footsteps of his father, he became court physician to Emperor Charles V of Spain, to whom he had dedicated the *Fabrica*. Once in service to the Emperor, Vesalius had little opportunity for furthering research or writing. Charles was a challenge to the skill of his physicians, with his bad combination of weak constitution, gargantuan appetites, "Hapsburg jaw," and gout. When the University of Pisa offered Vesalius the Chair of Anatomy, Charles refused to release him. In 1556 Charles abdicated in order to retire to a monastery. Vesalius was created Count Palatine, granted a pension, and transferred to the service of Philip II, the son of Charles V and successor to the throne. During this period Vesalius was sent to several other royal courts. When Henry II of France was injured while jousting, Vesalius and the famous French barber-surgeon, Ambroise Paré (1517–1590) were among the physicians assembled. Paré and Vesalius agreed that the wound was fatal. Another difficult case involved Don Carlos, son of Philip II of Spain. At the age of 17, while "hastily pursuing a wench, he fell down stairs and broke his head."[3] Vesalius found the Spanish doctors to be unbelievably bad and the people determined to use religious magic to save the prince. The remains of the Blessed Diego were placed near Don Carlos while 300 Spaniards paraded half-naked, flagellating themselves and praying for a miracle. The English ambassador was convinced nature had cured the Prince in spite of the surgeons' inconsiderate dealing.[4]

Vesalius could not be content in service to Philip II, known as a sullen and excessively pious man. With the Emperor totally controlled by his confessors, purely scientific studies were impossible. Soon Vesalius felt that the Inquisition was considering his revolutionary views rather suggestive of heresy. Yearning to return to the academic world, Vesalius hoped to recover his former position at Padua since his successor, Fallopius, had died in 1562. According to a persistant but doubtful tradition, Vesalius was forced to leave court and undertake a pilgrimage to the Holy Land as penance for doing a "premature" autopsy. The story of this "vivisection" has been attributed to Paré or Fallopius, either of whom might have been jealous of Vesalius' reputation. On the other hand, Vesalius might have used the excuse of a pilgrimage to escape the hostile atmosphere at Philip's court. In any case, he never returned from the pilgrimage and the exact cause and place of his death are uncertain.

The Fabric of the Human Body

Vesalius opened the *Fabrica* with a preface that amounts to a defense of science. He warned of the dangers that result from obstacles to the arts and sciences and the danger that comes from overspecialization. He included a liberal dose of flattery for Charles and the University of Padua, along with a

Lincoln Christian College

good word for the Senate of Venice for their liberality towards learning. Some of his arguments could well grace a modern research grant application.

The first book is devoted to the skeleton which supports the whole body and the second to the muscles which are the "workmen of motion." These studies are clearly the most successful aspects of sixteenth century anatomy. The third book describes the vascular and arterial systems. The venous system was still regarded as the most important, especially as it was routinely used for phlebotomy. Convinced that the function of the veins and arteries are quite obvious, Vesalius asserted that vivisection experiments were not needed to study these systems. Obviously, the veins were fashioned to carry nourishment. "The Creator made the veins for the prime reason that they may carry the blood to the individual parts of the body," wrote Vesalius, "and be like canals from which all parts suck their food."[5] Vesalius knew that the arteries pulsate and carry blood in the living animal, but he affirmed the common conviction that their function was to carry the innate heat and

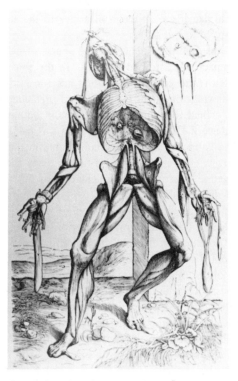

Illustration from the works of Andreas Vesalius. [From J. B. de C. M. Saunders and C. D. O'Malley (1950). *The Illustrations from the Works of Andreas Vesalius.* **Cleveland: World Publishing Co.]**

vital spirit. The fourth book was devoted to the description of the nervous
system. In keeping with Galenic physiology, the function of the nerves was
to carry the animal spirits. Descriptions of the abdominal viscera and generative organs in book five were quite good. The sixth book described the
thoracic viscera. In book seven, Vesalius described the brain, pituitary gland,
and the eye. A chapter on dissection in living animals concluded the book.
While Vesalius did practice vivisection on animals, he did not discover new
physiological truths from these studies. Generally he recommended such exercises for students to train their hands and senses for surgery and the closure
of wounds.

The *Fabrica* was a ponderous treatise intended for serious anatomists, but
Vesalius simultaneously brought out a condensed version of his masterpiece
entitled *Epitome* for the benefit of students. In 1555 Vesalius published a revised edition of the *Fabrica* and a reprint of the *Epitome*. The *Fabrica* and
its anatomical illustrations were plagiarized and reprinted without authoriza-
tion many times over during the life time of Vesalius and for several centuries
after his death. Such plagiarism, although offensive to the author, was a clear
sign of the superiority of his work.

Vesalius exerted a profound influence on the course of biological research,
yet his immediate impact was confined to Italy. In the conservative French
medical schools, Galen remained the dominant force for another century. In
Germany, the constant religious strife of the period precluded any serious
scientific progress in a field touching so intimately on the nature of man.
The intellectual heritage of the University of Padua was passed on from
Vesalius to Fallopius, Fabricius, and finally culminated in the revolutionary
work of William Harvey. It was Harvey who finally made the great step
forward from observation of structure to elucidation of function in his
analysis of the heart and circulatory system.

In contrast to his success with anatomy, Vesalius was not much more advanced than Aristotle and Galen in terms of physiology. Digestion was vaguely explained as the "cooking" of food in the abdominal cavity. The object of
respiration was to cool the blood. His views on reproduction and embryology
were similar to Aristotle's. Most surprising was his failure to appreciate the
function of the heart and circulatory system. Although he gave an exhaustive
description of the structure of the heart and dissected the arteries, veins, and
valves, he followed Galen's scheme of the function of these organs. He
searched diligently for the pores in the septum but, unable to find them, concluded that they either did not exist or were invisible. Although in the second
edition of the *Fabrica* he explicitly denied that any such channels could be
found to completely pierce the septum, he did not seize on this anomaly and
reject Galen's system, but merely wondered at the mystery.

It seems a defect in Vesalius that he did not go on to realize the physiological

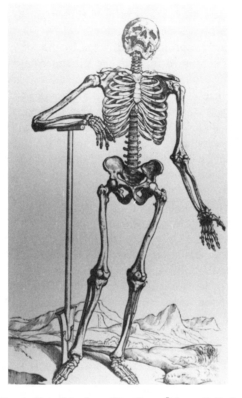

Illustration from the works of Andreas Vesalius. [From J. B. de C. M. Saunders and C. D. O'Malley (1950). *The Illustrations from the Works of Andreas Vesalius.* Cleveland: World Publishing Co.]

import of the dissections he so skillfully performed. Such an assessment underestimates the difficulty of the task. The problem of the circulation was too intimately tied to theological aspects of the distribution of "spirits" to be a safe research topic and Vesalius had already attracted the malevolent eye of the Inquisition. Yet it seems that although he possessed a powerful analytical mind, Vesalius lacked something in imagination and the appreciation of dynamic function.

Vesalius exhorted students and physicians to learn, not from Galen's anatomical books, but only from the completely trustworthy book of man. He denounced the "detestable procedures" by which dissection lectures were conducted. The ignorant technician dissected the cadaver while the learned professors sat "like jackdaws aloft in their high chairs, with egregious arrogance croaking things they have never investigated. . . ."[6] Anatomy was so wrongly taught in the schools that a medical student might learn more from a butcher than from his professor of anatomy!

Although anatomy was profoundly affected by the work of Andreas Vesalius, anatomists continued to "see" the structures described by Galen. Partly this was because human bodies were rare and anatomists often used the same animals Galen had used to clarify their studies of particular organs. Thus the five-lobed liver was still obvious in dogs even after Vesalius realized that it is not found in humans. Galen's *rete mirabile* was difficult to find in humans long dead, even for Vesalius. At an early period in his career, he kept the head of an ox or lamb handy to demonstrate this network of vessels which is found at the base of the brain in cattle but does not exist in man. Later he wondered at his own youthful credulity.

If it seems difficult to understand why more of Galen's errors were not thoroughly exorcised from the pages of anatomy, it might help to compare the Renaissance anatomist to a college student of today running a typical "cookbook" experiment in a laboratory course. The student does not expect to discover anything new and does the experiment merely to review or confirm some accepted fact or learn a standard procedure. It was in this same spirit that anatomists dissected in order to confirm their reading of ancient authorities.

Another Facet of Renaissance Science: Alchemy

Not all Renaissance biology was grounded in anatomical investigation. Indeed, another breed of investigators totally rejected anatomical research as irrelevant to understanding health and disease, life and death. These mystics and radicals immersed themselves in the ancient, obscure, and secretive philosophies and techniques of alchemy.

While chemistry as we know it today may be regarded as a young science, founded upon the atomic theory of John Dalton (1766–1844), chemical technologies have roots that go back to the earliest civilizations in their work with glass, metals, cements, perfumes, medicines, and alcoholic beverages. Ancient Sumeria had mastered techniques of creating glass "gems" and false gold. Ancient descriptions of chemical manipulations suggest that elaborate rituals were used to standardize processes which were but vaguely understood. Metallurgy was one of the most important of the roots of chemistry because it included the smelting of ores, interpreted as the transformation of ore to metal. Such technological achievements as making bronze, iron, and purifying gold and silver were of obvious economic importance and a source of great power to the societies that possessed them. Thus, the processes were often the closely guarded secrets of families or guilds.

The alchemists, who absorbed and transformed protochemical techniques and philosophy, are often dismissed as quacks, charlatans, or fools. Yet the alchemists deserve a place in the history of biology, for they were the first to attempt chemical analyses of living matter. This was not so much because they were rational scientists or nascent biochemists, but believing that the whole universe was alive, they did not limit their investigations to the inorganic world. Their systems were based on complex, obscure, and mystical religious and philosophical theories. Nevertheless, their techniques, including distillation and sublimation, were often quite sophisticated improvements of ancient methods.

According to the most fundamental alchemical traditions, the seeds of noble metals are contained in the base metals. If the right conditions and methods were known, these seeds could be encouraged to grow. In thinking that beyond the known gross qualities of common crude materials there were other hidden qualities, the alchemists were correct and in their search for these occult properties, they sometimes produced valuable products. For example, in trying to distill off the "essence" of wine, they discovered strong liquors (*aqua ardens*) which supposedly brought "increased awareness." The liquors and cordials prepared by alchemists were regarded as valuable medicines.

Alchemical philosophy was shared by many cultures, including ancient Babylonia, China, India, Greece, and the Moslem civilization. Alchemy did not originate as an empirical science, but as a means of manipulating the passions, marriages, growth, death, and transmutations of matter. Thus Mircea Eliade described alchemy as a "sacred science."[7] When alchemy gave way to a rudimentary chemistry, which was a secular experimental science, it lost its old "validity" and all reason for existing.

Alchemical tradition produced an explosion of creativity and buffoonery known as Philippus Aureolus Theophrastus Bombastus von Hohenheim Paracelsus (1493–1541). Mercifully, he is best known as Paracelsus, which presumably indicated one higher than Celsus. This strange man, who may have been the inspiration for Christopher Marlow's "Faust," could also be the rightful ancestor of our modern penchant for "better living through chemistry."

Paracelsus (1493–1541)

Little is known of the early life of Paracelsus, but legends and scandals abound. His father, Wilhelm von Hohenheim, illegitimate son of a noble family, was a physician at Einsiedeln. His mother seems to have been a nurse at the monastery hospital there. Rumor suggested that Paracelsus himself was the natural son of some nobleman. His mother seems to have suffered from some form of insanity and eventually committed suicide. From portraits and descriptions it seems that Paracelsus suffered from rickets in his youth, a not

uncommon disorder at the time. Another legend has it that he was a eunuch—
rendered so either by a band of drunken soldiers or during a hunt when attacked by a wild boar. Much better authenticated is the picture of Paracelsus
as generally drunk, dishevelled, and disputatious.

He is said to have entered the University of Basel at the age of 16, but to
have become a pupil of the Bishop Trithemius at Wurzburg shortly after that.
He had quickly tired of the scholastic dogmatism at the university, and preferred to study alchemy with the abbot who had set up a laboratory in his
monastery. To learn more of the secrets of metallurgy, Paracelsus became an
apprentice at the mines in the Tyrol. Sigismund Fugger, one of the owners of
the mines, was himself an ardent alchemist. In search of secret alchemical
lore, Paracelsus travelled through Germany, Spain, France, and possibly as far
as Moscow, Constantinople, Israel, and Alexandria. In his travels he visited

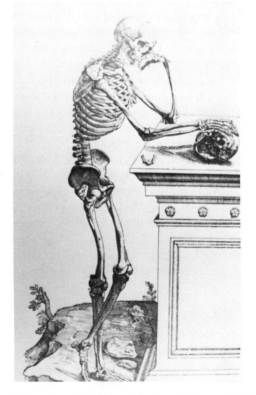

Illustration from the works of Andreas Vesalius. [**From J. B. de C. M.
Saunders and C. D. O'Malley (1950).** *The Illustrations from the Works of
Andreas Vesalius.* **Cleveland: World Publishing Co.**]

universities, absorbing what he could there, and associated with people believed to be experts in the "occult sciences"—witches, gypsies, executioners, barber surgeons, and teamsters. Although he probably did not formally get a degree in medicine, he enjoyed a successful, if precarious, private practice. In 1526 he returned to Switzerland and settled at Strasbourg where he cured a stubborn ailment of the printer, Johan Frobenius, a close friend of Erasmus. Their recommendations led to his appointment as Professor of Medicine and City Physician at Basel. As a medical practitioner he was remarkably successful. He produced dramatic cures with simple, cheap medicines. He ridiculed the pharmacists and other physicians as a "misbegotten crew of approved asses," and worse yet, he took away their business by undercutting their prices and curing their disillusioned patients.

At the University, Paracelsus enraged his colleagues when he ceremoniously initiated his lectureship by burning the works of Avicenna and Galen with sulfur and nitre. Thus he showed his contempt for ancient dogma and his faith in the new chemical medicine. Wearing the alchemist's leather apron rather than academic robes, he also broke with tradition by lecturing in the vernacular rather than in Latin. The opposition became so great that he was forced to leave Basel only two years later. Seemingly doomed to a wandering life, he would set up his alchemist's apparatus in a kitchen, produce marvelous cures, stir up new enemies, and then be forced to flee again. Finally the Archbishop Duke Ernst of Bavaria, a man also interested in alchemy, invited him to settle in Salzburg. When it seemed he could at last live in peace, he died suddenly in a mysterious but certainly unnatural fashion. His friends said he was the victim of a deliberate hostile attack, but his enemies said he met with an accident while drunk. Despite his turbulent life, Paracelsus was a prolific writer. Over 300 works, of which about one-third pertain to chemistry, were attributed to him.

Paracelsus has been called everything from genius to quack, from founder of medicinal chemistry to a bizarre footnote to the main march of the history of science. Although it is impossible to calculate the absolute worth and influence of Paracelsus, it is certain he was a marvelous combination of sorcerer and healer, magician and scientist, fraud and prescient promotor of chemistry. He claimed that "natural magic" would enable man to see beyond the mountains, hear across the oceans, divine the future, cure all diseases, make gold, and even create man. He himself claimed to have raised a homunculus which he incubated in a gourd without the help of any female. It was said that he had found the Philosopher's Stone, which he carried in the handle of his famous sword. After his death legends of his accomplishments grew even more inflated. Conrad Gesner called the followers of Paracelsus druids and magicians. John Donne wrote that Paracelsus carried out the devil's orders. Pilgrimages to his birthplace continued for many years. Even in the 1850s, during a cholera epidemic in Austria, there were visits to his tomb and appeals for his help.[8] Paracelsus has been called the "Luther of medicine." Just as Luther had burned the Papal Bull, so had Paracelsus burned Galen and Avicenna and opened the way for the reformation of chemistry.

Why do judgments of Paracelsus differ so greatly that his position in the history of biology and medicine is so ambiguous? Perhaps this is due to his being forced into an obscure and mystical idiom so different from our own that we cannot evaluate him in our own terms. For Paracelsus, natural magic was the key to understanding nature and her laws, which scholastic traditions could never teach him. While "natural magic" sounds foreign to the ways of science, at least it was an empirical art; unlike scholasticism, which remained a quagmire of useless, endless argumentation, it could lead to a new experimental science.

Alchemy and astrology were traditionally associated with medicine, which was itself more an art and occult science than a rational and scientific discipline. Astrology was commonly used in prognosis as well as the casting of horoscopes. But for Paracelsus, alchemy could serve humanity in a new way. First, alchemists could use their art to prepare new drugs, and second, alchemy could provide a new chemical analogy for physiological functions. Paracelsus held that life was a chemical process. While this is a commonplace concept today, at the time it was a strange and revolutionary idea. Nor did it enter the mainstream of physiology for a long time. Mechanical analogies became much more acceptable and the Paracelcians remained on the periphery of the field of science.[9]

New Diseases—New Drugs

Paracelsus approached problems of physiology and pathology as a chemist and rejected the traditional approach to disease. Unlike the academic physicians, he wanted to do away with the heavy emphasis on anatomy, which he felt was totally irrelevant to the disease process. Since he believed that life was a chemical process, it followed that disease was the result of a defect in body chemistry. Thus based on alchemical notions, his physiology was necessarily obscure and mystical.

The sad paradox of Paracelsus' struggle with ancient philosophies is that in attacking the old system he was forced to construct another to take its place. Not for lack of trying was he unsuccessful, but because of the thick cloud of alchemical mysticism that enveloped him. Despite his attempts to revise alchemical theory, the very heart of his beloved discipline was so mystical as to preclude any rational systematization. The old Greek elements were obviously unsatisfactory for developing a more modern chemistry. But despite the labors of the alchemists there was little chance of replacing them with anything much better at the time. Paracelsus taught that there were three chemical principles. Salt was the principle of solidity, sulfur the principle of inflammability and malleability, whereas mercury was the principle of fluidity, density, and that which was metallic in nature. Elucidating the various principles did not depend on mere speculation, but required empirical investigation through the use of the great analytical tool of the alchemist; the tool was fire.

According to Paracelsus, all physiological processes were fundamentally chemical transformations governed by the "archeus" or "internal alchemist" of the body. Disease was the result of some malfunction of the archeus and death its final loss. Paracelsus taught that diseases could be analyzed as one could analyze chemicals. He developed a scheme for the classification of diseases based on causes which, although heavily based on astrological concepts, had valuable derivatives. For example, he introduced the concept of metabolic diseases with his postulation of a group of "tartaric diseases," such as gout and arthritis. Gout involves local deposits of a metabolic product normally excreted. Thus, it is indeed body chemistry gone wrong. Paracelsus seems to have recognized the relationship between goiter and cretinism, and to have been a pioneer in the description of occupational diseases. Because he believed that diseases have specific causes, he demanded specific remedies. This was in sharp opposition to the prevailing "polypharmacy," which called for the use of complex mixtures of substances. Some mixtures, such as the popular Theriacum, might contain more than 60 ingredients. Apothecaries were also likely to lace their potions with noxious substances, such as dung and urine, and expensive exotics, such as "mummy powder," viper's flesh, and powdered unicorn horn.

By calling for the use of "pure" substances rather than complex and indeterminate mixtures, Paracelsus pointed the way toward a rational chemical pharmacology. In purifying drugs and poisons, alchemists were learning how tiny quantities of chemicals could exert great effects on the body. Yet their search for valuable remedies might be guided by astrological concepts such as the correspondence between the seven planets, the seven metals, and the parts of the body. The "doctrine of signatures" also provided clues in the search for new drugs. This meant that natural objects were stamped by the Creator with some sign of their use to man. A yellow medicine would be used for jaundice, whereas a heart-shaped leaf would suggest a remedy for heart disease. The Paracelsians adopted the motto of "violent remedies for violent diseases." Thus, large doses of mercury were preferable to the popular, but useless, "holy wood" in the treatment of syphilis.

Paracelsus and his followers introduced many important minerals into pharmacology. They favored the use of mercury, antimony, lead, sulfur, iron, arsenic, copper sulfate, and potassium sulfate. Luckily, many of their preparations caused the patient to vomit so quickly that there was no time to absorb a lethal dose. Two very important drugs favored by Paracelsus were laudanum and ether, which he called "sweet vitriol" (prepared from alcohol and sulfuric acid). Although Paracelsus recognized the narcotic properties of ether, through tests on man and chickens, its use was not taken up as an anesthetic until the 1840s.

Many of the ideas introduced by Paracelsus proved valuable in the development of a scientific pharmacology and in the study of physiology as a chemical process. Although his ideas were based on alchemical and astrological concepts now rigidly separated from respectable science, they are none the less impressive for their imaginative and prescient qualities. Paracelsus was to inspire other investigators whom we shall meet in later chapters. We should not leave Paracelsus without pointing out that, according to a Gallup Poll in 1975, about 22 percent of adult Americans still believe in astrology.[10]

Notes

1. W. Durant (1957). *The Story of Civilization.* Vol. 6. New York: Simon and Schuster, p. 159.

2. Durant, p. 160.

3. C. D. O'Malley (1964). *Andreas Vesalius of Brussels, 1514–1564.* Berkeley: University of California Press, p. 297.

4. O'Malley, p. 301.

5. M. Boas (1962). *The Scientific Renaissance, 1450–1630.* New York: Harper and Row, p. 148.

6. O'Malley, p. 319.

7. M. Eliade (1962). *The Forge and the Crucible.* New York: Harper and Row, p. 9.

8. H. M. Pachter (1951). *Paracelsus.* New York: Schumann, p. 4.

9. A. G. Debus (1966). *The English Paracelsians.* New York: Science History.

10. New York Times (Oct. 19, 1975); *Humanist 35* (1975).

References

Ashley-Montagu, M. F. (1955). Vesalius and the Galenists. *Sci. Monthly 80:* 230–239.

Belt, E. (1955). *Leonardo the Anatomist.* Kansas: University of Kansas Press.

Boas, M. (1962). *The Scientific Renaissance, 1450–1630.* New York: Harper and Row.

Brehaut, E. (1912). *An Encyclopedia of the Dark Ages: Isodore of Seville.* New York: Longmans, Green.

Bronowski, J. and Mazlish, B. (1960). *The Western Intellectual Tradition, from Leonardo to Hegel.* London: Hutchinson.

Byard, M. M. (1977). Poetic Response to the Copernican Revolution. *Sci. Amer. 236:* 121–129.

Chambers, M. M., ed. (1950). *Universities of the World Outside.* Wisconsin: George Banta.

Clagett, M., ed. (1969). *Critical Problems in the History of Science.* Madison: The University of Wisconsin Press.

Corner, G. W. (1927). *Anatomical Texts of the Earlier Middle Ages.* Washington, D.C.: Carnegie Institute of Washington.

Cushing, H. (1942). *Bio-Bibliography of Andreas Vesalius.* New York: Henry Schuman.

Daly, L. J. (1961). *The Medieval University.* New York: Sheed and Ward.

Debus, A. G. (1966). *The English Paracelsians.* New York: Science History Publications.

Durant, W. (1957). *The Reformation.* New York: Simon and Schuster.

Eliade, M. (1962). *The Forge and the Crucible.* New York: Harper.

Farber, E., ed. (1961). *Great Chemists.* New York: Interscience.

Florkin, M. (1949). *Biochemical Evolution.* Translated by Sergins Morgulis. New York: Academic.

Gumpert, M. (1948). Vesalius: Discoverer of the Human Body. *Sci. Amer. 178:* 24–31.

Holmyard, E. J. (1957). *Alchemy.* Harmondsworth, England: Penguin.

Lassek, A. M. (1958). *Human Dissection, Its Drama and Struggle.* Springfield, Ill.: Thomas.

Lind, L. R. (1974). *Studies in Pre-Vesalian Anatomy: Biography, Translations, Documents.* Philadelphia: American Philosophical Society.

Lind, L. R. and Asling, C. W., trans. (1969). *The Epitome of Andreas Vesalius.* Translated from the Latin, with introduction and anatomical notes. Cambridge: MIT Press.

MacCurdy, E., trans. (1958). *The Notebooks of Leonardo da Vinci.* New York: George Braziller.

McMurrich, J. P. (1930). *Leonardo da Vinci the Anatomist, 1452–1519.* Baltimore: Williams and Wilkins.

O'Malley, C. D. (1964). *Andreas Vesalius of Brussels, 1514–1564.* Berkeley: University of California Press.

O'Malley, C. D., and Saunders, J. B. de C. M., eds. (1952). *Leonardo da Vinci on the Human Body.* New York: Schuman.

Pachter, H. (1951). *Paracelsus*. New York: Schuman.

Pagel, W. (1958). *Paracelsus: An Introduction to Philosophical Medicine in the Era of the Renaissance*. New York: Karger.

Partington, J. R. (1961). *A History of Chemistry*. 2 vols. New York: St. Martin.

Randall, J. H. (1953). The Place of Leonardo da Vinci in the Emergence of Modern Science. *J. Hist. Ideas 14:* 191–202.

Reti, L., ed. (1974). *The Unknown Leonardo*. New York: McGraw-Hill.

Sarton, G. (1937). *The History of Science and the New Humanism*. Cambridge, Mass.: Harvard University Press.

Sarton, G. (1957). *Six Wings: Men of Science in the Renaissance*. Bloomington: Indiana University Press.

Singer, C. (1952). *Vesalius on the Human Body*. London: Oxford University Press.

Singer, C. (1957). *A Short History of the Anatomy from the Greeks to Harvey*. New York: Dover.

Singer, C., and Rabin, C. (1946). *Prelude to Modern Science*. Cambridge: Cambridge University Press.

Stevenson, F. G., and Multhauf, R. P., eds. (1968). *Medicine, Science and Culture*. Baltimore: Johns Hopkins Press.

Temkin, O. (1934). Karl Sudhoff, the Rediscoverer of Paracelsus. *Bull. Inst. Hist. Med. 2:* 16–21.

Temkin, O. (1952). The Elusiveness of Paracelsus. *Bull. Hist. Med. 26:* 201–17.

Temkin, O. (1965). Vesalius on an Immanent Biological Motor Force. *Bull. Hist. Med. 39:* 277–280.

Vallentin, A. (1938). *Leonardo da Vinci, the Tragic Pursuit of Perfection*. Translated by E. W. Dickes. New York: Viking.

Waite, A. E., ed. (1894). *The Hermetic and Alchemical Writings of Aureolus Philippus Theophrastus Bombast, of Hohenheim, Called Paracelsus the Great: Now for the First Time Faithfully Translated into English*. 2 vols. London: Redway.

Ziegler, P. (1969). *The Black Death*. New York: Harper and Row.

5

THE FOUNDATIONS OF A MODERN
SCIENTIFIC TRADITION

From Vesalius to Harvey: The Minor and Major Circulation

Renaissance artists and physicians made brilliant contributions to science. Certainly Leonardo da Vinci, Michelangelo, and Vesalius tried to examine the fabric of the human body with their own eyes and to understand the relationship of structure to function. Their work, with all its achievements and faults, points out the great problem of scientific research regarding living organisms. While accurate anatomical knowledge is absolutely necessary, it is not sufficient for progress in physiology. That is, the static aspects of living things can be understood through means of investigation that cannot tell anything significant about the dynamic aspects. The study of anatomy and the use of simple analogies, such as veins to canals, could not explain physiological function.

In addition to the purely technical problems of studying physiology, such as the lack of chemical knowledge, the difficulties were compounded by political and theological inhibitions. The mechanism of the human body was intimately connected with religious dogma. Books which were conceived of by their authors as purely scientific or medical could be denounced by the Catholic or Protestant censors as heretical. The very life of the author and publisher of such dangerous works could be in jeopardy.

The best illustration of progress from anatomical research to physiological insight is the story of the discovery of the circulation of the blood.

Blood and Life

During the Renaissance, biological knowledge increased as artists and scientists became more willing to abandon scholastic methods and learned to carry out original research. But to a large extent it is true that conceptually Renaissance science did not go very much beyond the theory and philosophy of Aristotle and Galen. Yet some of the new facts were so inconsistent with the old theories that by the seventeenth century a definite break with the old ideas was inevitable. Although the scientific revolution of the seventeenth century has mainly been studied in terms of astronomy and physics, biological research was also escaping from slavish deference to the past. In some ways, the application of the scientific method to the microcosm of man by William Harvey was more revolutionary than the study of the macrocosm by Galileo.

It is possible to imagine a people without an interest in astronomy, but impossible to imagine any uninterested in their own life's blood. Ancient and even modern emotional responses to blood involve images more powerful than its role as a physiological fluid, a liquid tissue of red cells, white cells, platelets, clotting agents, enzymes, and antibodies. People have used blood in sealing contracts, at weddings, and in pacts of brotherhood. Blood has been used in many ceremonies or fertility rites and as an amulet against disease. Many people shared the belief that the traits of the donor could be transmitted to the recipient through blood. Some tribes had the custom of drinking blood to assure youth and courage. Even scientists suggested that blood transfusions would change the character and personality of the recipient. Thomas Bartholin (1616–1680) reported that he had examined a young girl who displayed feline characteristics after drinking the blood of a cat.

Scientists and philosophers alike treated the blood as a most powerful fluid. Hippocrates and Aristotle shared the belief that the movement of the blood was fundamental to life processes. For Aristotle, the heart was the first organ of the body, being the seat of intelligence, and the organ that added animal heat to the blood. Pulsations in the blood system were the result of a kind of boiling up phenomenon when the blood in the heart met the "pneuma" drawn in by respiration. Galen promoted almost the same view and emphasized that the primary importance of blood was its role in the movement and distribution of "vital spirits." Philosophers were more concerned with the nature of these spirits than with the actual motion of the blood.

Galen had gathered together all that the ancients knew about the blood and vascular system and with his own experiments, observations, and philosophy synthesized a complete although inaccurate theory of the blood. Some of his insights and observations were valuable to Harvey, who respectfully quoted Galen and Aristotle at every possible point. Galen understood that both arteries and veins carry blood and that the arteries and left ventricle contained bright red blood. He believed that the color was acquired from the pneuma or vital spirits carried by the arterial system. The dark venous blood was formed from food carried from the intestines to the liver. Galen described the valves of the heart as fireplaces for heating the blood. The function of the lungs was to cool and ventilate the heart. According to Galenic physiology, the movement of the heart was identical to the movements of respiration. Although Vesalius had been unable to observe the pores in the septum that Galen's scheme required to get blood from the right to the left side of the heart, he had been unable to suggest any alternative mechanism for the movement of the blood.

Vesalius regarded the heart as of great importance because of its "agitative faculty." The task of the heart was to rekindle the natural heat and restore and nourish the spirits. He, too, assumed that the motion of the heart and respiration were identical. Breathing deeply attracted air through the mouth and nose as into a bellows. Part of the air went directly to the brain and the remainder to the lung. By a force peculiar to the substance of the lung, air was changed and adapted to the use of the heart. The best part of the air goes to the left side of the heart to form the vital spirit. The heart draws a large supply of blood from the right ventricle into the left ventricle where, by virtue of the substance of the heart, the combination of the steamy vapor of the blood and the air is transformed into vital spirit. The function of blood is to distribute the spirit. According to Vesalius, the respiration and the pulse have the same use, the dilation and constriction of the great artery of the heart.[1]

Vesalius, undisputably the greatest anatomist of the sixteenth century, was rather vague about the whole question of the distribution of the blood. He may have been dissatisfied with the prevailing theory, like his contemporary Michael Servetus (1511-1553), but not bold enough for a direct attack on the Galenic system and its "spirits." In the second edition of the *Fabrica* (1555) he dared to differ with Galen about the nature of the septum. He unequivocally stated that "the septum is as thick, dense, and compact as the rest of the heart. I do not see, therefore, how even the smallest particle can be transferred from the right to the left ventricle through it." Previously, he had searched for the pores and concluded that they must be invisible, although he believed that the blood "soaks plentifully through the septum." Most students would not be driven to challenge the ancient system from their reading of Vesalius, however. After all, when science was baffled, theology could always provide enlightenment of a sort. Vesalius concluded, "We are driven to wonder at the handiwork of the Almighty, by means of which the blood sweats from the right into the left ventricle through passages which escape human vision."[2]

Thus, even though the inaccuracies and contradictions of Galen's system were becoming obvious in the sixteenth century, anatomists still accepted Galen's views. The greatest obstacle to dismantling this increasingly dilapidated structure was the intimate relationship between Galen's system and the theological question of "spirits" or "souls." A direct attack on this problem required great courage. The first to do this was a man who spent his whole life struggling against entrenched authority and orthodox ideas and died a martyr to his principles. This was Michael Servetus.

Michael Servetus (ca. 1511-1553)

If any man ever died twice for his principles, it was Servetus. He so antagonized both Protestant and Catholic authorities that he was burnt in the flesh by
the Protestants and in effigy by the Catholics. He fell victim to the climate of
dogmatism and intolerance which permeated the Renaissance world, but is so
often forgotten in admiration of the great outpouring of beauty and creativity.
In challenging the theological orthodoxies of his day, Servetus also proved
that no further anatomical discoveries were needed to allow a clear-minded
man to see the pathway taken by the blood in the pulmonary or so-called
lesser circulation.

   Born in Tudela, Spain, Servetus studied law at Toulouse and served for a
time as secretary to Charles V, Holy Roman Emperor. Even in his youth
Servetus was too unorthodox for Spain. Traveling almost constantly, he supported himself by writing, editing, and translating while he studied at various
universities. His youthful work, *On the Errors of the Trinity,* aroused the ire
of both Protestant and Catholic censors. Servetus found refuge in Lyons and
studied with a physician. This aroused his interest in medicine. By 1536 he
was at Paris studying geography, astronomy, and medicine. Here he probably
met Vesalius and Ambroise Paré. Servetus was a pupil of Jean Fernel (1506-
1588), French court physician, mathematician, astronomer, and assistant to
Johann Guinther (1505-1574) of Andernach. Like Vesalius, he performed
dissections for the philologist. To support himself, Servetus lectured on geology,
astronomy, and astrology and also wrote a book about pharmaceutical products
called *On Syrups.* In 1538 his heterodox views on judicial astrology forced
him to leave Paris without a medical degree. He next practiced medicine at
Charlieu and Vienna.

   In Vienna, he resumed his writing and took up a peculiar correspondence
with John Calvin (1509-1564) in Geneva. Ostensibly asking for information,
Servetus attempted the theological re-education of Calvin. Servetus sent back
Calvin's *Institutio* with insulting annotations for Calvin's edification. At this
point Calvin wrote to a mutual friend that if Servetus ever came to Geneva he
would not get out alive. Servetus was denounced to the Inquisition for the
publication of another heretical book, called *On the Restoration of Christianity,*
which concluded with an attack on Calvin. Servetus managed to escape from
prison, but the trial was held in the absence of the defendant. He was condemned to be burnt in a "slow fire until reduced to ashes" along with all
copies of his book. Lacking the body, his persecutors burnt an effigy instead.[3]

   Within four months of his escape from death at the hands of the Catholic
Inquisition, Servetus was captured while in Geneva. Once again he was condemned to be burnt at the stake along with all copies of his book. Attempts
to mitigate the sentence to death by the sword or at least have the victim
burnt "with mercy" (meaning strangulation before the burning) were

unsuccessful. Having secured Servetus to the stake with an iron chain, the executioner set fire to the pile of green wood. Thus ended the tumultuous career of Michael Servetus who among other things had the clarity of mind to propose the correct course of the pulmonary circulation.

Since only three copies of the *Restoration* survived, it is unlikely that Servetus had much influence on anatomical thought. Even if anatomists knew of his work they were unlikely to admit any association with that ill-fated radical thinker. The *Restoration* is mainly a theological and mystically spiritualistic view of the relationship of God to the world and to man. However,

Michael Servetus. [From Bainton, 1958.]

Servetus included almost every conceivable field in his discussion—law, politics, astronomy, physics, and medicine. To know the Holy Spirit, he argued, one must understand the spirit of man. This required exact knowledge of the human body—its structure, function, and its moving and intangible components, which were the blood and the spirit.

In his discussion of human anatomy, Servetus mixed evidence from human dissection, citations from Galen, and his own heterodox theological considerations. Direct observation suggested that the septum did not have the pores Galen's scheme required. If these pores did not exist, then not much blood could pass from the right to the left heart. An acquaintance with anatomy showed that the pulmonary artery was very large and that blood is very forcefully expelled from the heart to the lungs. Certainly more blood is sent to the lungs than could be necessary merely for their nourishment. Servetus argued that blood must go to the lungs for aeration as well as for expulsion of soot. During passage through the lungs, the color of the blood changed. Galen was led into error, Servetus said, because he misunderstood the function of the lungs. Galen did not recognize the continuous flow of blood through the lungs, but assumed that if any blood returned to the heart from the lungs it was due to leakage. Aeration in Galen's scheme was the function of the left ventricle.

In accordance with the Renaissance view that all knowledge is interrelated, Servetus also used theological arguments which led to the conclusion that the soul is in the blood. It seemed very appropriate that the soul should be located in something that moves—not in the heart or liver or brain.[4] Unfortunately, Servetus did not follow his insight further to a full appreciation of the blood circulation. He was satisfied with having reconciled physiology and theology as to the idea of the unity of the spirit. Establishing his main point, he was not unwilling to go along with Galen's idea of the continuous synthesis of blood in the liver, nor did he preclude the possibility that some blood might pass through the septum.

Although Servetus studied at Paris at the same time as Vesalius, the famous anatomist was still very much under the influence of Galen at that time and remained more cautious than the fiery Servetus. Thus, it is unknown whether Servetus influenced other anatomists or even whether he carried out any particular experiments to support his theory. His career is of interest in revealing the dark underside of the Renaissance and showing that with knowledge available to a man who had completed only a limited study of medicine and anatomy, the path of the blood through the lungs was obvious.

Realdus Columbus of Cremona (1516–1559)

When Vesalius left Padua, his student, Realdus Columbus, was appointed as his deputy. Soon Columbus succeeded Vesalius as professor of surgery and anatomy. Displaying little reverence for his teacher, Columbus distinguished himself as one of the most vociferous critics of the *Fabrica.*

Columbus, son of an apothecary, served as apprentice to a surgeon for seven years before studying medicine and anatomy under Vesalius. Later Columbus denied he had learned anything from Vesalius. Soon the two became bitter enemies and Vesalius spoke of Columbus as an "uncultivated smatterer."[5] Indeed, although a clever man, Columbus did not have a good general education and even his Latin was faulty. Columbus used every opportunity in lecture and anatomical arena to ridicule his former mentor. In 1546, Columbus again assumed a professorship that was intended for Vesalius. The Chair of Anatomy at Pisa was vacant, but attempts to release Vesalius from his post at the imperial court failed. At the time, Pisa was not a prestigious university, so it is uncertain as to what caused Columbus to make this move. He remained at Pisa until 1548, when Pope Pius IV invited him to the chair of anatomy in the University at Rome.

Various deficiencies have been pointed out in the work and even character of Columbus, but it must be said that his description of the pulmonary circulation was the first clear-cut statement made by a prominent anatomist and thus likely to achieve a wide audience. In this sense, it was a major step away from the Galenic physiology of the blood and made the work of William Harvey possible. Columbus left only one medical treatise, *De re anatomica,* which was published by his children in 1559. This work has been denounced as an "almost barefaced imitation of Vesalius' *Fabrica.*"[6] Still, the work was clearly written and arranged, although it was not illustrated. There is abundant evidence of a boastful man, who exaggerated the number of dissections he had carried out, praised himself to excess, attempted to take credit for the discoveries of others, and lavished praise on those, like the Pope, who could advance his career. Some of the discoveries were original, such as the description of the lens of the eye and its correct position, but most important was his work on the blood system.

Unlike Servetus, Columbus made it clear that his idea of the lesser circulation was based on clinical observations, dissections, and experiments on animals. Columbus described the anatomy of the heart in detail and corrected many ancient errors. He denounced other anatomists for assuming the existence of a direct path from the right to the left ventricle. "But these make a great mistake," he wrote, "for the blood is carried by the artery-like vein to the lung and being there made thin is brought back thence together with air by the vein-like artery to the left ventricle of the heart. This fact no one has hitherto observed or recorded in writing; yet it may be most readily observed by anyone." Columbus went on to ridicule other authors as "not very wise" who wrote of "smoky fumes" being discharged in the lungs as if the heart were a chimney burning green logs. He boasted that he held a unique and true view of the situation, namely that the "vein-like artery was made to carry blood mixed with air from the lungs to the left ventricle of the heart. And this is not only most probable, but is actually the case; for if you examine not only dead bodies but also living animals, you will find this artery in all instances filled with blood." [7]

Although Columbus boasted of his originality and his daring in breaking away from the "dogmas of Galen," he did not go on to draw the logical conclusions from this important new fact. His general views on the blood are essentially Galenic. Even his claim to priority has been questioned on several accounts. In 1546 Servetus sent to Curio, a doctor in Padua, a manuscript copy of the *Restoration*. It has been suggested that Curio might have shown this to Columbus or Vesalius and that it resulted in the strong statement about the Galenic pores in the second edition of *Fabrica*. We do not know when Columbus first came to the idea of the pulmonary circulation. It has been suggested that one of the copies of the printed *Restoration* might have come into Columbus' hands in Rome, but Columbus may have been teaching his views on the pulmonary circulation as early as 1546.

Andreas Cesalpino (1519–1603)

Celebrated as the discoverer of both the minor and major circulation by some historians of medicine,[8] Cesalpino was a very learned man who combined a great reverence for the ancients, especially Aristotle, with an appreciation of the more modern aspects of natural history and medicine. From 1555 to 1592, he was Professor of Medicine and Botany at Pisa. In 1592 he was called to Rome to serve as physician to Pope Clement VIII and as Professor at the Sapienza University. Although his main scientific work was in botanical classification, Cesalpino also wrote a number of anatomical works. These have intimations of the theory of the circulation but are so obscure and tortuous in contrast to Harvey's work that they really bear no comparison. Excessively devoted to Aristotle, Cesalpino was particularly opposed to Galen and was ready to believe that whatever Galen said must be wrong. Thus, he welcomed discoveries which helped to emphasize Aristotle's idea that the heart was the most important organ of the body.

According to Cesalpino, "The lung and heart must be dilated at the same time and constricted at the same time; or the entrance of the spirits must take place while we breathe out." But, he continued, the facts prove otherwise, since we can "regulate our breathing by our will, but the beat of the heart is wholly beyond our power; and even when we are breathing involuntarily, breathing is in most cases slower than the pulse." From a consideration of the pulse of the arteries and the opening and closing of the "little membranes" whose function is to secure that "the orifices letting in do not lead out and that those leading out do not let in" he seems to have been close to recognition of the function of the circulatory system.[9] That is, implicit in his writing is the idea that during systole (contraction), the heart discharges blood into the aorta and pulmonary artery, and that during diastole (expansion) the heart receives blood from the vena cava and pulmonary vein.

In his *Medical Questions,* from considerations of blood-letting, he came close to the idea that blood flows from the arteries to the veins and from the veins to the heart. All who practiced blood-letting knew that veins swell on the far side of the ligature. But instead of working out the problem as Harvey did, he retreated into a digression on what Aristotle said about sleep, becoming involved with the movement of animal heat during sleeping and waking states.

Possibly, Cesalpino's ideas were based more on the spirit of controversy for its own sake than on either patient observation, experimentation, or real study of the phenomena in question. His anticipation of Harvey's theory, assuming such a claim is at all justified, seems to have made no impact on his contemporaries. His speculations were forgotten until historians disinterred them from the graveyard of premature and unproven theories. His botanical theories, on the other hand, were well received and influential.

Hieronymus Fabricius of Aquapendente (ca. 1537–1619)

Fabricius, born in the little Tuscan village of Aquapendente into a poor family, achieved distinction as an anatomist and a founder of scientific embryology. After medical studies at Padua, he practiced surgery and gave private anatomy lessons. In 1565, Fabricius was appointed to the chair of anatomy and surgery that had been vacant since the death of Gabriele Fallopio (1523–1562). Having served the University for 50 years, illness finally forced Fabricius to retire in 1613. Poor health did not prevent him from continuing to work, since he published eight new works and revised several previous works during his retirement.

The most important anatomical contribution by Fabricius was his discovery of the valves in the veins. Actually the venous valves were "discovered" by a number of sixteenth century anatomists. Yet it seems that none of these measured up to Fabricius in terms of the clarity and rigor of his work. While several anatomists had noticed the little membranes found in some veins, the real significance of these structures was first dimly perceived by Fabricius, and then with completeness by his student, William Harvey.

In 1603 Fabricius published the most complete account of the structure, position, and distribution of the valves of the veins in a brief work (only 24 pages) entitled *On the Valves in the Veins.* In attempting to investigate the entire venous system to discover which veins had the peculiar membranous structures,

Fabricius surpassed other anatomists. Progress towards a mechanical explanation of the circulatory system was stimulated by his use of a hydraulic analogy in speculations about the functions of the valves. Fabricius realized that
the valves opposed the flow of blood from the heart towards the periphery.
In his courses and public lectures he demonstrated the work of the valves in
living beings. If a ligature is thrown around an arm, little knots are seen
along the course of the veins. These swellings correspond to the location of
the valves revealed by dissection. A simple experiment involving an attempt
to push the blood past these knots towards the hand clearly demonstrated
that movement of the blood was prevented by the valves.

Despite this promising beginning, Fabricius failed to truly appreciate the
function of the valves. To Fabricius these structures seemed more like the
floodgates of a millpond, which regulate volume, than valves, which regulate
direction. "In order therefore that the blood should be everywhere distributed in a certain just measure and admirable proportion for maintaining the
nourishment of the several parts," declared Fabricius, "the valves of the veins
were formed." [10]

Because he could not free himself from old ideas about the veins and the
blood, Fabricius could not understand their true function. He believed that
the veins carried blood rich with nutriments but devoid of vital spirits away
from the heart for the use of all parts of the body. Thus, he was satisfied
that the function of the valves was to prevent the heart from sending an excess into the extremities by the veins.

Fabricius did not study the heart action himself, although he mentioned
the action of the heart in his work on respiration. Differences between his
concept of the mechanism of respiration and the teachings of Galen on that
subject were negligible. For Fabricius, the function of breathing was the generation of "spirits" and animal heat within the heart, which "burns as with a
flame." [11] Thus Galen still dominated the thinking of a man who had been
the favorite pupil of Falloppio, who in turn had been a favorite pupil of
Vesalius. Seemingly he was little influenced in his thinking by these, or by
Columbus or Cesalpino.

William Harvey (1578-1657)

It was William Harvey, a quiet and in many ways a conservative man, who succeeded in putting together a complete new theory of the uses and action of
the heart and the entire circulatory system. Harvey, born in Folkestone on
the south coast of England on the first of April in 1578, was the son of
Thomas Harvey, a yeoman-farmer. William, the eldest of what was called "a
week of sons," was the only one to choose the academic life. The other

brothers in this very close-knit family became merchants in London. Young Harvey showed a curiosity about the way living things work even while quite young. As a child he is said to have played with the hearts of animals obtained at the local slaughterhouse.

After receiving the Bachelor of Arts from Caius College, Cambridge, Harvey studied medicine at the University of Padua under the great Fabricius. Student life at Padua still included its elements of risk—outside of the classroom. During a brawl, Harvey's friend received a wound from a dagger which severed an artery. Ever observant, Harvey noted how the blood came out of the artery in spurts, very different from the way blood drained smoothly from a vein. Less dramatic episodes at Padua, such as the teaching of Fabricius, and the opportunity to attend anatomies also helped Harvey to see the problem of the circulation in a new way.

Having received the degree of Doctor of Medicine in 1602, Harvey returned to England where he soon became a prominent and well-to-do London physician. For his wife he chose Elizabeth Browne, the daughter of Lancelot Browne (d. 1605), physician to Elizabeth and James I. In rapid succession, Harvey was elected Fellow of the College of Physicians, appointed physician to St. Bartholomew's Hospital, Professor of Anatomy at the College of Physicians and Surgeons, and Physician Extraordinary to James I and Charles I. Harvey remained the faithful friend and physician of Charles I throughout the unhappy career of that monarch.

Harvey's notes for his first Lumlein Lecture show that by 1616 he had already arrived at an understanding of the blood circulation. This was 12 years before he published a complete account of his work in *De Motu Cordis* (1628). In this demonstration lecture Harvey announced the revolutionary new doctrine that blood is continuously transferred "through the lungs into the aorta as by two clacks of a water bellows to raise water." Application of a ligature proved that there is a passage of blood from the arteries to the veins. His studies demonstrated that "the beat of the heart produces a perpetual circular motion of the blood."[12]

The rigorousness of the proofs assembled in Harvey's tightly organized book of 1628 sets him off from all his predecessors. His achievements are even more remarkable when set against the turmoil through which Harvey lived. As physician to the king, Harvey frequently travelled with the court to continental Europe where he is believed to have discussed his discoveries with other physicians. Harvey remained dedicated to the king throughout the civil war and went to Oxford, where he obtained a professorship, when the court transferred there from London. The remarkable temperament of William Harvey is illustrated by the legend of how at the battle of Edgehill he sat at the fringes of the battle under a hedge, reading a book. When the Royalists were defeated and the unfortunate Charles lost his head, Harvey retired to live with his brothers in the country near London, rather than follow the king's associates into exile. During the war, Harvey's house in London was attacked and many of his manuscripts and collections were destroyed. Harvey left his fortune to the College at Oxford, which has an annual festival dedicated to him.

Harvey's practice deteriorated while his research progressed. When he published his book in 1628, he expected to meet considerable opposition. In the storm of reaction, some critics declared he must be mad, just as Vesalius had been villified when he published the *Fabrica*. Although he suffered great losses in his private practice, being the personal physician of the king and having some wealthy relatives helped him continue his research. Loyalty to the Stuarts also adversely affected Harvey's popularity. After the death of his king and then his wife Elizabeth, Harvey's own health deteriorated. He suffered from gout and probably helped to shorten his life with the peculiar remedies he inflicted on himself.

De Motu Cordis (1628)

Harvey's most significant work was his *Anatomical Treatise on the Movement of the Heart and Blood in Animals* (1628). Like Aristotle, whom he admired, Harvey asked seemingly simple but truly profound questions in the search for final causes. In thinking about the function of the heart and the blood vessels, he came to reject Galen's scheme and moved closer to Aristotle's idea that the heart is the most important organ in the body. Harvey asked many questions that other anatomists had dimly framed before him. He wanted to know why the two structurally similar ventricles of the right and left heart should have such different functions as the control of the flow of blood and of the vital spirits. Why should the artery-like vein nourish only the lungs, while the vein-like artery had to nourish the whole body? Why should the lungs need such an inordinate amount of nourishment? Why did the right ventricle have to move in addition to the movement of the lungs? But the unique question that Harvey asked was one that is astonishingly simple, even childlike. Furthermore, it is a question that lent itself to experimental and quantitative answers that demolished Galenic castles in the air once and for all. That question was, of course, *how much blood is sent through the body with each beat of the heart?*

During his student days and in the midst of a busy medical practice, Harvey pondered the question of the movement of the blood. In his last years he was to tell Robert Boyle that it was Fabricius' demonstration of the venous valves that helped him to think in terms of a circulation. In London, he participated in anatomies of six executed criminals per year. Observations on patients, such as the throbbing of an arterial aneurism, and the pulses on the wrists, temples, and neck suggested the action of a central pump. Harvey had been taught that there were two distinct kinds of blood: blood from the liver for nourishment of the parts by the veins, and blood from the heart for the distribution of vital spirits by the arteries. Although there certainly was a difference in color between arterial and venous blood, there was no major difference in any other parameter that Harvey could observe—such as taste.

Unlike most of his contemporaries, Harvey did not rest his case on supposition and logic alone. He was unique in applying experimental and quantitative methods to medical research. This was a remarkable accomplishment for biological science at a time when even in the physical sciences quantitation was still rare and unfettered speculation rampant. Because Harvey used only a hand lens, he was not able to directly observe the capillaries. Thus, the only weak point in his work was the nature of the connections between arteries and veins. Until the microscope was used in anatomical research by Leeuwenhoek (1632-1723) and Malpighi (1628-1694), the "anastomoses" which he had postulated remained hypothetical.

Modern in his approach to the problem, even if he was Aristotelian in his search for final causes, Harvey solved the question of the movement and purpose of the heart and blood by isolating the system of interest and refusing to speculate about nature in general. For his time, and even compared to many moderns, he demonstrated a remarkable degree of modesty and restraint. An approach to biology that reflected the increasingly popular mechanistic philosophy of the seventeenth century was most successfully used by Harvey, who applied a mechanistic model to what was essentially a mechanical problem. In contrast, attempts to apply mechanical principles to other systems, such as digestion, generation, and the function of the nervous system, were to remain unsuccessful. Until significant advances were made in chemistry and physics, such problems were insoluble.

Harvey used several modes of proof, some still valid and others which demonstrate that for all his genius, Harvey was deeply imbued with the spirit of his time. Though he transcended the limitations of Galen and Aristotle, he did not reject the philosophy and spirit of their work. With arguments based on dissection and vivisection, Harvey proved that, in the adult, all the blood must go through the lungs to get from the right to the left side of the heart. He proved that the heart is muscular and that its important movement is the contraction. Previously anatomists had taught that the heart was not a muscle and that dilation was the important function; movement had been ascribed to the chest and lungs rather than to the heart itself. In his first chapter, "Difficulties in Experiment," Harvey dismissed the work of anatomists who would study only one species, man, and that one only when quite dead. The movement of the heart at first seemed so rapid and complex that it "was only to be comprehended by God." But Harvey saw the advantage of using cold-blooded animals and dying mammals in which the heart beat is slower and easier to follow. Harvey compared the working of the heart and arteries to machinery acting wheel on wheel or a gun where one cannot see the rapid action of trigger, flint, steel, spark and powder, but one can study and understand the mechanism.

Harvey easily improved on previous studies of the pulmonary circulation. But he thought his views on the "quantity and source of the blood" were "so novel" that he feared he might have mankind at large for his enemies.[13] This radical step was the application of the idea of a continuous circulation through the whole body. In support of this new conception, Harvey turned to quantitative considerations. Although he did experiments aimed at getting an accurate measurement of the quantity of blood put out by the heart with each beat, he emphasized that exact measurement was unnecessary. Even the most cursory calculation would prove that the amount of blood pumped out of the heart per hour was so great that it exceeded the weight of the entire organism.

If the heart pumps 2 to 3 ounces of blood with each beat and beats some 65 times per minute, this would drive about 10 pounds of blood per minute or 600 pounds per hour. This is three times the weight of even a very large man. Such calculations were carried out for various animals, such as sheep and dogs. In all cases the amount of blood pumped by the heart in even a half-hour is more blood than is contained in the whole animal. One could exsanguinate experimental animals, as a butcher exsanguinates an ox, and easily prove this. Thus, the blood must move in a circle through the body. The movement of blood must be continuous, rather than a one-way trip of blood constantly synthesized from food by the liver.

Once this basic truth was grasped, many observations fell neatly into place. Every butcher knew, for example, that the heart must still be beating to drain out the blood of a slaughtered animal; he could do so in about one-quarter hour. At last scientists could explain what every butcher knew. Observations from bloodletting and ligatures provided further proof, while being explained in a rational manner for the first time. The purpose of the valves in the veins now became eminently clear—they controlled direction of blood flow and not simply volume.

Knowledge of the continuous circulation of the blood had immediate application to explaining many clinical observations. Circulation explained why poisons or infections at one site could affect the whole system, as could bites from snakes and rabid animals. Harvey even saw in his work an explanation for the way externally applied medicines could be absorbed and distributed. This included such quaint therapy as garlic applied to the soles of the feet.

Beyond any doubt Harvey proved that "the blood in the animal's body is impelled in a circle, and is in a state of ceaseless motion; that this is the act or function which the heart performs by means of its pulse; and that it is the sole and only end of the motion and contraction of the heart."[14] Harvey confined his explication to the mechanical aspects of the blood circulation. He was not drawn into arguments on the distribution of various kinds of spirits. These questions were merely acknowledged: "Whether or not the heart, besides propelling the blood, giving it motion locally, and distributing it to the body, adds anything else to it,—heat, spirit, perfection—must be inquired into by and by, and decided upon other grounds."[15] Harvey had discovered all that could be known of the blood system in terms of anatomy and hydraulics without a microscope. But the biochemistry of the system was beyond his abilities and very wisely he left such questions to the future.

Although violent controversies were sparked by Harvey's little book, his great influence was already apparent during his lifetime. Physicians attempted to draw practical benefits from his work and young scientists took inspiration from his methods. Medicines could be given by injection into the veins and would be distributed throughout the body. Attempts at blood transfusion quickly followed Harvey's work, and were sometimes fatal since nothing was known about blood types. The most puzzling aspect of Harvey's influence, was the tremendous stimulation of phlebotomy in the seventeenth century. Not even Harvey seemed to realize that the ancient practice was quite inconsistent with the continuous circulation of a limited quantity of blood.

The opposition to Harvey's new system began within weeks after publication of *De Motu Cordis* with a typical Galenist attack by James Primrose and

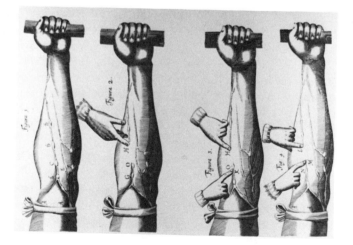

The valves of the veins, from Harvey's *De motu cordis,* 1628. [From J. F. Fulton and L. G. Wilson (1966). *Selected Readings in the History of Physiology.* Courtesy of Charles C Thomas, Springfield, Illinois.]

continued for decades. In 1649 Harvey published an answer to Jean Riolan (1571-1657) called *An Anatomical Disquisition on the Circulation of the Blood to Jean Riolan the Younger, of Paris.* Riolan had advanced the incredible but seemingly indestructible view that one should not admit Galen to be wrong. Even if dissection revealed differences, one must presume that Galen had been right and that nature had changed since he wrote. Another conservative French doctor, Guy Patin (1601-1672) of the Faculty of Paris wrote that Harvey's theory was "paradoxical, useless, false, impossible, absurd, and harmful."[16]

Until 1650, the conservative medical faculty of France rejected the new experimental science. Only heterodox thinkers like René Descartes were really enthusiastic about the new science, yet Harvey enjoyed support among many eminent physicians and anatomists. Having a clear understanding of the radical nature of his work, Harvey predicted that no one over 40 would understand it. He served as an inspiration to the young scientists who eventually founded the Royal Society in their attempt to develop a new experimental and practical natural science.

Harvey's originality in the discovery of the circulation is in no way compromised by the revelation that the idea of a circulation appeared in other cultures. In ancient Chinese writings there is a passage that seems to indicate knowledge of the continuous circulation of the blood. The *Neiching* or *Canon of Medicine* contains the statement that "the heart regulates all blood in the body. The river of blood flows in a circle and never ends." Furthermore, according to this work, which was composed about 2600 B.C., there are two kinds of blood.[17] Before rushing to the conclusion that the ancient Chinese discovered the circulation between the arterial and venous blood systems, we should remember that the Chinese did not distinguish veins from arteries. The two "forms" of blood correspond to the two principles of yang and yin, which are ubiquitous in all explanations of natural phenomena in classical Chinese philosophy. The flow is not circulation in the sense that Harvey meant but is more likely to be an analogy between the movement of this vital body fluid and the cycles of nature. Furthermore, dissection was not practiced by Chinese physicians and scientists.

Another mention of the minor or pulmonary circulation was made by a thirteenth century Arabian writer, Ibn an-Nafis, but this Arab manuscript was not known in the Western world and was not translated until the twentieth century.[18] Although interesting, such speculations are of a fundamentally different nature from Harvey's direct, experimental, modern scientific method.

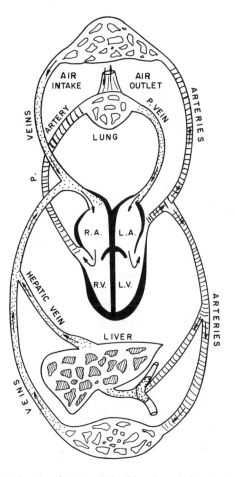

Diagram to illustrate the circulation of the blood as understood after Harvey's work. Compare this Figure with the illustration of Galen's scheme for the circulation of blood and spirits on page 67 of Chapter 3. [After Singer, 1959.] R.A. = right auricle, L.A. = left auricle, R.V. = right ventricle, L.V. = left ventricle, P. Artery = pulmonary artery, P. Vein = pulmonary vein.

Anatomy and Physiology After Harvey

Anatomists continued to pursue the study of gross anatomy with great enthusiasm throughout the seventeenth century. Many of the ducts and glands are named after the anatomists of this period, such as Wirsung's duct, Brunner's glands, Graafian follicles. The lymphatic system was discovered during Harvey's lifetime, but he showed little interest in it. Gasparo Aselli (1581-1626) had described the "lacteal veins," just before the publication of *De Motu Cordis*. But a complete description of this work would be a mere catalog of the parts of the body, revealing no fundamental new principles to make the study worthwhile.

A Different Kind of Lesson

Sanctorius (Santorio Santorio) (1561-1636)

It should be pointed out that other attempts to apply quantitative and mechanistic methods to physiological problems were dismal failures at the time. Sanctorius of Padua, for example, tried to study metabolism, then known as "invisible" or "insensible perspiration." Santorio Santorio has been called the founder of quantitative medical research. He was educated at Padua and practiced medicine in Venice after his graduation in 1582. In 1611 he was appointed to the Chair of Theoretical Medicine at Padua where he remained until his resignation in 1629. Thereafter he devoted himself to medical practice and research in Venice.

Sanctorius tried to quantitate his work on "insensible perspiration" with instruments that had only recently been invented, such as the clinical thermometer and pulse clock. For much of his own patient experimental career he sat on a balance—weighing himself and everything that went into or out of Sanctorius. He tried to quantitate variations of body weight after eating and drinking, while sleeping, at rest or exercise, in health and disease. He spent most of the day during his 30 years of experimentation in a chair suspended from a steelyard. In 1614 he published a book of aphorisms based on his work called *De medicina statica aphorismi*. The book went into at least 32 editions and was translated into several modern languages.

Sanctorius gave only vague indications of his methods in his *Medical Statics*. The only experiment illustrated is that in which Sanctorio himself was being weighed before and after a meal. Sanctorius told the reader how he would learn about a totally new field: "It is a new and unheard of thing in Medicine that anyone should be able to arrive at an exact measurement of insensible perspiration. Nor has anyone either Philosopher or Physician dared to attack this part of medical inquiry. I am indeed the first to make the trial, and unless I am mistaken I have by reasoning and by the experience of thirty years brought this branch of science to perfection, which I judged more advisable than to describe all the details of my inquiry."[19] Without much discussion of methods, Sanctorius gives his deductions as to what changes take

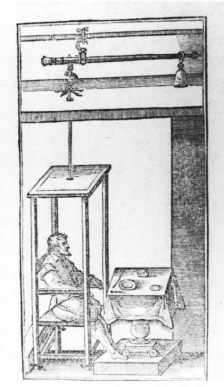

The weighing chair of Santorio Santorio. [From J. F. Fulton and L. G. Wilson (1966). *Selected Readings in the History of Physiology.* Courtesy of C Thomas, Springfield, Illinois.]

place in the body under various conditions. These deductions are based on knowledge of the amount of food and drink, and the sensible and insensible excreta of the body. In Sanctorius' opinion, these experiments had some practical applications to medicine. A man could take his meals in the balance and know when he had eaten just the right amount of food!

Some of Sanctorius' experiments indicate the difficulties of a premature attempt to solve biochemical problems. Such quantitative but meaningless experiments were still being carried out in the nineteenth century when biochemistry was just about to emerge as a true science. The great French physiologist Claude Bernard subjected such works to scathing criticism. Bernard tended to distrust the use of the statistical method in biological studies because it was often applied in too crude a form to give significant results. For example, a study of nutrition he cited consisted of a balance sheet of all the substances taken into a cat's body and excreted during eight days of feeding and 19 days of fasting. Kittens born on the seventeenth day were included in the calculations as excreta.

Notes

1. L. R. Lind and C. W. Asling (1969). *The Epitome of Andreas Vesalius.* Cambridge: MIT Press, pp. 58–59.

2. M. Foster (1970). *Lectures on the History of Physiology.* New York: Dover, pp. 13–14.

3. R. H. Bainton (1939). *Hunted Heretic: The Life and Death of Michael Servetus, 1511-1553.* Boston: Beacon, p. 164.

4. Bainton, pp. 118–127.

5. Foster, p. 26.

6. Foster, p. 27.

7. Foster, pp. 28–29.

8. A. Castiglioni (1969). *A History of Medicine.* New York: Aronson, pp. 436–440.

9. Foster, pp. 32–33.

10. Foster, p. 37.

11. Foster, p. 39.

12. W. Harvey. *Lectures on the Whole of Anatomy.* Annotated translation by C. D. O'Malley, F. N. L. Poynter, and K. F. Russell (1961). Berkeley: University of California Press, p. 191.

13. W. Harvey. *The Circulation of the Blood and Other Writings* (1907). London: Dutton, p. 55.

14. Harvey, *The Circulation,* p. 85.

15. Harvey, *The Circulation,* p. 39.

16. Castiglioni, p. 519.

17. I. Veith (1966). *The Yellow Emperor's Classic of Internal Medicine.* Berkeley: University of California Press, p. 34; K. Chimin Wong and Lien-Teh Wu (1936). *History of Chinese Medicine.* 2nd Ed. Shanghai: Nat. Quar. Serv., p. 35.

18. O. Temkin (1940). Was Servetus Influenced by Ibn an- Nafis? *Bull. Hist. Med.* 8: 731–734.

19. Foster, p. 146.

References

Adelmann, H. B. (1942). *The Embryological Treatises of Hieronymus Fabricius of Aquapendente.* Ithaca: Cornell University Press.

Bainton, R. H. (1960). *Hunted Heretic: The Life and Death of Michael Servetus, 1511-1553.* Boston: Beacon.

Bayon, H. P. (1939). William Harvey, Physician and Biologist: His Precursors, Opponents and Successors, Part IV. *Annals of Science 4:* 65–106.

Bonelli, M. L. R., and Shea, W. R. (1975). *Reason, Experiment, and Mysticism in the Scientific Revolution.* New York: Science History.

Bridges, J. H. (1892). *Harvey and His Successors.* London: MacMillan.

Castiglioni, A. (1920). *La Vita e l'opera di Santorio Santorio.* Bologna: Coppelli.

Castiglioni, A. (1947). *A History of Medicine.* New York: Knopf.

Chauvois, L. (1957). *William Harvey.* New York: Philosophical Library.

Clendening, L. (1933). *The Romance of Medicine.* New York: Garden City.

Fermi, L., and Bernardini, G. (1961). *Galileo and the Scientific Revolution.* New York: Basic Books.

Foster, M. (1970). *Lectures on the History of Physiology during the Sixteenth, Seventeenth, and Eighteenth Centuries.* New York: Dover.

Franklin, K. J. (1961). *William Harvey, Englishman, 1578-1657.* London: Macgibbon and Kee.

Franklin, K. J. (1933). *De Venarum Ostiolis 1603, of Hieronymus Fabricius of Aquapendente (1533?-1619),* Springfield, Ill.: Thomas.

Gasking, E. (1970). *The Rise of Experimental Biology.* New York: Random House.

Harvey, W. (1628). *Exercitatio Anatomica de Motu Cordis et Sanguinis in Animalibus.* Translated by C. D. Leake (1930). Springfield: Thomas.

Harvey, W. *Lectures on the Whole of Anatomy.* Annotated translation by C. D. O'Malley, F. N. L. Poynter, and K. F. Russell (1961). Berkeley: University of California Press.

Keele, K. D. (1965). *William Harvey: The Man, the Physician and the Scientist.* London: Nelson.

Keynes, G. L., ed. (1949). *Blood Transfusion.* Bristol: Wright.

Keynes, G. L. (1949). *The Personality of William Harvey.* Cambridge: Cambridge University Press.

Keynes, G. L. (1966). *The Life of William Harvey.* Oxford: Clarendon.

Landsteiner, K. (1962). *The Specificity of Serological Reactions.* Revised Ed. New York: Dover.

Meyer, A. W. (1936). *An Analysis of the De Generatione Animalium of William Harvey.* California: Stanford University Press.

Marcus, R. B. (1965). *William Harvey, Trailblazer of Scientific Medicine.* London: Chatto and Windus.

O'Malley, C. D. (1953). *Michael Servetus.* Philadelphia: American Philosophical Society.

Payne, J. F. (1897). *Harvey and Galen.* London: Oxford University Press.

Power, D'A. (1897). *William Harvey.* New York: Longmans, Green.

von Sachs, J. (1906). *History of Botany, 1530-1860.* Oxford: Clarenden.

Sigerist, H. E. (1933). *The Great Doctors: A Biographical History of Medicine.* Translated by E. and C. Paul (1958). Garden City: Doubleday.

Singer, C. (1925). *The Evolution of Anatomy.* London: Paul, Trench, Trubner.

Titmuss, R. M. (1971). *The Gift Relationship: From Human Blood to Social Policy.* New York: Random House.

Underwood, E. A. (1963). The Early Teaching of Anatomy at Padua, with Special Reference to a Model of the Padua Anatomical Theatre. *Annals of Science 19:* 1-26.

Veith, I. (1966). *The Yellow Emperor's Classic of Internal Medicine.* Berkeley: University of California Press.

Webster, C. (1976). *The Great Instauration: Science, Medicine and Reform, 1626-1660.* New York: Holmes and Meier.

Whitteridge, G. (1971). *William Harvey and the Circulation of the Blood.* London: MacDonald.

Willius, F. A., and Keys, T. E. (1961). *Classics of Cardiology.* 2 vols. New York: Dover.

6

SCIENCE AND CHANGE: SCIENTIFIC SOCIETIES

The Search for Order

Europe was torn by great political, social, and spiritual unrest during the period called the Age of Reformation. The seventeenth century was the stage for the final acts of the Protestant Revolution and Catholic Reformation. In England, it was the age of Elizabeth, Shakespeare, Milton, Bacon, Boyle, and Newton. Despite or perhaps because of the turbulence of the times, remarkable scientists were produced all over Europe, including Descartes, Leibniz, Pascal, Kepler, William Gilbert, and Giordano Bruno. Revolutionary new concepts initiated by Copernicus and Vesalius culminated in their seventeenth century counterparts, Galileo and Harvey. The year 1600 saw the last burning for a scientific heresy when the unfortunate heterodox thinker Giordano Bruno died at the stake.

Because powerful and despotic rulers came to control newly centralized governments, the period from the late fifteenth to the late eighteenth century has been called the "Age of Absolutism." But the search for order and rigid system permeated spheres other than political; theology, philosophy, and science also suffered an excessive respect for rigidity and organization. Most universities remained under the control of religious authorities. For admission, all professors and students had to take the official religion. Censorship, which could be secular as well as ecclesiastical, was almost ubiquitous. However, the codes of the censors in both Protestant and Catholic countries were so diverse that some authors were able to publish in one state works that would be smuggled into more conservative countries. As we shall see, private academies were the only institutions that served to encourage free thought and stimulate experimental sciences.

Certainly, the physical sciences advanced most as first Galileo and then Newton transformed mechanics and astronomy. On the whole, biology was a

137

very backward member of the family of sciences. Only Harvey's great work on the circulation of the blood was in the first rank of scientific discoveries. Still, the great explorations of this period brought back many new species of plants, animals, drugs, and foods to Europe, providing material for future progress in botany and zoology.

The Academy Movement

The purpose of this chapter is to sketch the background and philosophical orientation of the intellectuals associated with the rise of the new scientific institutions and the scientific revolution rather than to provide a catalog of the societies variously formed and dissolved during this period.

As we have seen, the individuals drawn to the study of science have been associated with various professions. The pre-Socratics, philosophers rather than scientists, interested in practical matters as well as natural explanations of the cosmos, were independent scholars and not attached to any "scientific" or educational institution. After Socrates, natural philosophers were concerned with ethics rather than pure science. In Alexandria, science was not only encouraged, but a major institution was devoted to its development. Scholarly work in the Middle Ages was generally confined to the monasteries before the universities began to assume this role. During the Renaissance, significant contributions to biology flowed from the work of artists, but physicians like Vesalius and Harvey, associated with the courts of powerful rulers or the universities, were the main protagonists in our history of biology.

By the seventeenth century, membership in a novel institution became the sign and symbol of science. Called the "scientific society," or "academy," the institution is in some sense a resurrection of Plato's school, but with a new purpose and mission. Practical and experimental goals superseded abstract and philosophical pursuits.

Eminent philosophers of science such as René Descartes (1596-1650) and Francis Bacon (1561-1639), who will be discussed in more detail later, provided much of the philosophical rationale for the new institution, while the great accomplishments of Harvey, Galileo, and others provided the methods and interest in experimentation that animated the academy movement.

New tools as well as new ideas and institutions appeared during the seventeenth century. Valuable scientific instruments included the telescope, pendulum, thermometer, barometer, hydrometer, air pump, watch spring, and, especially important to biology, the microscope. A salient feature of the multiplication of tools, instruments, and machines was the realization that it was possible, proper, and even necessary to design special instruments for scientific investigations.

Perhaps there has never been any period in which science had a greater impact on prevailing concepts of the earth, the heavens, and the nature of human beings than the sixteenth and seventeenth centuries. The structure and mechanism of the human body were being revealed just as the "eternal and unchanging" heavens were being radically rearranged. Parallels between the microcosm of man and the macrocosm went beyond the nonsense of astrology; the blood travelled continuously in a circle around the body, just as the earth moved in a circle around the sun. Harvey delighted in comparing the heart to the sun and the sun to his king.

The new experimental science became intimately associated with the academies rather than the prestigious, long-established universities. This raises an obvious question: why did the universities fail to serve this purpose? The generation of new institutions is never a simple process, but then as now the structure and function of the university differed from the scientific society. Furthermore, at that time the difference in philosophy was profound. The pedagogical purposes of the university encouraged formal methods, based on the teacher-student relationship, which are not conducive to new ideas. On the whole, universities, despite the achievements of certain individuals, continued to stress the ancient, respectable subjects where authoritative texts and recognized scholarship existed. Universities generally regarded their mission as the preservation and transmission of the past to the new generation rather than the encouragement of novel ideas and techniques.

Unlike the universities, the academies served as a setting where talented and enthusiastic individuals interacted as equals. Demonstrations and experiments stimulated new enquiries by bringing together a critical mass of intellectuals in a stimulating environment. The exchange of information was enhanced by the regular publication of journals and the transactions of the societies. Competition for awards and prizes encouraged the search for new facts and ideas and led to study and critical appraisal. Academies could also serve as safe places to exchange new thoughts when they succeeded in obtaining government charters and protection. Such institutions could also attract rich sponsors who might donate their own "curiosity cabinets" and funds for collections and apparatus.

Francis Bacon (1561-1639)

In spite of his great enthusiasm for the scientific method, Bacon made no direct contributions to scientific knowledge. However, he did influence the philosophy and institutions that directed the course of science. Like Pliny, he was all too enthusiastic an amateur to be a critical judge of facts and fancies. Yet Bacon was remarkably prescient in realizing that science could play a major and very valuable role in society. He devoted himself to defining the

methodology of science, suggesting means of insuring its application, and providing encouragement and direction for the new scientific enterprise he predicted: understanding and controlling nature.

As the son of a prominent government official, the Keeper of the Great
Seal, Francis Bacon lived in comfort and wealth until his seventeenth year.
When the death of his father forced him to work for a living, he became obsessed with the quest for financial security to allow him the leisure time
needed for his scientific and intellectual interests. Much energy was diverted
from his beloved scientific projects to the scramble for political power, public
office, and wealth. Bacon became entrapped in the complexities of seventeenth century politics. Educated at Trinity College, Cambridge and Gray's
Inn, London, he first served with the English Ambassador in France. He took
his seat in Parliament in 1584 but was frustrated in attempts to gain public
office during the reign of Queen Elizabeth. Under James I, Bacon finally
approached his goal. He became Knight (1603), Member of the King's Council
(1604), Solicitor-General (1607), Attorney-General (1613), Keeper of the
Great Seal (1617), and finally Lord High Chancellor (1618). In 1621, the unfortunate Bacon was involved in a bribery scandal, convicted of corrupt
practices, fined, impeached, and imprisoned in the Tower "at the King's
pleasure." Although King James I remitted the fine and terminated the
prison sentence after four days, Bacon's public career was ruined. Disgraced
and depressed, he spent the last five years of his life completing and enlarging
his essays.

In his writings, Bacon outlined grandiose schemes for the review and analysis of all branches of human knowledge, an encyclopedia of the crafts and experimental facts, predicted that a great new synthesis would emerge from this
grand accumulation of knowledge, and that the sciences would be directed
and harnessed to provide for human welfare, comfort, and prosperity. *The
Advancement of Learning,* published in 1605, became a popular work of inspiration to later generations of scientists. His major project, called the *Great
Instauration of Learning,* was published in part in 1620 but was never
completed.

In time, the "Baconian method" became synonymous with the "scientific
method." Even Charles Darwin imagined that he proceeded entirely on
Baconian principles. Bacon's method (the inductive method) involves the exhaustive collection of particular instances or facts and the elimination of
factors which do not invariably accompany the phenomenon under investigations. The foundation of this method was Bacon's idea that hidden among
the confusing facts known to us are causes or forms. The task of science was
to analyze experience "as if by a machine" to arrive at true conclusions, by
going from less to more general propositions. From "Tables and Arrangements
of Instances," hypotheses would be derived (like "Boyle's law," for example).

Bacon distrusted the deductive method which proceeded from the more to the less general and utilized intuitive thinking rather than the strictly mechanical sifting of facts which Bacon advocated.[1]

Bacon hoped to point the way to "a true and lawful marriage between the empirical and the rational faculty, the unkind and ill-starred divorce of which has thrown into confusion all the affairs of the human family."[2] He believed that the crafts had developed because many individuals contributed to their growth, but that philosophy had been stifled in its progress because the accepted philosophical systems were the products of only one mind. Instead of adding to human understanding, followers of the great thinkers had surrendered their intelligence to past masters of philosophy. The separation of philosophy from its roots in experience led to the decline of philosophy from the time of its first proponents, the pre-Socratic Greeks. Bacon rejected the scholasticism of the universities and launched open attacks on Aristotle and Plato. Theology, he argued, should be separated from science and philosophy. Since his new philosophy was anti-idealist and pro-materialistic, he favored the philosophies of the pre-Socratics, especially Democritus and Lucritius. Bacon favored a reunion of the craft tradition with natural philosophy. From such a fertile union would come understanding and control of nature along with valuable inventions to improve the human condition.

Science would be the basis of a new education, according to Bacon, but the universities could not provide the technical and utilitarian education that was required to unite the crafts and the sciences. Therefore new institutions where science and technology could be united and redirected were needed.

Financial aid for scientific research was a high-priority item in Bacon's schemes. Properly supported, science could create an international government and a brotherhood of experts. Scientists and technologists would serve as the supervisors and reformers of a world made as beautiful as Eden before the Fall of Man.

Bacon believed that assembling an encyclopedia about six times as great as Pliny's *Natural History* would equip him to explain all natural phenomena. James I, who supposedly fell asleep reading Bacon's *Novum Organum,* ignored his requests for the support needed to carry out this grand project. William Harvey dismissed Bacon's work with the remark, "He writes philosophy like a Lord Chancellor." But Thomas Jefferson's trinity of portraits of the world's greatest men consisted of Newton, Locke, and Bacon. Ironically, Bacon, who earnestly if ineffectually struggled to fulfill his vision of a life dedicated in service to humanity, has lately become the target of a modern wave of anti-rationalism and anti-science.[3]

Bacon made some clever observations and predictions. He noticed the fit between the continents of the old and the new world. Recognizing the importance of varieties of plants and animals in nature, he predicted that "by art" man would be able to produce useful new species. Little that was known to the seventeenth century could be credited to the sciences, which were then "merely systems for the nice ordering and setting forth of things already invented" rather than "methods of invention of directions for new works." Progress in science would depend not on "superinducing and engrafting of new things upon old," but by erecting new foundations for science.

To reassure the theologians who feared that science might subvert the authority of religion, Bacon portrayed science as the "most faithful handmaid" of religion and "the surest medicine after the word of God . . . against superstition. . . ." Bacon even offered a prayer "that things human may not interfere with things divine" and that when the mind was purged of "fancies and vanities" it would remain "subject and entirely submissive to the divine oracles" and "give to faith that which is faith's."[4]

Like Pliny, whom he resembled in some ways, Bacon believed he fell a martyr to scientific curiosity. He died in April 1626 of a cold he caught conducting an experiment to determine the effect of stuffing a chicken with snow.

René Descartes (1596–1650)

Another influential philosopher of science was René Descartes. In many ways he was the opposite of Francis Bacon, the "philosopher of industrial science." Descartes, a talented mathematician honored as the inventor of analytic geometry, advocated the deductive, mathematical approach to science. Bacon, who favored fact gathering and experiment, tended to distrust mathematics and deductive logic, which, he believed, had dominated scientific thought for too long.

Descartes came from a wealthy French family. His father was a Counsellor in the Parliament of Rennes, but his mother died when he was born, and as a sickly youth with a "pulmonary weakness" he seemed destined for an early grave. Descartes was educated in the Jesuit school of La Fleche, where he became interested in mathematics and dismissed the rest of the teachings as mere quibbling. Thereafter Descartes worked as an engineer, dabbled in astrology, and associated with mathematicians and natural philosophers. Apparently restless and dissatisfied with these activities, Descartes enlisted in the army of the Prince of Orange.

During a lull in the warfare in the Netherlands, Descartes set off seeking diversion in Germany, where he joined the Bavarian service in 1619. Here while sitting in a "stove" (or warm room), he was overwhelmed with an emotional and intellectual experience that left him with a profound philosophical insight. His *Discourse on Method* (1627) gave his discovery to the world. Having sold his estates so that he could live and study without working, Descartes toured Switzerland and Italy and settled in Paris for some time. In

1629, realizing that his ideas were too unorthodox for France, he moved to Holland, where he remained for the next 20 years. During this time he studied mathematics, optics, the theory of music, astronomy, meteorology, and anatomy.

When Galileo was censored by the Inquisition for his heretical support of the Copernican system, Descartes abandoned his plan of publishing his own book based on similar ideas. Descartes believed that the Copernican doctrine was so intimately connected with every part of his *Treatise of the World* that he "could not disconnect it without making the remainder faulty." Thereafter, fear of persecution for heresy influenced much of his conduct, including perhaps the destruction of some of his writings and the suppression of other works during his lifetime.[5]

Even while living in Holland, Descartes was apprehensive about the opposition to his published works, especially certain essays in which he attempted to prove the existence of God. Despite his habitual caution, his views proved too unorthodox and charges were filed against him. He fled to the Hague and appealed to the Prince of Orange for help. Finally Queen Christina of Sweden, who was very interested in science and mathematics, invited Descartes to join her court at Stockholm. After five months of trudging through the cold at five A.M. to give Christina her lessons, Descartes contracted pneumonia and died.

So pervasive was Descartes' influence on science and philosophy that many contemporaries of Newton thought of themselves as the disciples of Descartes. Like Bacon, whose work he had read and criticized, Descartes was an amateur scientist, generally correct in his broad premises but often wrong in the particulars as we shall see in our discussion of physiology (Chapter 9). Unlike Bacon, Descartes placed a priori principles first and subordinated his observations and experimental findings to them. He too had a grand scheme and task for natural philosophy. An examination of method was a primary part of his plan to use the mathematical method in developing a general mechanical model of the workings of nature. Although experimentation had a role in Descartes' system, it was definitely a minor one. He believed that experiments should serve as illustrations of ideas that had been deduced from primary principles or should help decide between alternative possibilities when the consequences of intuitive deduction were ambiguous. It would be wrong to leave Descartes without remembering his most famous axiom or first principle, "I think, therefore, I am." He also said, "Give me motion and extension, and I will construct the world."

Obviously, neither Descartes nor Bacon alone could have served as a complete guide for the development of science and technology. Christiaan Huygens (1629–1695) recognized this when he remarked that Descartes had ignored the role of experimentation, while Bacon had failed to appreciate the role of

mathematics in scientific method. A synthesis of the two approaches was
really needed, or the admission that there is no one scientific method.
Plodding, dull fact finding and flashes of intuition have served scientists in
their actual work, no matter what formal doctrine of method they professed
to follow.

Like Bacon, Descartes rejected the conceit that science should be "pure"
and snobbishly useless. In particular, he believed that the study of biological
subjects would contribute to medicine and ultimately to human welfare. We
shall deal with particular aspects of Descartes' work in the chapter on physiology since he played a major role in the development of the mechanistic
philosophy.

The Academy Movement in Italy

While the most famous and long-lasting of the scientific societies were the
Royal Society of London and the Académie Royale des Sciences of Paris,
Italy, which had been the heart of the scientific and artistic renaissance, was
also the site of the earliest academies. The *Academia Secretorum Naturae* was
established in the 1560s and met at the house of its leader, Giambattista della
Porta (1538-1615) in Naples. This group took the important step of requiring
members to present a new fact in natural science as the price of admission.
By encouraging the collection of new facts and the performance of experiments
at meetings, the Academy generated some 20 books of results. Unfortunately,
suspicion that the natural philosophers were dabbling in witchcraft led to the
rapid demise of the society.

Another group called the Academy dei Lincei (Academy of the Lynx Eye)
met in Rome from 1603 to 1630 under the patronage of Duke Federigo Cesi
(1585-1630), a rich and talented experimentalist. Among the 32 members
were Galileo, della Porta, and Nicolas Fabri de Peiresc (1580-1637), a Frenchman interested in Galileo's work and the exchange of scientific information
throughout Europe. Galileo's "microscope" was developed in this academy,
as were other important instruments and publications. Again, a valuable experiment in scientific institutions was destroyed prematurely. The membership
was split by the condemnation of the Copernican theory in 1615 and finally
disbanded with the death of the duke in 1630.

The Florentine Accademia del Cimento (Academy of Experiments) encouraged the cooperative efforts of natural philosophers. The group included
disciples of Galileo, as well as Giovanni Borelli (1608-1679), a mathematician
particularly interested in muscle physiology, and Francesco Redi (1621-1691),
well known for his experiments on spontaneous generation. With the patronage of the Medici brothers, Grand Duke Ferdinand II and Leopold, the
Academy assembled a well-equipped scientific laboratory which encouraged

experimentation and discussions. The *Report of Experiments* (1667) included
the first published descriptions of important scientific instruments such as the
barometer, thermometer, and hydrometer. The group had been meeting as early
as the 1640s, but the Academy ended abruptly in 1667. Although the reasons
for the demise of the society are obscure, Leopold's appointment as a cardinal
that year may have been the major cause.

These early Italian societies pursued experimental science and generally
avoided speculative or controversial areas. This was not because "science"
had become well separated from philosophy and religion at this stage, but because natural philosophers, for their own protection, had to avoid dangerous
topics. While this was unfortunate for the development of scientific theory, it
did encourage a period of experimentation and work on scientific instruments
which would be of great value to later scientists.

Miscellaneous Academies

Academies developed in many European countries and in the United States of
America, although none of these matched the prestige of the Royal Society of
London and the French Academy of Sciences.

English Academies

Because of the political disruption in England during the seventeenth century
(already discussed in terms of its effect on the career of William Harvey), the
academies got off to an uneven start in England. Although the Royal Society
was not officially founded until chartered by Charles II in 1662, it grew out
of informal meetings of scientists going back to the 1640s in London and
Oxford. As early as 1616, Edmund Bolton had attempted to establish a
formal organization and had actually gotten a favorable reception from King
James I and Parliament, but the death of the king caused the project to lose
all momentum.

In the 1640s a group of young scientists began meeting regularly in London as
the Philosophical College. The members included John Wilkins, John Wallis, Samuel Foster, Jonathan Goddard, George Ent, and a little later, Robert Boyle and
William Petty. The meetings were held at the homes of the members, at the Bull
Head Tavern in Cheapside, or at Gresham College in Bishopsgate.

Soon after Cromwell seized Oxford a Parliamentary Commission was set up
to reform the university (1648). John Wilkins, appointed Warden of Wadham
College, soon attracted important students to Oxford and founded a scientific
club called the Philosophical Society, which met until 1690. The balance of
power, and university appointments, changed again in 1660 when Charles II

was restored to the throne and the center of scientific activity in England returned to London. After a lecture by Christopher Wren at Gresham College
in November 1660, the assembled scientists formally declared interest in establishing a "College for the promoting of Physico-Mathematical Experimental
Learning." The ever-present John Wilkins was elected chairman and, of the
many persons interested in science, 41 were chosen for the official project.
The king gave his verbal approval via the courtier Sir Robert Moray, who was
made president of the assembly. A formal charter was granted two years later
incorporating the institution as The Royal Society for the Improvement of
Natural Knowledge. Thomas Sprat, although not himself associated with any
particular scientific finding, was enthusiastic about the new experimental
method and wrote the *History of the Royal Society* in 1667. The very valuable secretary of the Society, Henry Oldenburg (1615–1677), was not a
scientist either, but was a promoter of the experimental method who carried
out a massive correspondence to share his enthusiasm for science and the
Royal Society with scientists and societies all over Europe.

Francis Bacon, ignored during his lifetime, became the vivifying spirit of
the Royal Society. Some of the fellows tried to carry out policies and
projects he had suggested, such as "histories" of the crafts and commerce.
Robert Hooke (1635–1703), the Curator of Experiments for the Society, reflected the Baconian principle that truth and utility are the same thing in his
description of the "noble matters" pursued by the Royal Society: "The Improvement of Manufactures and Agriculture, the Increase of Commerce, and
the Advantage of Navigation." While pursuing the useful arts, manufactures,
and inventions, members of the Society pledged to avoid meddling with
grammar, rhetoric, logic, politics, morals, metaphysics, or divinity. The rules
which the society prescribed for philosophical progress required "avoiding
Dogmatizing, and the espousal of any Hypothesis not sufficiently grounded
and confirm'd by Experiments."[6]

Although the Royal Society did not emphasize biological studies, many of
the inventions and experiments performed by members were of great importance to progress in biology. Robert Boyle (1627–1691) is best remembered
as a reformer of chemistry, but his experiments on respiration and combustion,
using the air pump, were important in physiology. In addition to experiments
in physics and chemistry, some members of the society carried out studies of
the blood system which continued and refined the work of William Harvey.
Many experiments were conducted on the intravenous injection of food and
drugs, and blood transfusions between man and beast were attempted.

One of the most versatile, hard-working, and underpaid scientists of all
time was Robert Hooke, who served as Curator of Experiments for the Royal
Society. Hooke worked as assistant to the wealthy Robert Boyle, designing
and improving many experimental instruments such as the air pump, balance,
watch spring, and universal joint. As Curator, Hooke was responsible for

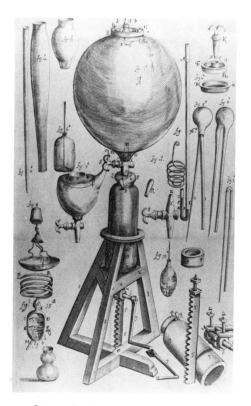

Boyle's first air pump. [From J. F. Fulton and L. G. Wilson (1966). *Selected Readings in the History of Physiology.* **Courtesy of Charles C Thomas, Springfield, Illinois.**]

weekly demonstrations for the edification of the members. He carried out experiments described by Vesalius, Antony van Leeunwenhoek, and original experiments of his own design. Although mainly remembered as a physicist, he made important contributions to biology, particularly through his microscopic studies, experiments on the properties of air, and investigations into the relationship of air, respiration, and combustion. Hooke served as joint secretary after the death of Henry Oldenburg in 1677. Because of his conflict with Newton, Hooke became almost a forgotten man in the history of science.

The French Academy of Science

Like the Royal Society of England, the Royal Academy of Sciences in France originated from a tradition of informal scientific meetings—but the two institutions developed quite differently thereafter. While the Royal Society of London remained quite independent of government control, the French institution became intimately linked to the State and so dependent on royal patronage that the academicians owed their appointments and salaries to the government.

Early in the seventeenth century, numerous groups of intellectuals met privately in and around Paris to explore the new science then permeating Europe. Peiresc was the motive force behind one such group. Another group revolved around Marin Mersenne (1588–1648), a writer and popularizer of science who enjoyed a far-flung correspondence with men such as Galileo, Descartes, and Hobbes. Although such groups enjoyed a voluntary, idiosyncratic, and spontaneous character, they suffered from chronic financial difficulties and instability. Ultimately, serious scientists turned to the State for the money and stability they needed. Finally, on December 22, 1666 the new Academy of Sciences held its first official meeting in the private library of Louis XIV. The academy owed its official status to Jean-Baptiste Colbert (1619–1683), advisor and minister to Louis XIV, who persuaded the king to charter and patronize the new scientific society. During the next half-century the academy grew into what Roger Hahn calls a "new organ of government" and "the means by which the French virtuosi were metamorphosed into professional scientists." [7] In 1699 the academy received a set of regulations dealing with its duties and relations to king and state, membership and procedures. Luxuriously accommodated by the king at the Louvre, the academy was clearly a creature of the state.

The French Academy differed fundamentally from the English. Charles II had given the Royal Society a charter and his blessing, but really nothing else. The Paris Academy was founded as a group of about 20 members who were appointed and salaried by the State. Thus, the Paris Academy was the first in Western Europe to turn scientific inquiry into a profession financed and controlled by the government. There were certainly many advantages, at first. The academy could hold meetings and conduct experiments in the Royal Library. Melchisedec Thevenot (1620–1692), director of the Royal Library and patron of the microscopist Swammerdam, was an influential member. Colbert, a man with a keen interest in pure and applied science, allowed the academy wide latitude and governed the academy only through general directives. The academy had the resources to pay salaries, set national standards, establish regulatory power over various industries and inventions, and serve government as a "think tank" on technical matters.

Bacon's writings influenced the projects adopted by the academy in its first phase. "Histories" and catalogs were planned and initiated. Just as in England, the influence of Baconian utilitarianism soon declined, whereas Descartes' philosophy of science became a major influence. However, unlike the Royal

Society, the academy was tied by a strong umbilical cord of funding to the objectives of the state. Colbert was succeeded by the Marquis de Louvois, the war minister. Not at all ambivalent about the proper purpose of the academy, he believed that scientific research should be devoted to useful ends in the service of the king and the state. Unfortunately, this narrow minded man saw the needs of king and state in singularly practical and even trivial terms. He ordered scientists to study problems connected with the royal fountain and games of chance enjoyed by the court. In 1693 Pontchartrain succeeded Louvois and appointed his nephew, Bignon, as director of the academy. The academy was increased to 70 members in hierarchically arranged positions, salaried and controlled by the state.

The French Academy was most important from the time of Louis XIV to Napoleon. During these turbulent times the interaction between the Academy and French society generated different tensions, conflicts, and political pressures. These rendered the academy at times very powerful, but subsequently historical forces crippled, dissolved, and ultimately reshaped the institution. As part of the Old Regime, the academy became rigid and smug, boasting an elitism that produced many enemies. Forced to satisfy a dual allegiance to the state and to science, the academy ultimately found that serving two masters generated more friction than progress.

The French Revolution destroyed the old system. Some of the academicians died in the Reign of Terror and others were forced to emigrate. Eventually the institution was reorganized on a more egalitarian basis. The academy had helped to develop high quality science in eighteenth century France and had transformed science from a pastime to a profession. After the revolution, the National Institute of Arts and Sciences replaced the academy. Although the new institute was to serve as proof of French cultural achievements and useful discoveries, almost half the members were to be representatives of scientific fields. All members of the institute, salaried by the government, were expected to serve as government advisors and functionaries. The government kept the institute under close control and emphasized the utilitarian role that was the price of government support. Compared to the independent scientific and professional societies developing in other countries, the French Institute became an obsolete organization and a conservative force which helped to sap the vigor and undermine the great promise of French science.

Scientific Societies in the United States

Even during the early colonial period, science was a subject of interest in America. The English colonies were particularly influenced by the Royal Society. Several Americans were elected Fellows of the Society. During the

Civil War in England, Governor John Winthrop invited the Society to move to Connecticut. In 1663 Winthrop was elected a Fellow of the Society and served, thereafter, as chief correspondent of the Royal Society in the Western World. Cotton Mather maintained a correspondence with the Royal Society in 1756. Increase Mather had established a scientific society in Boston as early as 1683 and modeled it on the Royal Society. This became the Boston Philosophical Society. Unfortunately, not as much care was taken as in England to exclude disputes over religion and politics, and these soon destroyed the Society.

Among the Founding Fathers, Benjamin Franklin (1706-1790) and Thomas Jefferson (1743-1826) were particularly active in bringing science to America. Franklin established a group called the Junto, which met weekly in Philadelphia. This club, established to discuss natural history and philosophy, eventually evolved into the American Philosophical Society, the oldest existing learned society in the United States. The American Academy of Arts and Sciences, incorporated in Boston in 1780, was the second major learned society established in the United States. Members of the Adams family were closely associated with the academy in its inception and development. Among its first members were George Washington, Benjamin Franklin, Thomas Jefferson, Alexander Hamilton, James Madison, and John Adams. Various special and regional clubs formed and dissolved, but only the American Philosophical Society and the American Academy of Arts and Sciences evolved into permanent national learned societies. Engendered in an age permeated by Baconian utilitarianism, the American societies also served the spirit of "cultural nationalism" of the new republic.[8] Like the Royal Society, the American academies remained essentially independent of university or government ties.

Government sponsorship of scientific societies and research is a quite recent development in the United States. Following World War I, various societies were grouped into national "councils" or "institutes": the American Council on Education, the National Academy of Sciences, National Research Council, the American Council of Learned Societies, and the Social Science Research Council.

The roots of the National Academy of Sciences go back to two national societies founded in 1840: the National Institution for the Promotion of Science, and the American Society for Geologists and Naturalists. (In 1848 the latter became the American Association for the Advancement of Science.) Scientists and government officials called for an academy to supplement the existing societies and serve as a guide for public policy concerning scientific matters. In 1863, Congress chartered the National Academy of Science to act as the official adviser to the federal government on scientific and technical questions. According to the charter "the Academy shall, whenever called upon by any department of the government, investigate, examine, experiment, and report upon any subject of science or art, the actual expense of such investigations, examinations, experiments, and reports to be paid from appropriations which may be made for the purpose, but the Academy shall receive no compensation whatever for any services to the Government of the United States."[9] This institution has been the subject of recent controversy reminiscent of tensions generated by the relationship between government and science in the French Academy.[10]

Although the history of the National Institutes of Health and the National Science Foundation is beyond the scope of this book, some reference to these organizations is necessary. The National Institutes of Health, established in 1930, evolved from inauspicious beginnings in 1887 as a laboratory in Staten Island for bacteriological studies which was part of the United States Public Health Service. The National Science Foundation, an institution originally dedicated to the support of pure research, was not created until 1950. Obviously quite different from the scientific societies generated by the first scientific revolution, these organizations illustrate the continuing interaction between science and society as manifested in the evolution of new institutions.

Notes

1. Francis Bacon (1870). *The Works of Francis Bacon.* Collected and edited by J. Spedding, R. L. Ellis, and D. D. Heath; new edition in 14 vols; reprinted 1968, New York: Garrett; especially vol. IV, *The New Organon.*

2. Bacon, IV-19.

3. T. Roszak (1973). *Where the Wasteland Ends.* New York: Anchor; especially Chapter 5.

4. Bacon, IV-52, 48, 89, 20.

5. R. Descartes (1664). *Treatise of Man.* Translation and commentary by T. S. Hall (1972). Cambridge, Mass.: Harvard University Press, p. xiii.

6. R. Hooke (1665). *Micrographia.* Reprint 1961. New York: Dover; see Hooke's letters to the king and Royal Society (no page numbers).

7. R. Hahn (1971). *The Anatomy of a Scientific Institution: The Paris Academy of Sciences, 1666–1803.* Berkeley: University of California Press, p. 4.

8. See B. Hindle (1956). *The Pursuit of Science in Revolutionary America, 1735–1789.* Chapel Hill: University of North Carolina Press.

9. J. C. Kiger (1963). *American Learned Societies.* Washington, D. C.: Public Affairs, p. 124.

10. P. Boffey (1975). *The Brain Bank of America.* New York: McGraw-Hill.

References

Abbot, E. A. (1885). *The Life and Works of Francis Bacon.* London: MacMillan.

Andrande, E. N. da C. (1954). Robert Hooke. *Sci. Amer. 191:* 94–98.

Bacon, F. (1870). *The Works of Francis Bacon.* Reprint 1968. Collected and Edited by J. S. Spedding, R. L. Ellis, and D. D. Heath. 14 vols. New York: Garrett.

Bates, R. S. (1945). *Scientific Societies in the United States.* New York: Wiley.

Beck, L. J. (1952). *The Method of Descartes; A Study of the Regulae.* Oxford: Clarendon.

Bloland, B. C., and Bloland, S. M. (1974). *American Learned Societies in Transition: The Impact of Dissent and Recession.* New York: McGraw-Hill.

Boffey, P. (1975). *The Brain Bank of America.* New York: McGraw-Hill.

Brasch, F. E. (1931). The Royal Society of London and its Influence upon Scientific Thought in the American Colonies. *Sci. Monthly 33:* 337–355, 448–469.

Bronk, D. W. (1975). Marine Biological Laboratory: Origins and Patrons. *Science 189:* 613–617.

Bronk, D. W. (1975). The National Science Foundation: Origins, Hopes, and Aspirations. *Science 188:* 409–414.

Bush, D. (1939). *The Renaissance and English Humanism.* Toronto: University of Toronto Press.

Cameron, H. C. (1952). *Sir Joseph Banks.* London: Batchworth.

Crowther, J. G. (1960). *Francis Bacon: The First Statesman of Science.* London: Cresset.

Descartes, R. *Treatise of Man.* Translated by Thomas Steele Hall (1972). Cambridge: Harvard University Press.

Descartes, R. (1637). *Discourse on Method: Optics, Geometry, and Meteorology.* Translated by P. J. Olscamp (1965). Indianapolis: Bobbs-Merrill.

Drake, S. (1966). The "Academia dei Lincei." *Science 151:* 1194–1200.

Éspinesse, M. (1956). *Robert Hooke.* Los Angeles: University of California Press.

Farrington, B. (1949). *Francis Bacon, Philosopher of Industrial Science.* New York: Schumen.

Fay, B. (1932). Learned Societies in Europe and America in the Eighteenth Century. *Amer. Historical Review 37:* 255–266.

Hahn, R. (1971). *The Anatomy of a Scientific Institution: The Paris Academy of Sciences, 1666–1803.* Berkeley: University of California Press.

Haldane, F. S. (1905). *Descartes: His Life and Times.* London: Murray.

Hall, A. R., and Hall, M. G., eds. (1967). *The Correspondence of Henry Olden-burg*. Madison: University of Wisconsin Press.

Hartley, H., ed. (1960). *The Royal Society: Its Origins and Founders*. London: Royal Society.

Hartley, H. (1960). The Tercentenary of the Royal Society. *Amer. Sci. 48:* 279–299.

Herrnstein, R. J., and Boring, E. G., eds. (1965). *A Source Book in the History of Psychology*. Cambridge: Harvard University Press.

Hindle, B. (1956). *The Pursuit of Science in Revolutionary America, 1735-1789*. Chapel Hill: University of North Carolina Press.

Kearney, H. F. (1971). *Science and Change, 1500-1700*. New York: McGraw-Hill.

Keeling, S. U. (1968). *Descartes*. London: Oxford University Press.

Keynes, G. L. (1960). *A Bibliography of Dr. Robert Hooke*. Oxford: Clarendon.

Kiger, J. C. (1963). *American Learned Societies*. Washington, D.C.: Public Affairs.

Kraus, M. (1942). Scientific Relations Between Europe and America in the Eighteenth Century. *Sci. Monthly 55:* 259–272.

Kronick, D. A. (1952). *A History of Scientific and Technical Periodicals*. New York: Scarecrow.

Lange, E. F., and Buyers, R. F. (1955). Medals of the Royal Society of London. *Sci. Monthly 81:* 85–90.

Lyons, Henry (1944). *The Royal Society, 1600-1940*. Cambridge: Cambridge University Press.

Mason, S. F. (1966). *A History of Science*. New York: Collier.

Martin, D. C. (1960). The Tercentenary of the Royal Society. *Science 131:* 1785–1790.

Matzke, E. B. (1943). Concepts of Cells Held by Hooke and Grew. *Science 98:* 13–14.

Merton, R. K. (1938). Science and Technology in 17th Century England. *Osiris IV:* 360ff.

Moe, H. A. (1960). Tercentenary of the Royal Society. *Science 132:* 1816–1822.

Montagu, B. (1844). *The Works of Francis Bacon*. Philadelphia: Carey and Hart.

More, L. T. (1944). *The Life and Works of the Honourable Robert Boyle*. London: Oxford University Press.

Oleson, A., and Brown, S. C. (1976). *The Pursuit of Knowledge in the Early American Republic: American Scientific and Learned Societies from Colonial Times to the Civil War*. Baltimore: Johns Hopkins University Press.

Ornstein, M. (1936). *The Role of Scientific Societies in the Seventeenth Century*. Chicago: University of Chicago Press.

Porta, della G. B. (1658). *Natural Magick*. New York: Basic Books. Reprinted 1957.

Power, H. (1664). *Experimental Philosophy*. London: Martin and Alestry. Reprinted 1966. New York: Johnson Reprint.

Redi, F. (1668). *Experiments on the Generation of Insects* (1668). Translated by M. Bigelow (1909). Chicago: Open Court.

Rosenberg, C. E. (1976). *No Other Gods: On Science and American Social Thought*. Baltimore: Johns Hopkins Press.

Roszak, T. (1973). *Where the Wasteland Ends*. New York: Doubleday.

Roth, L. (1937). *Descartes' Discourse on Method*. Oxford: Clarendon.

Scott, J. F. (1952). *The Scientific Work of Rene Descartes, 1596–1650*. London: Taylor and Frances.

Smith, A. G. R. (1972). *Science and Society in the 16th and 17th Centuries*. New York: Harcourt, Brace, Jovanovich.

Spedding, J. (1878). *An Account of the Life and Times of Francis Bacon*. Boston: Houghton, Osgood.

Sprat, T. (1667). *History of the Royal Society*. Facsimile of 1667 edition, edited with critical apparatus by J. I. Cope and H. W. Jones. St. Louis: Washington University Press. (1958)

Stimson, D. (1968). *Scientists and Amateurs: A History of the Royal Society*. New York: Greenwood.

Thorndike, L. (1923). *A History of Magic and Experimental Science*. New York: Macmillan.

Thorndike, L. (1929). *Science and Thought in the Fifteenth Century*. New York: Columbia University Press.

Thornton, J. L., and Tully, R. I. J. (1959). *Scientific Books, Libraries, and Collectors*. London: Library Association.

Wolf, A. (1935). *A History of Science, Technology and Philosophy in the 16th and 17th Centuries*. New York: Macmillan.

7

MICROSCOPES AND
THE SMALL NEW WORLD

As we have seen in the previous chapter, during the seventeenth century many instruments fundamental to modern experimental science were either invented or first put into use on a significant scale. Although the barometer, thermometer, pendulum clock, and air pump are important devices, the microscope and telescope, which gave man new access to the world of the infinitely small on earth and the infinitely large cosmos, undoubtedly created the greatest intellectual impact. Certainly the microscope was the most important of all instruments for progress in biology, literally creating the fields of cytology, histology, and microbiology.

Roots of Microscopy

Lenses were used in ancient times, but their application to scientific and medical problems is comparatively recent. The Assyrians are known to have used lenses and Ptolemy of Alexandria wrote a treatise on optics dealing with indices of refraction. The ruins of Nineveh, Pompeii, and Heraculaneum contained lenses that seem to have been ground and polished. The Emperor Nero is said to have watched performances in the arena with the aid of a jewel having curved facets which he held up to one eye. Although it is possible that the jewel contained concave facets to correct shortsightedness, there are few other reports of such uses of lenses until the thirteenth century.

According to Pliny, Roman doctors used convex lenses as a way to converge the rays of the sun, i.e., as a burning glass. "I find that among doctors it is thought that nothing is better when any part of a body is to be burned," he wrote, "than to use a crystalline sphere placed in the sun's rays." However, the voluble Pliny does not mention the use of lenses for aiding human vision. Seneca (4 B.C.–65 A.D.), a Roman philosopher of the first century, wrote: "Letters, however small and indistinct, are seen enlarged and more clearly through a globe or glass filled with water." Since he also noted that fruits in water seem to be enlarged, he may have failed to appreciate the effect of the sphere and merely believed that water itself caused apparent magnification. Towards the end of the thirteenth century some unknown and probably illiterate artisan invented spectacles—lenses placed close to the eyes to correct the farsightedness common in middle age. The inventor was probably a glazier, who would have made ornaments and glass discs for windows. A passage in a Florentine manuscript dating back to 1299 praised the use of spectacles. The writer stated: "I find myself so pressed by age that I can neither read nor write without glasses which they call spectacles, newly invented, for the great advantage of old men when their sight grows weak."[1] Roger Bacon wrote about lenses and light and in his description of the properties of lenses mentioned their magnifying properties. Craftsmen continued to make lenses to correct vision and found that the curve had to be more accentuated as people grew older. These lenses were all of the convex or converging type and could not help victims of myopia. It was not until 1568 that the use of concave lenses to correct nearsightedness was first mentioned.

Even the limited use of spectacles must have occasioned a profound effect on attitudes towards human limitations and debilities. Certainly, spectacles made it possible for scholars and copyists to continue their work—but what was more fundamental, they accustomed men to the idea that certain physical limitations could be transcended by the use of new technologies.

The first to show a real appreciation of the way convex lenses produced a magnified image was Ibn al Haitham, known to Europe as Alhazen (962–1038). His treatise was translated into Latin in the twelfth century as the *Opticae Thesaurus Alhazeni Arabis* and was published four centuries later. Alhazen dealt with experimental means of proving various propositions concerning the action of lenses. He also included an anatomical description of the eye. Although Alhazen clearly knew that an image would appear enlarged if viewed through a sphere when the eye and the object lie on an extension of the same diameter, he did not suggest any practical applications of this phenomenon.

In 1589 Giovanni Battista della Porta (1543–1615) published an encyclopedic work on popular science in 20 books under the collective title of *Magia Naturalis* (*Natural Magick* in the English translation). One book was called *Of Strange Glasses* and dealt with some practical uses of lenses to correct defects of human vision. According to della Porta,

Concave Lenticulars will make one see most clearly things that are far off; but Convexes things neer at hand; so you may use them as your sight requires. With a Concave you will see small things afar off very clearly; with a Convex, things neerer to be greater, but more obscurely; if you know how to fit them together you shall see both things afar off and things neer at hand, both greater and clearly. I have much helped some of my friends who saw things afar off, weakly; and what was neer, confusedly, that they might see all things clearly.[2]

Certainly, spectacle makers had a good deal of empirical knowledge of optics in the sixteenth century and the making of lenses was an established industry.

Complex Lens Systems

Although Galileo is sometimes cited as the inventor of the microscope and telescope, these instruments seem to have originated with some rather obscure Dutch spectacle makers, Hans and Zacharias Janssen of Middleburgh. About 1590, Zacharias Janssen created a crude compound microscope by combining a concave and a convex lens at the ends of a tube which was about an inch in diameter and one-and-a-half feet long. There is no evidence that the Janssens ever made any significant observations with their magnifying device. Another Dutchman, Cornelius Drebbel of Alkmaar (b. 1572), who devised a better instrument at about the same time, seems to have done more than the Janssens to make the compound microscope known to scientists. When Galileo heard of the device, he prepared lens systems for himself. Although he did describe the complex eye of an insect as seen under the microscope, Galileo's major interest was of course in the companion invention, the telescope. While sparing "neither pains nor pence" to perfect the telescope, Galileo left the development of the microscope to others.

Even with magnification no greater than 10-fold, the first microscopes provided exciting new visions. Creatures just barely visible to the naked eye were found to be complex and bizarre monsters when studied in detail. Galileo said that his microscope made flies look as big as lambs. The ubiquitous flea was a popular object of microscopic study. Indeed a common name for the instrument was *vitrum pulicare,* or flea glass.

Not all scholars and scientists were ready to accept the new instruments. Many of Galileo's contemporaries refused to look through his telescope. Trusting the rational more than the empirical, they feared that man's fallible senses could only be mislead by such unnatural devices. The microscope was similarly distrusted. The very life of George Stiernhielm, a Swedish poet who amused himself with scientific experiments, was placed in jeopardy by the use of magnifying lenses. A Lutheran clergyman who had been persuaded to look at a flea through the magnifying glass was so frightened by the unnatural dimensions of the creature that he denounced Stiernhielm as a sorcerer and an atheist. Without the intervention of Queen Christina, the poet might have been burnt as a witch.

Microscopy in the Seventeenth Century

With the new magnifying glasses, seventeenth century scientists assembled a
remarkable body of observations and experiments, beginning with the work of
Francesco Stelluti in 1625. His strikingly detailed illustrations of the parts of
the bee, obtained at magnifications of five and 10 diameters, were published
by the Academy of the Lynx Eye. Giovanni Borelli (1608–1679), better known
for his studies of muscles, turned magnifying lenses on objects as varied as the
blood of a nematode worm, textile fibers, and spider eggs. Athanasius Kircher
(1602–1680) turned the microscope to the study of putrefaction and disease.

From these tentative beginnings, a group of five men born between 1628
and 1641, working quite independently of each other, went on to establish a
highly developed level of microscopic investigation. Indeed, the work accomplished by Leeuwenhoek, Malpighi, Hooke, Swammerdam, and Grew was
the best to be done until the middle of the nineteenth century. After looking
at the work of these classical microscopists, we shall return to a discussion of
the technical problems of microscopy and the means developed for its
improvement.

Because the life and work of Antony van Leeuwenhoek are certainly the
most interesting, he will be discussed first and in most detail.

Antony van Leeuwenhoek (1632–1723)

Leeuwenhoek was the greatest amateur scientist and microscope maker of the
seventeenth century. His admiring biographer, Clifford Dobell, called him
the "only earnest microscopist in the whole world" and a man with "no
rivals and hardly a single imitator."[3] Antony's father was a basket maker and
his mother was the daughter of a brewer. The family lived in Delft, Holland,
a town famous for Vermeer, beer, and china. Leeuwenhoek's father died
when Antony was only six. At the age of 16, Antony was apprenticed to a
linen draper in Amsterdam. He returned to Delft six years later, bought a
house and shop and engaged in a remarkable number of jobs ranging from
keeping his own dry goods store to bookkeeper, cashier, surveyor, and Cham-
berlain to the Sheriff of Delft, with duties ranging from care and cleaning of
City Hall to wine gauger and tax collector. These municipal sinecures apparently took little of his time, the bulk of which seems to have been devoted to
his hobbies of glass blowing, lens grinding, and fine metal work. Leeuwenhoek
became so wrapped up in his scientific experiments that he was regarded as
crazy by his neighbors and accused of neglecting his family. Perhaps the last
charge was true, since he lost two wives and only one of some six children
survived him. Yet Leeuwenhoek himself lived, as recorded on his epitaph, for
"ninety years, ten months, and two days."

It is difficult to imagine a more unlikely candidate for best microscope maker and user of the seventeenth century—a period not lacking in great minds. Leeuwenhoek was a totally self-educated scientist with no formal training and no knowledge of Latin or any language other than Nether-Dutch. Thus, he could not read the works of the learned natural philosophers. This was probably an asset since it kept him remarkably free of the preconceptions and philosophical dogma that clouded the minds of his more educated contemporaries. Ingenious, patient, dexterous, and immensely curious, Leeuwenhoek produced very small biconvex lenses of short focal length. With these tiny, high-powered lenses, Leeuwenhoek's simple microscopes were superior to contemporary optical systems where multiple lenses invariably produced multiple aberrations and artifacts.

Although Leeuwenhoek was an uneducated draper, he had friends in the world of art and science. Regnier de Graaf (1641–1673), the well-known anatomist, wrote to the Royal Society in 1673 about the great work of Antony van Leeuwenhoek. A long series of letters to the Royal Society, some 372 communications in Nether-Dutch, was edited and translated into some 120 extracts in English or Latin and printed in the *Philosophical Transactions*. Leeuwenhoek's letters contained accounts of scientific observations, experiments, remedies for hangover, and what he had for breakfast.

In 1680, Leeuwenhoek was elected a Fellow of the Royal Society. Although he treasured this honor, he generally found fame an annoyance, especially when it brought visitors to bother him. When he died, more than 400 microscopes and magnifying glasses were found. The Royal Society was bequeathed a cabinet with 26 instruments and extra lenses. The optical properties of some of these lenses were measured. Magnifying powers of from 50- to 200-fold were found, but unfortunately these instruments were lost within 100 years of Leeuwenhoek's death.

With no apparent scientific plan, Leeuwenhoek allowed his insatiable curiosity to determine the course of his research. Crystals, minerals, plants, animals, water from different sources, scrapings from his teeth, saliva, seminal fluid, even gunpowder were examined under his lenses. Unfortunately, Leeuwenhoek was a secretive man and kept his method for seeing the "smallest animalcules" for himself alone. He apparently was jealous of his own methods and contemptuous of others who were lacking in curiosity and persistence. Thus, when asked why he did not teach his methods to others he said: "To train young people to grind lenses, and to found a sort of school for this purpose, I can't see there'd be much use . . . because most students go there to make money out of science, or to get a reputation in the learned world. But in lens-grinding, and discovering things hidden from our sight, these count for nought. . . . And over and above all, most men are not curious to know. . . ."[4]

Leeuwenhoek obviously took great joy in his work, but he also took care to make it as quantitative as possible. Although he was ever the enthusiastic

Antony van Leeuwenhoek studying the small new world through one of his microscopes. Note that the microscope—actually a tiny lens mounted between metal plates—is brought close to the eye for observation. (© 1959, Parke, Davis and Company.)

amateur, he was not a sloppy dilettante. There were no standard units of measurement available for microscopic work, so he used simple objects for comparison—such as hair, or grains of sand, and stated the measurement in terms of fractions or multiples. For some objects he made comparisons to "biological units" such as the eye of a louse.

In the course of his long and vigorous career, Leeuwenhoek made so many discoveries it is impossible to cover them in any detail. He is most famous for his studies of microorganisms and blood circulation, but he also made studies of the life history of insects (such as fleas, aphids, ants), rotifers, digestion, and the histology of striated muscle, dental bone, and optic lenses. Leeuwenhoek confirmed Malpighi's discovery of blood capillaries, described the red blood corpuscles in fishes, frogs, and birds, and the disc-shaped corpuscles in man and other mammals. In demonstrating quite rigorously that the capillaries link the arteries and veins, Leeuwenhoek helped to complete

Harvey's work. He carefully noted that the circulation in the capillaries is dependent on the heart beat. Leeuwenhoek did not passively observe, but attempted to analyze what he had seen. He found that the passage of blood from the arteries into the veins in tadpoles took place in blood vessels so thin that only one corpuscle could be driven through at one time. From his observations of capillaries in the tails of tadpoles, he ventured the opinion that the same thing must occur in human bodies but could not be seen because of the thickness of our skin. He estimated that "the corpuscles or globules that make the blood red, are so small that ten hundred thousand of them are not so big as a grain of course sand. . . ." This report was not the result of a single set of observations, but was quite reproducible as Leeuwenhoek assured the reader: "The observations told here have not been made once, but they have been resumed repeatedly, giving me much pleasure, and every time on different tadpoles. . . ."[5]

In 1677 Johan Ham seems to have made the first observation of sperm in the semen of a man with gonorrhea. He attributed the "animalcules" he saw in the semen to the disease. The idea that sperm were a sign of disease proved very persistent, despite the observation of Leeuwenhoek that these entities were found in the semen of dogs, rabbits, and men. Coming close to a recognition of the significance of sperm, Leeuwenhoek imagined he could see female and male producing sperm. His studies of fertilization in fish and frogs noted the association of sperm with the egg. (However, Leeuwenhoek held incorrect spermist-preformationist views, which we shall discuss in Chapter 8.)

It took science over a century to catch up with this self-educated microscopist. Work on sperm, fertilization, and development became hopelessly muddled by more educated but less acute observers. Leeuwenhoek's discoveries of parasites and bacteria were fundamental to the development of medical microbiology, but were not put to good use until the end of the nineteenth century.

Marcello Malpighi (1628–1694)

Quite appropriately, Malpighi was born in the year that *De Motu Cordis* was published. Harvey had made the existence of the capillaries a logical necessity, but the microscopist turned this into an experimental reality. Second only to Leeuwenhoek for his breadth of interests and enthusiasm for microscopic studies, Malpighi was a pioneer in embryology, plant anatomy, and comparative anatomy as well as histology.

Malpighi studied Aristotelian philosophy at Bologna, where he met Bartolomo Massari (1603–1655), a professor of anatomy and founder of the *Coro Anatomico,* a scientific academy which conducted dissections and discussions of the new natural philosophy. Recognizing Malpighi's ability, Massari allowed him to use his private library and had him elected a member of the academy. Eventually Malpighi married Massari's sister.

When his parents died in the epidemic of 1649, Malpighi decided to become a physician to support the rest of his family. Having completed his studies of medicine and philosophy, Malpighi practiced medicine in addition to his position as lecturer in logic at the University of Bologna. Soon he had established a dangerous reputation for his anti-Galenist, proexperimental ideas. In 1656 the Chair of Theoretical Medicine was created for him at the University of Pisa. Although the intellectual climate was an improvement, the hot, humid weather wreaked Malpighi's health. Nevertheless, biology advanced when Malpighi joined Francesco Redi, Giovanni Borelli, and other members of the *Academi del Cimento*. In 1667 Henry Oldenburg, Joint Secretary of the Royal Society, invited Malpighi to communicate his studies to the society. One year later Malpighi was elected a Fellow. After years of wandering from one university to another, Malpighi settled in Bologna. Unfortunately, in 1684 his house burned down, with the loss of many microscopes, books, and research notes. When Innocent XII became Pope, he requested that Malpighi become his personal physician. Reluctantly Malpighi went to Rome in 1691. Three years later he died of apoplexy.

Malpighi was a founder of microtechnology for both animal and plant materials. Among his most important studies were those of: the capillaries and circulation of the blood; the fine structure of the lungs and kidneys; the cerebral cortex; plant microanatomy; and invertebrate biology, particularly the structure and natural history of the silkworm in its evolution from egg to pupa.

Malpighi's fundamental studies of the lung were begun in September of 1660 and completed the summer of 1661. Little was known of the structure of the lung at the time. It was merely described as a fleshy organ of a porous parenchyma, where the smallest divisions of the blood vessels and the ramifications of the windpipe were lost. "Parenchyma" at the time was used to describe tissues which were not obviously composed of fibers, but seemed mainly porous. The term goes back to Erasistratus, who used it to mean something poured from the veins.

Anatomists assumed that blood and air were simply poured into the parenchyma of the lung and mixed freely in the spaces of this porous material. In studies of lung tissue from dogs, Malpighi found that the lung was composed of membranous vesicles filled with air. The relationship between the air sacs and the minutest blood vessels was left unclear in these studies. When he turned to the study of the frog, he discovered that the blood did not leak out of its proper vessels into the air spaces, but went from artery to vein through minute "tubules" which we call capillaries. While the heart continued to beat, he could actually see the blood moving through the capillaries. The substance of the lungs was revealed as a network "where certain bladders and sinuses are bound together." By washing out the blood, inflating the lungs through the windpipe and drying them, Malpighi was able to trace a network of thin-walled bladders which were connected to the windpipe by very fine ramifications. So thorough was his work that he felt he might have "destroyed almost the whole race of frogs."

The greatness of this work lies not simply in the patient morphological analysis, but in the elegant hypotheses about the relationship between the minutest elements of the tissue and the function of the organ. "Observations by means of the microscope," he wrote, "will reveal more wonderful things than those viewed in regard to mere structure and connexion." He predicted that physiologists would transcend the prevailing mechanical explantions of function.

Despite the crudity of his instruments and what he called his own clumsiness, through "both reason and sense, aided by research" he linked studies of structure with new theories of function. "I never reached my idea of the structure of the kidneys by aid of books," he wrote, "but by the long, patient, and varied use of the microscope. I have gotten the rest by deductions of reason, slowly, and with an open mind, as is my custom."[6] There is a sense of modernity in this work. The use of stains, and mercury and wax injections were introduced in his pioneering histological studies. While the techniques are certainly primitive (ink and other readily available liquids were his stains), Malpighi, unlike Leeuwenhoek, tried to tell his reader exactly the method by which he proceeded, the preparation of tissue, the kind of lenses, and the means of illumination to use. With these methods Malpighi explored many other tissues, always trying to trace the smallest elements of the tissue under study.

Robert Hooke (1635-1702)

As Curator of the Royal Society Robert Hooke was responsible for the demonstrations performed at the weekly meetings. Although his reputation suffered because of his controversies with Newton and other scientists, Hooke was probably the seventeenth century's greatest inventor and designer of scientific instruments. His contributions to astronomical instrumentation were particularly valuable, but he also made many suggestions for improving the efficiency of the microscope and the methods of illumination. This is not surprising because the study of light was one of his main theoretical preoccupations.

In 1663 Hooke was required by the Royal Society to undertake a program of microscopical investigations of such vigor that he "bring in at every meeting one microscopical observation, at least." Hooke was ready with observations of various insects, such as fleas, lice, and gnats. He also demonstrated different types of hair, the point of a needle, textiles, the edge of a sharp razor, molds, and moss. These microscopic studies were quite difficult, even painful, for Hooke, who suffered from an eye disorder.

Hooke's *Micrographia*, published in 1665, contains many important descriptions of things observed through the microscope and techniques associated with the instrument. Four major ideas were elaborated in this work. First, Hooke described the colors of membranes and thin plates of mica, showing how light patterns vary with the thickness of the plates. He suggested an undulatory theory of light, comparing the waves in water to the spreading of light vibrations. Second, he described experiments to show that combustion depends on air and is similar to respiration. Hooke also described his observations of fossils and a correct theory of their origin. Finally, in his studies of the structure of cork he observed the empty little structures which he called "cells." Although Hooke's most famous observation was of cells in cork, he also described the contents of live plant cells: "for in several of those Vegetables, whilst green, I have with my Microscope, plainly enough discovered these cells or Pores filled with juices, and by degrees sweating them out."[7]

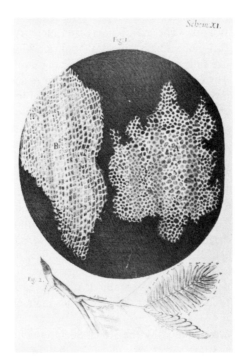

Illustration of molds from *Micrographia,* **1665. [From R. Hooke's** *Micrographia* **(1961). New York: Dover.]**

Illustration of molds from *Micrographia,* 1665. [From R. Hooke's *Micrographia* (1961). New York: Dover.]

Jan Swammerdam (1637–1680)

Like Leeuwenhoek, Jan Swammerdam was one of the great Dutch scientists produced in that country's Golden Age. In outlook, training, and methodology these two differ in everything but their genius. Swammerdam was more methodical, even at times obsessively so. Yet his life was the more disorderly and his genius almost wasted. Using the microscope as a tool rather than an end in itself, Swammerdam became the first true entomologist and one of the great comparative anatomists of the seventeenth century.

Jan Swammerdam's father was a prosperous apothecary who was interested in natural science and had a fairly good collection or "museum" which his son catalogued (published anonymously in 1679). Nevertheless he was determined that his son should take religious orders, even though Jan wanted to study natural history. As a compromise, Jan was allowed to study medicine at Leyden. Here he found Nicolaus Steno (1638–1686) and Regnier de Graaf (1641–1673) among his fellow students. Friendships with Steno and Thevenot, the king's librarian, proved invaluable to Swammerdam. Friends provided encouragement and even financial support to the often troubled, sickly, and impoverished naturalist. Interested in research, not medical practice, Jan was constantly at odds with his father who seems to have been a stingy, surly old man who had the unreasonable idea that his 30-year-old son should practice medicine and support himself rather than waste his time in studies of insects.

While a student, Swammerdam gained fame for his fine work in microdissection and microscopic investigations. Continuing his life-long studies of insects, Swammerdam also dissected animals and humans. Friends at court obtained special permission for him to autopsy criminals. Through such work, Swammerdam discovered the values of the human lymph system. Important observations appeared in his graduation thesis on respiration (1667). Proof that the lung of a newborn mammal would float in water if the animal had begun to breathe by itself but sink if respiration had not been initiated had important medical and legal ramifications. Swammerdam also worked on clinical thermometers, hermaphroditism, and fertilization in snails.

In 1669 Swammerdam published a general history of insects written in Dutch. Unfortunately, during the course of his research he contracted malaria which undermined his health for the rest of his life. When his father sent him to the country to improve his health so that he would be able to practice medicine, Swammerdam instead used the time for his classic research on the mayfly.

The prodigious amount of work carried out by Swammerdam in his short, unhappy, and embittered life is truly remarkable. In the course of his battles with his family over money, and the debilitating effects of malaria, he seems to have become mentally unstable and to have suffered a progressive decline in his intellectual powers. Desperate for money, he tried to sell his collections and even his microscopes. Eventually he became enmeshed in religious mysticism. Although Steno's attempts to convert him to Catholicism were unsuccessful, Jan became a follower of Antoinette Bourignon, a Flemish mystic and religious fanatic. Swammerdam abandoned science as an unworthy activity. In 1675 he wrote: "I am now resolved to addict my thoughts more to love the Creator of these things, than to admire him in his creatures."[8] At the age of 43, worn out by illness, hard work, religious excess, and quarrels with his sister over the family inheritance, he died. His zoological museum included 3000 species of insects, carefully dissected and displayed. His heirs demanded 5000 florins, but finding no buyers, broke up and dispersed the invaluable collection.

The bulk of Swammerdam's work was unknown to his contemporaries and remained unpublished until 1737. The manuscripts were left to his friend

Thevenot, who planned to have them published, but died before the project was even begun. After a precarious existence as private manuscripts, Herman Boerhaave (1668-1738) of Leyden purchased the papers and published them in 1737-1738 as two folio volumes in Dutch and Latin. The plates were engraved from Swammerdam's own fine drawings. Although published 100 years after the birth of the author, the *Biblia naturae* (*Bible of Nature*) was the best study of insect microanatomy available in the eighteenth century.

Working with tiny insects, Swammerdam developed a very fine dissection technique and skill in the construction of his instruments. He designed a simple dissecting microscope which had two arms; one held the object under study and the other held the lens. Both arms had coarse and fine adjustment. Like Leeuwenhoek, he had many different microscopes for particular purposes. His dissecting instruments were of the finest construction. For some purposes he used glass tubes drawn out as fine as hairs and scalpels ground to the proper degree under a magnifying glass. Fine scissors were used, in preference to knives and lancets, because these did not disturb the structural relationships of his experimental subjects quite as much. Fine glass tubes were used to inflate tiny vessels in the subjects or for the injection of colored liquids, ink, or wax. Swammerdam used wine and turpentine for preservatives and in even more ingenious ways. He was able to remove obstructions, such as the fat body of insects, by dissolving them away with turpentine or other solvents. Swammerdam was so willing to put himself into his work that in studies of the louse he let the louse bite his hands to observe the operation of the mouth parts.

Swammerdam's work on insects was influential in establishing preformation as the favorite theory of generation in the eighteenth century. His treatise on the mayfly was not just a scientific account of the insect but a deeply felt reflection of Swammerdam's religious commitment. Knowledge of the brief life of the mayfly might "give us wretched mortals a lively image of the shortness of this present life, and thereby induce us to aspire to a better."[9] With this "insecto-theology" he tried to establish an ethical system to harmonize biological and religious sources of knowledge.

The *Bible of Nature* was the first great scientific study of insect forms, transformations, and classification. The work also contains observations of the Crustacea, Mollusca, and the metamorphosis of the frog. The illustrations are remarkably detailed and accurate. Insect "evolution," which in this context means "metamorphosis," was revealed as a carefully predetermined sequence of events.

Very critical of Harvey's epigenetic theory of development and his casual acceptance of spontaneous generation of lower creatures, Swammerdam supported preformationist theory. The insect did not undergo any true transformation according to Swammerdam, but grew from tiny parts already present. In support of this theory, he was able to dissect out the adult insect under the pupal skin. Swammerdam applied his principle of evolutionary development not only to insects but to other animals and plants as they advanced from egg to frog or from bud to flower. Studies of the frog included aspects of generation, blood circulation, and muscle physiology. Under generation, he gave a detailed comparison between the development of man, insect, and frog. He assumed moulting occurred in all, that is, the new parts develop under the old skin which is then shed. Like the insect, the tadpole hides its new limbs under the skin; a sudden transformation makes them visible.

Microscopists like Leeuwenhoek and Swammerdam proved that the "lower creatures" were complex and interesting and that their life cycle conformed to rules rather than chance. Both were mistakenly advocates of preformationist doctrines, but partially based this on their opposition to spontaneous generation. They argued that even "simple" creatures were magnificently and lavishly gifted with complex parts when viewed under the microscope and did not arise from mud and slime via spontaneous generation, but only from parents like themselves.

Nehemiah Grew (1628–1712)

Although primarily remembered for studies of plant anatomy which led to Rudolf Camerarius' (1665-1721) discovery of plant sexuality, Nehemiah Grew also carried out detailed studies of the comparative anatomy of animals. Nehemiah was the only son of the Reverend Obadiah Grew, an anti-Royalist clergyman who lost his position after the civil wars, when Charles II was restored to the throne. After graduating from Penbroke Hall, Cambridge in 1661, Grew studied medicine at the University of Leyden. The University enjoyed a great reputation and furthermore, it was expedient for the Grews to leave England at the time. Having received the M.D. degree in 1671, Nehemiah practiced medicine in a small town but devoted the better part of his energies to studies of plant anatomy. Eventually he was able to settle in London where in addition to his medical practice he carried out his researches on plants and animals.

Essentially theological impulses seem to have motivated his scientific studies. His research was guided by the belief that animals and plants must possess similar structure because they are both "Contrivances of the same Wisdom." An essay on this theme was read to the Fellows of the Royal Society in 1671. When Henry Oldenburg died in 1677, Grew became Joint Secretary of the Royal Society with Robert Hooke. He edited the *Philosophical Transactions* for a year, resigning in 1679 because of a time-consuming but lucrative medical practice. Grew's treatise, *The Comparative Anatomy of Stomachs and Guts begun,* appeared as an appendix to a catalog of specimens in the museum

of the Royal Society. Quite modest about his work, Grew similarly entitled
part of his great work on plants, *The Anatomy of Vegetables begun* (1682).
This study was the first part of *The Anatomy of Plants,* which included *The
Anatomy of Roots, The Comparative Anatomy of Trunks,* and a fourth part,
dedicated to Robert Boyle, *The Anatomy of Leaves, Flowers, Fruits and
Seeds.* This work helped to revive the science of botany which had stagnated,
except for herbals of dubious value, since the time of Theophrastus. Grew
studied the vascular tissue of plants, noted its tubular nature, and discussed
its different distribution in roots and stems. Although his discussion was
probably more correct than that of Malpighi, Grew modestly acknowledged
Malpighi's priority in the discovery of the vascular system of plants.

Grew came close to discovering the sexual reproduction of plants. He
speculated that flowering plants exhibit sexuality, but never tested his idea
that the flowers carried the sexual organs. Although he recognized the pistil
as the female part, he was uncertain about the purpose of the stamens which
he called the "attire" of the flower. He noted that the pistil and stamens
were sometimes present in different flowers and sometimes in the same flower.
This suggested that some flowers were hermaphroditic like snails. In 1694, 12
years after publication of the collected *Anatomy of Plants,* Rudolph Camerarius described plant sexuality based on anatomical and experimental studies.

Comparative Anatomy of Stomachs and Guts begun, was the first zoological
book with the phrase "comparative anatomy" in the title and the first book
to deal with one system of organs by the comparative method. Viewing man
as the standard, Grew compared all other animals to the human type. Grew
also attempted to develop a classification of the "quadruped" based on the
anatomical and physiological characteristics revealed by studies of the gut.
Approaching more physiological ideas, Grew argued that the morphological
significance of any organ could not be grasped until its function had been considered. This was illustrated by a discussion of the food eaten by the animal
and the type of gut required to process different foods.

Much of Grew's work was at the microscopic level, indicating that there
was still much work to be done in gross anatomy after the introduction of the
microscope. By studying freshly killed animals, Grew was able to see the
peristaltic action of the gut. At a finer level, he studied the villi of the
mucosa. Patient and thorough in his research, Grew dissected the guts of 40
different species of birds alone. Various species of fish were also examined,
though clearly he found the higher mammals most worthy of study. As with
most medically trained anatomists, he regarded the human being as the highest
type and the standard against which all other species must be judged.

Malpighi's first sketch for his *Anatomy of Plants* was sent to the Royal
Society which later published the full work. Curiously, it was presented and
read before the Society on December 7, 1671, the same day on which Grew

presented the Society with his printed book *The Anatomy of Vegetables begun.* The order to print Grew's book had been given at a meeting of the Society the previous May. Michael Foster in comparing the two studies finds Grew's work to be "a sound piece of honest, arduous labour" while that of Malpighi "shines . . . with the light of genius, and is richer than the other in philosophic insight." [10] But Grew was remarkable for his hard-headed, empirical search for facts. Indeed, Conway Zirkle says that his work on plants was unsurpassed for over a century. [11] A strong theological interest pervaded his life and came out in his last work, *Cosmologica Sacra* (1701), in which he tried to demonstrate the wisdom of God in the excellence of His creations.

Technical Problems of Microscopy

The work of the five classical microscopists discussed in this chapter was the best done until the middle of the nineteenth century, when serious problems with the lens systems were finally solved. Chromatic and spherical aberrations were the most serious problems, although poor-quality glass—cloudy and with bubbles—was also troublesome. As more convex lenses were fashioned to increase magnification, they become more like prisms which separate light into different wavelengths. Naturalists brought enough imagination to their work without the addition of a rainbow of color. Newton himself studied the problem of chromatic aberration and pronounced it insoluble. Fortunately, the great Newton was not worshipped in the same manner as Aristotle and his followers were not willing to allow science to stagnate on this account.

One ingenious answer to Newton was that the human eye is not troubled by chromatic aberrations, although it is a lens system. This challenge was issued by David Gregory (1661–1708), a professor of mathematics who served at Edinburgh and Oxford. Newton was correct in the sense that chromatic aberration is an insoluble problem when single lenses are used with ordinary light. Other scientists suggested that the problem could be avoided by using monochromatic light or by constructing doublet or triplet lenses with different indices of refraction. The way to avoid chromatic aberration was either to change the light or the lens.

Achromatic lenses were finally constructed by John Dollond (1706–1761), a London resident of Huguenot descent. Dollond had been trained as a silk weaver and was, like Leeuwenhoek, a self-educated amateur scientist. His eldest son, Peter, was in business as an optician. John joined his son's business in 1752 and became interested in the problem of making achromatic lenses. Using a combination of crown glass and flint glass, he made an achromatic lens. The silk weaver turned optician won the Copley Medal of the Royal Society in 1758 for a telescope using his combination lens. In 1761 he was elected a Fellow of the Royal Society and appointed optician to the king.

In the 1830s improved compound microscopes, with ocular and objective spaced by a body tube, were on their way to becoming standard equipment in biological research. The color fringes around the outside of the optical field were eliminated by apochromatic objectives with better correction for color aberration.

Giovanni Battista Amici (1786–1863)

In 1827 Giovanni Battista Amici demonstrated the advantages of corrected lenses. Amici's career included contributions to both astronomy and microscopy. He studied at Bologna and was Professor of Mathematics at Modena between 1815–1825. Later he became astronomer to the Grand Duke of Tuscany and Director of the Observatory at the Royal Museum in Florence. The reflecting microscope which he invented about 1827 contained a three-inch concave mirror between objective and eyepiece–all of which were mounted in a horizontal tube. In this microscope the mirror magnified the image without causing chromatic aberrations. With this improved instrument, Amici made some of the earliest observations on the growth of the pollen tube. Although he came very close to the discovery of fertilization in flowers, the subject was not completely clarified until further refinements in microscopes and sample preparation in the 1880s and 1890s.

Immersion Microscopy

The maximal resolution of the optical microscope was eventually attained with the aid of the "immersion principle." Although not widely used and developed until the nineteenth century, it had been suggested by Robert Hooke for use with simple lenses as early as 1679. Though the English microscopists valued high resolution and great magnification, they did not take up the immersion principle as quickly as did continental scientists. Amici is often credited with priority in the introduction of immersion lenses in the nineteenth century, but Sir David Brewster (1781–1868) actually deserves that honor for work done in 1812. Attempting to develop an achromatic lens, Brewster suggested that the front element of a microscope's objective lens could be immersed in the liquid in which the object of study was mounted. These ideas were elaborated in his work *A New Treatise on Philosophical Instruments,* published in 1813. Diverse objects such as aquatic plants and animals, shells, and unpolished minerals were studied by immersion microscopy.

The fluid in which the object was immersed could be carefully selected to eliminate chromatic effects and the dispersion that occured when light rays traveled through media of differing refractive index. Oils with indices of refraction close to that of glass would keep the light rays in the same plane— going from glass to oil and back to glass. In addition to immersion microscopy Brewster may have been first to use filters to get monochromatic light (ca. 1836).

Working quite independently, Amici developed the immersion principle, not so much out of concern for the problem that Brewster was working on, but rather to diminish the loss of light in high-power lens systems and improve definition. The scientific world and the public had the opportunity to become acquainted with this advance in microscopy at the Paris Exhibition of 1855. Amici used various fluids, including glycerine and mixtures of oils—but mainly water because the oils often attacked the cements commonly used in the mounts at that time. The Amici system had various defects and probably the resolving power was not as good as the best of the "dry" objectives of the time.

There was a great deal of opposition to the use of immersion lenses in England. Probably the rich amateurs who were very influential in the 1860s were to blame for rejection of this kind of microscopy. English fashion leaders could get good results with their "dry" microscopes by using better illumination than the Continental scientists needed for their simpler immersion apparatus. American instrument makers departed from the British on this point and soon were manufacturing high-quality immersion lenses.

Ernst Abbe (1840–1905) and Carl Zeiss (1816–1888)

Further advances in microscopy resulted from the fruitful collaboration of Ernst Abbe, a Professor of Physics, and Carl Zeiss, supplier and repairer of instruments for the University of Jena. Abbe improved immersion microscopy with his "homogeneous immersion system," which eliminated refraction at the front of the lens and limited the loss of light by reflection. More than 300 different liquids were examined for their refractive indices and dispersive powers before cedar wood oil was chosen as the best medium.

Although Zeiss was attempting to develop high-quality optical instruments, he lacked the theoretical background to carry out elaborate computations concerning the properties of lenses. Abbe and Zeiss formed a good team, one with expertise in science and mathematics to complement the practical experience of the other. Their attempts to develop achromatic compound lenses were frustrated by the limited range of optical glasses then available. Zeiss and Abbe persuaded Otto Schott (1851–1935) to experiment with new kinds of glass. This resulted in useful borate and phosphate glasses which had properties quite different from crown and flint glass. Abbe developed the apochromatic objective in 1886, improved the oil-immersion microscope, and developed oil-immersion objectives with at least eight component lenses. The substage condenser was developed to complement his high-power objectives (1872).

Phase-Contrast Microscopy

Fritz Zernike (1888–1966), Professor of Physics at the University of Gronin-
gen, modified the light compound microscope so that it would be possible to
see living objects which cannot be distinguished in the ordinary light microscope. Mainly interested in optics, Zernike published a theoretical treatment
on the Foucault knife-edge test for testing the accuracy of grinding of mirrors
for telescopes in 1934. It was this work which led to the discovery of the
phase-contrast microscope.

Phase-contrast microscopy makes it possible to distinguish any structure in
a microscopic preparation that differs in refractive index from the surrounding
matrix. Biological research was particularly aided by this technique, because
it became possible to study the structure of still living cells which would be
colorless and transparent in ordinary light microscopy. In favorable cases this
technique obviated the need for dehydration, fixation, staining, and other
techniques which kill the specimen and introduce artifacts. Zernike won
many honors, including the Nobel prize in 1953, for this great contribution to
biological studies.

The Limits of Light Microscopy

Since the 1880s there have been many improvements in the optical microscope and, as we shall see, along with new methods of sample preparation,
these led to great advances in our knowledge of the fine-structure of complex
organisms and the workings of microorganisms. After correction of many of
the technical difficulties of optical microscopy, one problem put a limit on
the usefulness of the instrument. That is the size of the light rays themselves.
Objects smaller than the wavelength of light cannot be observed with such
"coarse" waves. To obtain higher magnification and resolving power rays
shorter than ordinary light must be used.

One approach to obtaining these objectives was the use of ultraviolet microscopes which could double resolution. Since glass filters out the short ultraviolet rays, quartz-transmitting or glass-reflecting lenses had to replace the
standard lenses. In theory, this would have given double the resolution or-
dinarily obtained. Among many practical problems, foremost is the fact that
the human eye is not sensitive to light in the ultraviolet range. Originally the
final image was recorded on a photographic plate and the final prints were
studied, but more recently such microscopes have been used with television
cameras. Fused quartz lenses were used in the original ultraviolet microscopes, but synthetic fluorite can also be used. Just as normal glass lenses
cannot be used, substitutes for the normal mounting media and immersion
oils were required. Water, castor oil, or glycerine jelly were found to be

appropriate. The coverslip and slide for the specimen must also be made of fused quartz. Thus, the use of this technique is not only very complex, it is also quite expensive. Once the electron microscope was developed, the ultra- violet microscope became limited to very specialized studies. It can be used with television cameras and receivers to observe living microorganisms in color. It can be used with monochromatic ultraviolet light of 260 mμ to study nucleic acids which absorb strongly at that wavelength. Caspersson and others used this approach to the study of nucleic acids from the late 1930s. Nucleic- acid-rich materials can be located within the cell and studied during cell divi- sion. In combination with a spectophotometer, the amount of nucleic acids in cells can be measured by this technique.

At present, ultraviolet (or deep blue light) is used in fluorescence micro- scopy. Compounds in the specimen absorb energy from the short-wave energy source and re-emit it in the visible portion of the spectrum as fluorescence. This is often a valuable tool for medical research, such as studies of antibody production using fluorescent dyes coupled to antigens or to antibodies.

The Electron Microscope

Biological studies have been profoundly affected by the availability of the electron microscope—an instrument perhaps 100 times as acute as the light microscope. One comparison that points up the difference is that trying to see an influenza virus with the light microscope is like trying to pick up a needle with a steam shovel (Hawley, 1946). Although the basic discoveries that made this instrument possible were recorded by physicists as early as 1858, the first successful use in America was as recent as 1938. The work of Sir William Crookes, J. J. Thomson, and William Konrad Roentgen is beyond the scope of this book, although their studies of cathode rays led to a new understanding of atomic theory and eventually the electron microscope.

No one individual can really claim credit as the inventor of electron micro- scopy, although this did not inhibit Reinhold Rudenberg from filing patents claiming this honor in 1937 in Germany and the United States. A prerequisite was the perfection of the vacuum tube, used in radio and other electronic devices as well as for electron microscopy. Here Crookes, Geissler, and Braun were important. Ordinary lenses obviously cannot be used for electron beams since they do not penetrate glass. Therefore, the big problem in using electron beams was how to focus them. Hittorf had shown in 1869 that a magnetic field would deflect cathode rays. This meant that magnets could be used to focus electron rays, for just as light rays are bent when passing through a glass lens, the electron stream traversing a magnetic field is similarly affected. After the pioneering work of Busch in electron optics, a group of workers in

Berlin began the systematic study of the possibilities of electron lenses about 1928. By 1932 Max Knoll and Ernst Ruska, German physicists, had built a model which they called a "super-microscope." They were able to obtain magnifications of about 400-fold, but most of their images were so out of focus that they were of no practical use.

Although it seemed that the curious electron microscope might never be a practical scientific instrument, physicists persevered. Rudenberg obtained German patents in 1931 and, after emigrating to America, was granted several patents in 1936–1937 which did anticipate some features later used in electron microscopes. In 1932 Ladislaus Marton at the University of Brussels built a microscope which he planned to use in the study of bacteria. Because of the fuzzy details and fading of the image, his final results were no better than those from ordinary microscopes. Ruska continued his work and by 1933 had built the instrument that may be regarded as the true ancestor of current models. Pictures at a resolution of about 500 Å were obtained. The rather humble objects of study were a piece of aluminum foil and bits of cotton fiber—which were carbonized by the intense electron beam. In 1935, Friest and Muller improved the microscope and produced photographs of the legs and wings of a housefly at a resolution of about 400 Å, which was about five times better than with the light microscope. The Berlin firm of Siemans and Halske continued to improve the instrument and by 1938 was obtaining impressive results. In 1940 von Ardenne produced the first treatise on the application of electronic principles to microscopy. Soon several teams were competing to make the instrument a practical part of science.

In Toronto, Professor E. F. Burton put James Hillier and Albert Prebus to work on the problem. Working under difficult and dangerous conditions with 30,000 volts produced numerous accidents, but a microscope was developed which, though crude, produced impressive images.

Dr. Vladimir K. Zworykin, associate director of electronics research at RCA Laboratories invited Dr. Marton to come to RCA from Brussels. He intended to construct an instrument that could be produced on a commercial basis and adapted for industrial use. Marton was primarily interested in the study of biological materials with electron microscopy. In 1934 he published the first electron micrograph of a biological specimen. This was a 15-μm thick section of the leaf of the sundew, *Drosera,* impregnated with osmium salts. It was soon apparent that the major problem was going to be preparing specimens in a way that would exploit the possibilities of the new tool. In Britain, Professor L. C. Martin of Imperial College, London led the search for a better electron microscope. The E.M.I. was the first electron microscope made by a commercial firm, but World War II soon arrested further progress.

Although biological research has flourished with the electron microscope, the timing of its development was such that the early use of the instrument was nearly confined to the study and development of war materials—cements, metallurgy, synthetic rubber, and catalysts.

Before leaving the story of this instrument, we should compare the capabilities of the eye, the optical microscope and the electron microscope. These comparisons can only be approximate because the actual limitations of each system vary greatly with particular equipment and conditions. The human eye can distinguish no objects smaller than 1/250 of an inch or 10^6 Å. The light microscope can do about 500 times better, since it can distinguish objects of 2000 Å (1/125,000 of an inch). It can enlarge objects about 2500-fold. The electron microscope can enlarge about 40,000-fold, but it is possible to obtain magnification of 80,000 to 100,000 diameters. The electron microscope can distinguish particles of about 20 Å, although even 2- to 3-Å resolution can be obtained.

In taking the story of the microscope from its beginnings with the Dutch spectacle makers, we have concentrated on the "brass and glass" aspects of microscopy. Later we shall see how developments in sample preparation and staining interacted with improvements in the microscope to advance biological research.

Notes

1. S. Bradbury (1967). *The Evolution of the Microscope.* Oxford: Pergamon, pp. 3–5.
2. Giambattista della Porta (1658). *Natural Magick.* Reprint 1957. New York: Basic Books, p. 368.
3. C. Dobell (1960). *Antony van Leeuwenhoek and His "Little Animals."* New York: Dover, p. 52.
4. Dobell, p. 325.
5. L. Clendening (1960). *Source Book of Medical History.* New York: Dover, pp. 219–220.
6. Clendening, pp. 210–212.
7. R. Hooke (1665). *Micrographia.* Reprint 1961. New York: Dover, p. 116.
8. F. J. Cole (1975). *A History of Comparative Anatomy.* New York: Dover, p. 273.
9. Cole, p. 278.
10. M. Foster (1970). *Lectures of the History of Physiology.* New York: Dover, p. 93.
11. N. Grew (1682). *The Anatomy of Plants.* Reprint 1965. New York: Johnson Reprint, p. xviii.

References

Adelmann, H. B. (1966). *Marcello Malpighi and the Evolution of Embryology.* 5 vols. Ithaca: Cornell University Press.

Allen, R. R. (1940). *The Microscope.* New York: Van Nostrand.

Auerbach, F. (1919). *Ernst Abbe.* Leipzig: Akademische Verlagsgesellschaft.

Bradbury, S. (1967). *The Evolution of the Microscope.* Oxford: Pergamon.

Bradbury, S., and Turner, G. L. E., eds. (1967). *Historical Aspects of Microscopy.* Cambridge: Royal Microscopal Society.

Casida, L. E. (1976). Leeuwenhoek's observations of Bacteria. *Science 192:* 1348-9.

Cassedy, J. H. (1976). The Microscope in American Medical Science, 1840-1860. *Isis 67:* 76-97.

Clay, R. S., and Court, T. H. (1932). *The History of the Microscope.* London: Griffin.

Cohen, B. A. (1937). On Leeuwenhoek's Method of Seeing Bacteria. *J. Bacteriol. 34:* 343-5.

Cole, F. J. (1937). Note on Swammerdam's Home. *Annals of Science 2:* 236.

Cole, F. J. (1937). Leeuwenhoek's Zoological Researches. *Annals of Science 1:* 1-46, 185-235.

Cole, F. J. (1949). *A History of Comparative Anatomy, from Aristotle to the Eighteenth Century.* London: MacMillan. (Reprinted by Dover Publications, 1975.)

Dobell, C. (1932). *Antony van Leeuwenhoek and his "Little Animals."* New York: Harcourt, Brace. (Reprinted by Dover Publications, 1960.)

Espinasse, M. (1956). *Robert Hooke.* Berkeley: University of California Press.

Foster, M. (1901). *Lectures on the History of Physiology during the Sixteenth, Seventeenth, and Eighteenth Centuries.* Cambridge: Cambridge University Press. (Reprinted by Dover Publications, 1970.)

Gage, S. H. (1964). Microscopy in America. *Trans. Amer. Mic. Soc. 83:* 9-125 (supplement).

Gunther, R. T. (1930-38). *The Life and Work of Robert Hooke.* 5 vols. Oxford: Oxford University Press.

Hawley, G. G. (1946). *Seeing the Invisible: The Story of the Electron Microscope.* New York: Knopf.

Hooke, R. (1665). *Micrographia.* (Reprinted by Dover Publications, 1961).

Hooke, R. *The Diary of Robert Hooke.* Edited by H. W. Robinson (1935). Reprinted 1961. London: Taylor and Francis.

Hooke, R. *The Posthumous Works of Robert Hooke.* New York: Johnson Reprint (1969).

King, H. C. (1955). *The History of the Telescope.* Cambridge: Sky Publishing.

Klencke, H. (1860). *Swammerdam.* Leipzig: Costenoble.

Leeuwenhoek, A. van. (1680–95). *Arcana Naturae Delphis Batavarum.* Reprinted 1966. Bruxelles: Culture et Civilisation.

Leeuwenhoek, A. van (1941). *The Collected Letters of Antoni van Leeuwenhoek.* Amsterdam: Swets and Zeitlinger.

Lindberg, D. C. (1976). *Theories of Vision from al-Kindi to Kepler.* Chicago: University of Chicago Press.

Miall, L. C. (1912). *The Early Naturalists: Their Lives and Work* (1530–1789). London: MacMillan.

Oliver, F. W., ed. (1913). *Makers of British Botany: A Collection of Biographies by Living Botanists.* Cambridge: Cambridge University Press.

Porta, della G. B. (1658). *Natural Magick.* New York: Basic Books. Reprinted 1957.

Sachs, J. von (1906). *History of Botany, 1530–1860.* Oxford: Clarendon.

Schierbeek, A. (1967). *Jan Swammerdam 1637–1680, His Life and Works.* Amsterdam: Swets and Zeitlinger.

Schumann, W., ed. (1962). *Carl Zeiss.* Berlin: Rütten and Loening.

Singer, C. (1959). *A History of Biology to about the Year 1900.* New York: Abelard-Schuman.

Swammerdam, J. (1680). *Ephemeri Vita: of the Natural History and Anatomy of the Ephemeron.* Translation and preface by E. Tyson. London: Faithorne, 1681.

8

THE GENERATION GAP:
PROBLEMS IN GENERATION,
REPRODUCTION, AND DEVELOPMENT

Problems in Generation

The term *generation* has become so obsolete and archaic that many specialists in the fields which grew out of studies in generation are totally unfamiliar with the original meaning of the term. "Generation" originally encompassed some of the most exciting aspects of modern biology—genetics, development and differentiation, embryology, and regeneration of parts. In its classical usage the term would have been restricted to apply to the coming into existence of new individual organisms, animals or plants, regardless of the methods involved.[1] Although the higher animals were known to have sexual methods of reproduction, lower forms of life seemed to arise through spontaneous generation from mud or slime, while plants arose from seed or through vegetative methods.

Scientists had many questions about generation and elaborated colorful and peculiar theories which were more or less satisfying to different periods. But interest in development and reproduction—of human beings and of useful animals and plants—predates science and philosophy. Before the development of any scientific theories, man was tending crops, developing domesticated animals, cross-pollinating plants, and using castration of both humans and animals to make them more useful and pliable in temperament. Natural philosophers looked at the body of information built up through centuries of experience and produced theories which proved they had not assimilated all the empirical observations at their disposal. For example, scientists did not appreciate plant sexuality until the work of Rudolf Camerarius (1665-1721), although the date palm had been cross-pollinated for centuries and the ratio of female to male trees standardized. If we look at all the questions that generation dealt with, it is understandable that the basic unity of life, in terms of the cell and the sexes, was obscured by the baroque profusion of different forms and their mysterious methods of reproduction and development.

First, it seemed strange and wonderful that offspring are always similar to but usually not identical to the parents. Why should they be the same and why should they differ? What causes births that deviate from the normal range and what produces strange and monstrous births? In sexually reproducing beings, what were the contributions of the two sexes to the offspring? At the earliest known stages, all organisms were simple, perhaps even unrecognizable. What caused them to reach their adult form? How was it possible for some species to regenerate lost parts while others were permanently mutilated by even trivial accidents?

Although the greatest minds in biology pondered these problems, until the late nineteenth century no explanation satisfying to modern biologists appeared. There was no scarcity of theories before that time and proponents of each theory were generally ready and eager to do battle with holders of other, to our minds, equally unsatisfactory viewpoints. Looking at the history of generation, we can see the problems of scientific progress more clearly than in any other branch of biology. From Hippocrates to the present, scientists looked at the same materials but drew very different conclusions. In retrospect, it seems that students of generation were faced with an insurmountable theoretical problem until the elaboration of cell theory and its application to development.

The story of generation shows so clearly that a scientist's implicit assumptions can be just as important in determining his approach to problems as his explicit scientific theories and the instruments, techniques, and model systems available. When scientists perceived all living things as composed of organs, they could not understand the earliest stages of development. This is the barrier over which scientists inevitably stumbled before cell theory was accepted. Their assumptions about the organization of living things militated against their understanding of the very earliest stages in the development of organized creatures. As Einstein said, "It is the theory which decides what we can observe." Not until the work of William Harvey do we see the beginnings of a modern approach to the riddle of generation.

Embryology in Antiquity

The philosophical framework forged by the ancients was still in place during the time of William Harvey and exerting a profound influence on the spirit and content of seventeenth century work. Thus, to understand Harvey's ideas we must review the work of the ancients. Some fragments from the writings of Alcmaeon show that the chick egg had already been chosen by scientists in the sixth century B.C. as a model system for investigating embryology. However, the first proposal for systematic embryological studies appeared in the Hippocratic writings. Certainly Hippocrates was interested in the development of human embryos, but his observations during the practice of gynecology and obstetrics would have been very limited. In *De Natura Pueri,* the developing chick egg is described as an object of systematic research:

> Take 20 eggs or more, and set them for brooding under two or more hens. Then on each day of incubation from the second to the last, that of hatching, remove one egg and open it for examination. You will find that everything agrees with what I have said, to the extent that the nature of a bird ought to be compared with that of man.[2]

Not only is the suggested program systematic, but there is the recognition that different life forms can be compared, if proper caution is exercised. This seems like a very promising beginning, but there is little evidence that after the path to systematic embryological research was pointed out it was really used.

Preformation, Epigenesis, and Aristotle

Aristotle pioneered the scientific study of reproduction and development. Since he used mode of reproduction and stage of development at birth for his classification of animals, studies of these problems were fundamental to all of Aristotelian natural science. One of the great debates that runs through the study of generation was initiated by Aristotle, who stated the alternative models of preformation and epigenesis as modes of development. Because these terms are such an important part of the story of generation, they should be defined at the outset. Preformationist theories hold that a miniature individual is present in the egg or sperm and grows to its adult form after the proper stimulus. According to the theory of epigenesis, the organism begins as an undifferentiated mass and goes through various steps and stages during development in which new parts are added. Aristotle favored the latter theory which remained most popular until the triumphs of mechanistic philosophy rendered the preformationist model more satisfying to scientists.

According to Aristotle, the female merely contributed unorganized "matter" to the new individual while the male contributed the principle of "form." When the "movements" proper to form cannot get control over matter, a deformity or monstrosity develops. Such problems are not at all rare, since Aristotle regarded the female as a kind of "deformity" characterized by the deficiency of heat. Like Hippocrates and Alcmaeon, Aristotle studied the developing chick egg. Despite his great thoroughness in many other aspects of his work, he apparently failed to carry out a systematic study of this system. Indeed, Harvey was of the opinion that Aristotle's discussion of the developing chick was based on only three inspections.

While these studies were important in initiating experimental embryology,

Aristotle's major influence was as a theoretician of development. His theory of causes and ideas about the soul, the nature of semen, the conceptual mass, the primacy of the heart, the roles of the two sexes, epigenesis, and fetal nutrition formed the basis of embryological thought for some 2000 years. Indeed, very little in the way of better work was done during that long period.

Galen

While Galen often harangued his readers about the importance of direct observation and the comparative approach, he too failed to carry out systematic embryological observations. Galen described the fetal membranes, the blood vessels of the uterus, the fetal organs, and the foramen ovale, including the way it closes over after birth. In *On the Semen,* Galen disputed Aristotle's idea that the female does not form any counterpart to the male semen. According to Galen, the testes muliebres secrete "semen" into the horns of the uterus, where the male and female sexual products mix with each other and with blood furnished by the mother. From this mixture the fetus was formed. Galen also disagreed with Aristotle about the function of the testes. While Aristotle believed they were merely weights, Galen argued that they play an important part in the formation of semen. As evidence, he provided a description of the effects of castration.

In his treatise "On the Formation of Fetuses," Galen described the order of the formation of the parts of the developing fetus. He believed that the liver formed before heart and brain so that the fetus lived first like a plant, being nourished under the influence of the vegetative soul. After formation of the heart, the fetus lived like an animal and the heart served as the source of heat. The brain was formed last, since the fetus does not hear, taste, smell, engage in any voluntary action, nor have any intellectual faculties. Thus, in accordance with Galen's teleological principle, the fetus does not need a brain until it is ready to be born.

Embryology and the Scientific Revolution

For original embryological studies, the period from the death of Galen (200 A.D.) to the late fifteenth century is a period devoid of interest. Although some scholars of the Middle Ages had begun to suspect that "Aristotle did not reach the limit of wisdom," even "radicals" of the sixteenth and seventeenth centuries hardly emerged from his shadow.

In his discussion of fetal anatomy, Andreas Vesalius perpetuated many ancient errors and introduced some original blunders. Not surprisingly, many of his errors concerning the anatomy and physiology of the human fetus are pure Galenism. Opportunities for original work on human development were so rare that Vesalius limited his discussion of observational embryology to the dog. Unfortunately, he could learn about the fetus only from dissections of animals and from ancient texts.

Realdo Columbus, best known for his discovery of the pulmonary circulation, was also interested in human embryology. Although Columbus dissected a fetus he believed to be about one month old, he could not study other phases of human development because of the insufficient supply of aborted fetuses. Nevertheless, he criticized other anatomists who had "unblushingly" applied what they imperfectly observed in the lower animals to the human fetus. However, many of his own observations were unreliable. For example, he stated: "Testes have been formed in women to produce semen and I can testify that I have sometimes observed thick white semen in them. The semina of both sexes are united in the uterus, whatever Aristotle may say. If the semen of the male preponderates, a male will be conceived and vice versa."[3]

Having criticized Aristotle, Columbus was secure enough to dismiss as fables the Hippocratic idea that the fetus suckles in the womb and the Aristotelian notion that the fetus is nourished by menstrual blood. Columbus stated that the fetus is nourished through the umbilical vein by a very pure and perfect blood. The term "placenta" was introduced by Columbus. Several other sixteenth century anatomists made interesting observations, but few saw embryology as a separate and distinct field, worthy of systematic study.

The work of Hieronymus Fabricius shows just how difficult it was for even the best trained Renaissance anatomists to cast off ancient dogma and observe embryological structure and development with their own eyes. Yet Fabricius can be regarded as the founder of scientific embryology. Indeed, Adelmann regards his embryological studies as of greater anatomical importance than his work on the venous valves.

In the anatomical theater of Padua, the first of its kind, Fabricius demonstrated various aspects of embryology to the crowds of students who flocked to his lectures. Fabricius was interested in comparative as well as human anatomy since the comparative approach seemed the best way to discover the means of action or function of the parts. During his course of 1576, Fabricius dissected three cadavers, one of which was a woman who had died in labor. He demonstrated the foramen ovale, the anatomy of the fetal horse and sheep, and closed a long discussion of development with the vivisection of a pregnant ewe. In addition to many books on anatomy, Fabricius published two important embryological works: *On the Formed Fetus* ("De formato foetu") and *On the Development of the Egg and the Chick* ("De formatione ovi et pulli"). The illustrations are remarkable, especially considering that Fabricius worked without the benefit of magnifying lenses.

Fabricius illustrates the tensions in the sixteenth century mind. Every page he wrote reflects the conflict between observation and authority. He must

have found it terribly painful to accommodate the evidence of his own eyes
to the philosophy of Aristotle and Galen that provided the framework for his
every thought. While clearly struggling for originality Fabricius quoted
Aristotle on almost every page. Sometimes the citation allowed him to refute
the ancient philosopher, nevertheless it seems that Fabricius could not take a
step without tripping over the corpus of Aristotle. In the section "The Uses
of the Semen of the Cock," Fabricius explained that the semen is not only
white and cold, but also foamy, spirituous and airy, which means it contains
much innate heat. Fabricius makes this point very clear:

> Indeed, this heat is lodged in a cold substance, because if it had been
> placed in a warm one it would easily be dissipated and pass away before
> the animal was formed from it, and this the coldness prevents.[4]

Such passages can easily make sixteenth century science seem ludicrous,
but accounts of errors and defects must not outweigh the importance of such
pioneering work. A balanced account must note that Fabricius did observe
the germinal disc, although he incorrectly assumed the chick formed from the
chalazae. Although Fabricius believed that the bursa of the hen was a recepta-
cle for semen, he must be given credit for discovering it. He did come closer
than any of his contemporaries to an appreciation of the role played by ovary
and oviduct in forming the egg. His illustrations of the daily progress of development in the chick embryo were the best up to that time. Also, by coin-
ing new words to describe more clearly what happened in development he
aided further work. Most important, as in his study of the circulatory system,
he inspired his pupil, William Harvey.

William Harvey and the Generation of Animals

Although Harvey's work *On the Generation of Animals* did not appear in print
until 1651, he had been working on the problem for many years. Many historians see Harvey's work on the circulation as a great triumph, but dismiss
his work on generation as a failure of his old age—almost as a case of scientific
senility. But work on generation was of profound importance to Harvey as an
experimental scientist and natural philosopher. More than any other aspect of
his work it demonstrates the keenness of his observation and experimentation
and the insurmountable barriers posed by lack of reliable instrumental and
theoretical guides.

Harvey set out to provide experimental proof for the venerable Aristotelian
theory of epigenetic development. In doing so, he came close to demonstrating that the central dogma on which all prevailing theories of generation rested

was absolutely wrong. The generally accepted theory of human development in the seventeenth century had the virtue of being logically consistent and the minor defect of being completely wrong. From the time of Aristotle, the embryo had been regarded as a combination of matter from the female and form from the male which "precipitated out" in the uterus immediately after mating. Blood from the female nourished the embryo and provided additional matter for growth during pregnancy.

While striving for truth through observation and experimentation, Harvey modestly admitted following in "the footsteps of those who have already thrown a light upon the subject." As his leader among the ancients he acknowledged Aristotle; Fabricius he took as his "informant of the way."[5] Generally, Harvey used the developing chick egg and deer from the Royal Park provided by the king for Harvey's experiments. As in the study of the heart and blood, beginning with respect for Aristotle and ancient philosophy, Harvey's experimental genius wrecked the rotten foundations of prevailing theory. But in this case he did not have the apparatus or theoretical framework to provide a satisfactory new solution to the problem.

While Harvey began with an admirable experimental program, his observations led him to certain misconceptions. Since he found nothing resembling the development of the hen's egg associated with the "female testicles" (ovaries), he concluded that they played no important role in generation. Searching for the seminal mass from which the embryo was supposed to originate, Harvey examined the uterus of numerous does during and after the rutting season. A careful series of dissections revealed that until about six or seven weeks after the rutting season nothing substantial could be found in the uterus. This was a surprise to Harvey and anyone else who would examine his results. But Harvey could find nothing to account for the "matter" necessary for the embryo—either as semen from the male or blood from the female.

Since Harvey's work on generation was confined to sexually reproducing organisms, his discoveries had no effect on the question of spontaneous generation of certain lower creatures. While Fabricius had concluded that the majority of animals arise from eggs, Harvey extended the idea to all animals as epitomized in his dictim: "Omnia ex ovo." This dictum was not rigorously grounded in experimentation, but was based on his assumption that all animals arise from "a certain corporeal something having life in potentia, or a certain something existing per se." This something he called an "egg," but he misinterpreted the mammalian "conceptus" as an ovum. The "egg" in Harvey's "omnia ex ovo" was actually the embryo large enough to be seen without magnifying lenses.

According to Harvey, both male and female contributed to the new animal. Yet Harvey accepted the ancient idea that some creatures developed by spontaneous generation, probably by means of "seeds" provided for them "without

Frontispiece of Harvey's *De generatione animalium.* [From J. F. Fulton and
L. G. Wilson (1966). *Selected Readings in the History of Physiology.* Courtesy
of Charles C Thomas, Publisher, Springfield, Illinois.]

any distinction of sex at all." Thus Harvey failed to build up a complete new
theory of generation and totally failed to come to grips with the question of
spontaneous generation. Despite these deficiencies, Harvey's work was an advance over that of his predecessors. Before any progress could be made in the
study of generation, the old misconceptions had to be thrown into relief and
experimental methods taken up consistently and objectively. Harvey contributed much that was fundamental and, in view of the lack of lenses and microscopes, carried out remarkable experiments. While Harvey's "egg" was only a
metaphysical concept, in time it became the fertilized ovum, real and visible
for those who followed Harvey.

Careful studies of the developing chick egg convinced Harvey that generation proceeded by epigenesis—the gradual addition of parts. Ironically, Harvey
stood almost alone as an advocate of epigenesis during the late seventeenth
century and for most of the eighteenth century. The mechanistic view of the
living world, which his own work on the blood circulation had done so much
to support, led to a preoccupation with performation. The preformation
model seemed to fit the mechanistic mode of explanation better than the
vague and purely descriptive epigenetic model.

Preformation Theories

Although preformationist theories enjoyed a prominent place in eighteenth century science, the philosophical roots of preformation go back to Democritus, Empedocles, Plato, Lucretius, Seneca, and the Church Fathers. Indeed Seneca expressed the concept quite explicitly:

> In the seed are enclosed all the parts of the body of the man that shall be formed. The infant that is borne in his mother's womb hath the roots of the beard and hair that he shall wear one day. In this little mass likewise are all the lineaments of the body and all that which Posterity shall discover in him.[6]

It is even possible that if Aristotle had seen sperm in the seminal fluid, he would have been the authority behind preformation.

During the seventeenth century, several naturalists claimed to have seen the rudiment of the chick embryo in the egg before incubation. As early as 1625, Joseph of Aromatari made this claim and also revived Empedocles' idea that the plant is already present in the ungerminated seed. In the chick egg observers found a complete and pulsating heart, although the blood, being colorless at such early stages, was most difficult to see.

Ironically, the microscope helped convert many zealous experimenters away from the epigenetic views of William Harvey to unfounded preformationist theories. The very instrument that would have enabled Harvey to complete his work on the circulation and possibly his work on generation, in the hands of men of less genius, became the tool that forged "scientific preformationism." Great pioneers of microscopy, such as Malpighi and Swammerdam initiated this reversal. Marcello Malpighi's work on that classical system—the developing chick egg—emphasizes the paradox that proponents of each philosophy were able to study the same system and come to diametrically opposite conclusions. Thus, although in general development seemed to occur gradually, Malpighi believed that some parts were present in the chick egg at the beginning of the incubation period. Although the heart did not start beating until about 40 hours of incubation, it seemed to exist fully formed before that.

Although Malpighi had the advantage over Harvey in working with the microscope, his equipment was not good enough to observe the details of very early development. Indeed, he often complained about the defective magnifying glasses that hampered his work. The microscope that he used was not much better than the toys sold for the amusement of precocious children today. Failure of intuition probably hampered Malpighi almost as much as poor-quality tools when he made one of the most famous mistakes in the history of embryology. Malpighi claimed to have seen the preformed chick in an egg that had not been incubated by any hen. But an egg warmed by the summer sun of Italy did not constitute a valid example of an unincubated egg. This single observation led to one of the most famous "half-baked ideas" associated with Malpighi.

Despite the poor quality of his microscope and his own misconceptions as to the meaning of what he saw, Malpighi revealed details of earlier stages of the embryo than had never been published before. These include his description of the vascular area, the development of the heart and gill arches, the dorsal folds, the developing brain, the mesoblastic somites, the amnion, and the allantois. Although Malpighi himself never used the term "preformation" and avoided dogmatic statements on the mechanism of generation, the emergence of preformation as a scientific rather than philosophical theory can be dated from the publication of Malpighi's papers.

By 1688, in public dissection lectures, Jan Swammerdam had shown that an insect larva, pupa, and imago may exist simultaneously at one stage of the life cycle. Because these forms could be dissected out of each other like a nest of boxes within boxes, Swammerdam concluded that no new parts were formed during insect metamorphosis. On the contrary, he held that the perfect and complete insect is present in some form during the entire life cycle. Swammerdam buttressed his arguments from insect metamorphosis with studies of tadpole development. The limbs can be dissected out from under the skin of the tadpole some time before they would normally emerge. Swammerdam ridiculed Harvey's studies of metamorphosis and sarcastically labeled his lengthy work on generation as one with almost as many errors as words. Although his studies of insect metamorphosis were intriguing and more skillful than Harvey's, Swammerdam exerted a pernicious influence on the science of embryology.

Nicolas de Malebranche (1638–1715), a French philosopher, priest, and follower of Descartes, reformulated Swammerdam's vague preformationist views into a philosophical scheme of an endless series of embryos within each other—like a nest of boxes. In this manner, the pious philosopher generalized the theory to cover both plants and animals. Preformationist science became a prop for the theological dogma that Adam and Eve were quite literally ancestral to all human beings. Although preformation theory impeded progress in embryology, it had one virtue. Preformationists generally rejected spontaneous generation since it conflicted with their concept of development. However, the boxes-within-boxes theory allows only one parent to serve as the source of the preformed individual. The egg was well known for many species and seemed the reasonable vehicle of heredity. The unformed liquid semen of the male still was accorded the critical role of providing the "generative principle" necessary for the initiation of development. But when the microscope revealed great numbers of active little animals in the semen, some scientists rejected ovist preformation and contended that the new individual was present in the sperm.

Ovists vs. Spermists

While some microscopists suggested that sperm were either associated with disease or were themselves permanent parasites, Anthony van Leeuwenhoek believed they were a normal part of male development. Although his statements about generation were not always consistent, he did claim to have seen two different kinds of sperm, which obviously represented "boy" or "girl" miniatures. Leeuwenhoek's studies of the "little animals of the sperm" were of a fundamental nature. Johan Ham, a medical student at Leyden, first noticed spermatozoa in August 1677. His uncle, a professor of medicine, took Ham to see Leeuwenhoek. Confirming Ham's observations, Leeuwenhoek studied the appearance, motility, survival, and other characteristics of the "little animals of the sperm." These observations were sent to the Royal Society in November 1677 and published in a Latin translation in the *Philosophical Transactions* of 1679.

Leeuwenhoek's attachment to spermist preformation is paradoxical since it was Leeuwenhoek who discovered parthenogenesis in aphids. Failing to find any males among an entire generation of aphids, he carefully studied the females and, while trying to find eggs, dissected out tiny individuals just like the parent. By 1699 he began to suspect that males might not exist. Although he concluded that aphids were uniquely able to produce young without the male sex, he remained a spermist. His studies of the female aphids should have led him to an appreciation of the female contribution, but instead he elaborated an ingenious scheme to explain away the aphid problem. He suggested that the aphid was actually a spermatic animalcule and that the female sex was really the one missing in these peculiar creatures.

Although some outstanding scientists and philosophers attempted to define a logically consistent doctrine of preformation, many of lesser genius contributed such ludicrous details that the doctrine eventually collapsed under the weight of absurdities. Some scientists argued that the whole human race and all human parasites had been present in the ovaries of Eve, mother of the race. Niklaas Hartsoeker (1656–1725) produced a famous sketch of the homunculus, a little person stuffed into the sperm. Aspiring to be more "quantitative," Hartsoeker calculated the size necessary for a rabbit large enough to incorporate all rabbits at the beginning of time. Hartsoeker was not alone in his foolishness, for N. Andry de Boesregard even saw (ca. 1650) sperm displaying behaviors characteristic of the type of animals to which they would give rise. While some animalculists based their ideas on pure speculation and theological dogma, others attempted experimental work—even if they did not know which end of a microscope to look through. Combining imagination with spherical and chromatic artifacts, investigators produced peculiar drawings and elaborate descriptions of the sperm and the earliest stages of development.

Sketch of the sperm containing the miniature individual or homunculus as described by Hartsoeker.

Ovist Preformation

Two of the most prestigious of the proponents of ovist preformation were
Albrecht von Haller (1708–1777) and Charles Bonnet (1720–1793). Von
Haller is best known as a physiologist, but his studies of the developing chick
egg convinced him that the embryo could be found in the egg, along with all
the essentials for maturation, before the egg emerged from the chicken. Von
Haller did not carry out many purely embryological studies himself, but encouraged Charles Bonnet in his work and in the expression of the ovist viewpoint. Leadership by such prominent scientists helped focus attention on the
egg as the source of the preformed individual.

The struggle between ovists and spermist was carried out more in print than
in the laboratory. Illustrious scientists such as von Haller, Reaumur, Bonnet,
and Spallanzani overwhelmed the nonentities who speculated about the
"vermicelli spermatici" or little worms of the sperm.

Portrait of Albrecht von Haller. [From A. Meyer (1939). *The Rise of Embryology*. Stanford, California: Stanford University Press.]

Charles Bonnet (1720–1793)

Bonnet was born in Geneva, Switzerland, to a family of wealthy Huguenots who had been forced to leave France because of religious persecution. He studied first for a career in law, but became devoted to the natural sciences, especially entomology, influenced largely by R.A.F. de Reaumur (1683–1757). A serious eye disease forced Bonnet to give up experimental work, but because of his wealth he was able to spend the rest of his life on purely speculative and theoretical work. Extremely pious, like his colleague von Haller, Bonnet labored so greatly to harmonize natural philosophy with religious dogma that even his scientific papers tended to sound like sermons. After an initial loose attachment to a vaguely epigenetic view of development, Bonnet became convinced that generation involved preformation. In this he was influenced by his own research and his correspondence with von Haller. While preformation was generally a philosophical principle, Bonnet worked out the details of a scientific "encapsulation" theory via the experimental approach.

The best argument for preformation, ovist style, grew out of Leeuwenhoek's observation that certain insects could reproduce without fertilization—without males. While studying the reproduction of aphids, Bonnet found that females which hatched during the summer gave birth to live offspring without fertilization. In the autumn the new generation of males and females mated and the females laid eggs. By careful segregation, Bonnet was able to grow 30 generations of virgin aphids by parthenogenesis. Bonnet worked out in great detail many other peculiarities of insect reproduction and metamorphosis. According to Bonnet, every female contained within her the "germs" of all creatures that would originate from her. Therefore, the first female of every species contained within her ovaries miniatures of all future individuals of that species in the form of "germs." When provided with its proper nutrition, the germ was stimulated into growth. Male semen served this purpose, and was needed to initiate growth, while at later stages of development the mother provided nutrition. To cope with special cases and new discoveries, the theory had to become ever more complex and convoluted.

For example, while the "germs" in higher animals must be confined to the reproductive organs, in lower animals the germs had to be scattered through the body. Bonnet had experimented with polyps and hydra in 1741 and had also confirmed the ancient observation that certain creatures can regenerate lost parts or form complete new individuals from segments. Scattering the germs about to save the theory created a difficulty in terms of the relationship of body and soul, which troubled the very pious Bonnet. If the soul were to be always one and indivisible, what happened to the soul when new individuals were regenerated from bits and pieces of the original?

There was some doubt in Bonnet's mind as to whether the "germ" was actually the particular individual or merely determined the species. Eventually he adopted the idea that the germ "carries the original imprint of the species, and not that of the individual. It is a man, a horse, a bull, etc., in miniature, but it is not a *certain* man, a *certain* bull." Individual variations were caused by external factors, such as the structure of the mother's body and nutritional status during development.

This theory could not be disproved simply through microscopic investigations in which the miniature individual was not found, because, Bonnet explained, it was impossible to actually see the miniature organism within the germ. Growth was a purely mechanical process whereby nutriments were absorbed into the appropriate parts of the expanding organism. Although there seem to be many difficulties with this theory, Bonnet's work was very well known and greatly admired by Lamarck, Cuvier, and many others. Indeed, Rádl, who finds Bonnet's ideas "tedious and insipid," nevertheless believes that they influenced August Weismann.[7] Yet even Bonnet was forced to admit that preformation theory was "one of the greatest triumphs of rational over sensual conviction."[8]

Lazzaro Spallanzani (1729-1799)

Spallanzani was one of the great figures in eighteenth century physiology, a founder of experimental biology, and foremost authority on fertilization phenomena. Although he originally studied law at Bologna, his cousin, Laura Bassi (1711-1778) aroused his interest in science. (Laura Bassi was a remarkable woman who held a chair of mathematics at Bologna, a very rare distinction for a woman at the time.) Although Spallanzani also took religious orders and became the "Abbe" Spallanzani, he held various university appointments. Versatile, energetic and ingenious, Spallanzani taught mathematics, physics, philosophy, and Greek. His last appointment came in 1769 at the summons of Maria Theresa to the Chair of Natural History at the University of Pavia, where he remained until his death. In addition to his teaching, Spallanzani made many trips to Switzerland, the Adriatic, and Asia Minor to explore their natural history. Spallanzani investigated many fields including digestion, blood circulation, regeneration, spontaneous generation, reproduction of plants and animals, respiration, the senses of bats, the zoology of sponges, and the electric action of the *Torpedo.*

Lazzaro Spallanzani. [From W. Bullough (1960). *The History of Bacteriology.* **Courtesy of Oxford Press.]**

In view of his experimental genius, Spallanzani's unswerving support for ovist preformationism is paradoxical. In retrospect, his studies of sperm should have served as an empirical foundation for a correct theory of reproduction and development. Nevertheless, his reputation had much to do with the predominance of ovist preformationism in the eighteenth century. Spallanzani was drawn into the study of the sperm as part of his continuing debate with Buffon and Needham about the nature of infusoria and spontaneous generation. When Needham criticized Spallanzani, Bonnet urged him to take up a systematic study of the sperm. At first, Spallanzani was reluctant and restrained because of his respect for Buffon. Nevertheless his preliminary investigation revealed obvious deficiencies in the work of the famous French naturalist. (The views of these scientists on spontaneous generation will be discussed in Chapter 10.)

Amphibians proved to be an excellent choice for studies of generation and regeneration. Microscopic studies of the regeneration of tail, limbs, and jaws were carried out with remarkable qualitative purpose and quantitative rigor. Attempts to change regeneration by variations in temperature and nutrition were of a remarkably reproducible character. From work in regeneration, Spallanzani was led backwards in the life history of amphibia to their point of origin, which he concluded was undoubtedly the egg. Spallanzani's studies of the development of amphibian eggs gave him what he considered sufficient proof for ovist preformation. Frog eggs inside the female body were found to increase in size. For Spallanzani growth could only be explained if the frog were already present in the egg. Therefore, the tadpole was already present in the egg—somehow coiled up and concentrated, ready to unfold itself in the presence of the fecundating liquid of the male.

Although he believed the egg contained the preformed individual, Spallanzani did not ignore the sperm. Beginning microscopic studies of frog sperm in 1771, he proved his results were generally valid, rather than a peculiarity of species by observations of the sperm of other amphibia, fish, dog, ram, bull, horse, and man.

Ideas about fertilization were still very vague in the eighteenth century. Indeed, the mammalian ovum was not discovered until 1828 and the union of sperm and egg nuclei was not seen until the end of the nineteenth century. Some naturalists believed that the egg had nothing to do with fertilization, while others denied the existence of the sperm or explained them away as a kind of parasite like other infusoria. Many investigators denied that external fertilization ever occurred. "Never, in any living body," declared Linnaeus, "does fecundation or impregnation of eggs take place outside the body of the mother." Those who had studied the semen of the male argued about which part represented the fertilizing power—liquid, sperm, some incorporeal agent or "aura seminalis." Even Harvey and Fabricius had believed in some active nonmaterial agent in semen for fertilization. Harvey likened this invisible power to that of a magnet.[9]

Spallanzani was able to provide rigorous proof that fertilization in frogs was external and that the idea of the aura seminalis was incorrect. Observing the mating behavior of frogs, he did experiments in which the female was killed while the eggs were being emitted. Eggs that had been discharged and had come in contact with the semen developed normally, but eggs dissected from within the body of the female did not develop. To confirm his hypothesis that fertilization was external and contact with seminal fluid necessary for development, Spallanzani designed special tight-fitting taffeta pants for his frogs. Dressed in these unique costumes the frogs attempted to mate as usual. Although many eggs were discharged none of them developed. When some of the eggs were mixed with semen that had been retained by the trousers, normal development took place. Eventually semen was collected from the seminal vesicles and carefully "painted" onto eggs. Eggs which were not mixed with semen disintegrated, while the treated eggs began to develop into tadpoles. Thus Spallanzani established a method of artificial fertilization.

Bonnet was very excited about this work and suggested various exotic hybrids. Spallanzani eventually carried out artificial insemination of salamanders, frogs, toads, silkworms, and finally a dog. It is worth noting that he was not the first to do artificial insemination: Arab horse breeders had done this for centuries. Charles Bonnet wrote to Spallanzani in 1781: "I do not know, but one day what you have discovered may be applied in the human species to ends we little think of and with no light consequences." As early as 1790, John Hunter (1728-1793) very discreetly supervised artificial insemination for a linen draper with a rather peculiar problem, enabling his wife to bear a child.[10]

Bonnet, who doubted that the aura seminalis was a sufficient stimulus, suggested that Spallanzani use his artificial fertilization system to test the question of which portion of the semen caused fecundation. This led to an ingenious series of experiments. First Spallanzani placed eleven grains of semen in a watch glass. He then fixed 26 eggs to a small watch glass by means of gluten, inverted the smaller watch glass over the larger and waited. Although the eggs were quite moist, and measurement proved that one grain and a half of semen had evaporated, the eggs did not develop when placed in water. To test whether the remaining semen was still effective, he applied some of it to other eggs. These did develop. Such experiments proved to Spallanzani that "fecundation in the fetid toad is not the effect of the aura seminalis, but of the sensible part of the seed."[11] On the suggestion of Bonnet, Spallanzani tried to replace semen with other agents, such as blood, blood extracts, electricity, vinegar, wine, urine, lemon and lime juices, oils, etc. All such attempts were unsuccessful in stimulating development. In attempts to remove the fecundating power of semen, he found dilution, exposure to a vacuum, cold, and oil ineffective, while heat, evaporation, wine, or passage through filter paper removed the capacity for fertilization.

In retrospect, Spallanzani's most striking experiments involved the filtration of semen, which produced a liquid portion devoid of fertilizing power and a thick residue containing sperm. When removed from the filter paper and suspended in water, this material was capable of fertilizing the eggs. Yet Spallanzani concluded that it was the liquid remaining in the filter paper with the sperm that accomplished fertilization rather than the sperm per se. How could Spallanzani come to such a conclusion? Other experiments had proved to him that the "spermatic worms" were not responsible for fertilization. He had tested liquid that was apparently free of sperm and found it caused eggs to develop. He also found that semen diluted in urine or vinegar could cause eggs to develop although it apparently killed the sperm. Convinced that the sperm were real animals, like infusoria, Spallanzani concluded that the seminal liquor was their natural habitat. It is possible that Spallanzani's observations of development with "sperm-free fluids" were the result of artificial parthenogenesis; alternatively, the concentration of sperm may have been very low.[12] However, it should be noted that Spallanzani usually applied the semen with a brush or needle. In 1910, Bataillon succeeded in producing artificial development of frog eggs by pricking them with a glass needle or micropipette.

Spallanzani believed that spermatozoa were genuine parasites passed on from one generation to the next during intercourse or within the uterus during pregnancy or even through the milk. After all, Buffon and Bonnet had reported finding "spermatic animals" in female semen or blood. Given this body of misconceptions, it is probable that if Spallanzani could have seen fertilization take place he would have taken it as confirmation of his parasite theory. Assuming that the seminal fluids were necessary for the development of the egg, these parasites might penetrate the egg, find their way into the tiny preformed genital organs, and rest there until puberty unmasked the disease state. According to this theory, all men would be afflicted with "spermicitis" and the transmission of sperm could be regarded as a veneral disease.

Much to his credit, Spallanzani preferred experimentation to the elaboration of pompous theories. He noted that the resemblance of children to both parents raised a question concerning the preexistence of the germ in the female. Unwilling to define a function for the spermatic animals he had so carefully studied, he concluded that the question transcended the sphere of human knowledge.

A Challenge to Preformation Theories

Casper Friedrich Wolff (1733–1794)

Even as preformationists were assembling what they regarded as incontrovertible proofs of their theory, Caspar Wolff was introducing a new approach to the analysis of development and differentiation. Wolff, born in Berlin, was the son of a master tailor. He studied at the Collegium Medicochirurgicum at Berlin and later at Halle. Along with his medical studies he absorbed large doses of philosophy from Christian Wolff, professor of mathematics and philosophy, and author of *Thoughts on God, the Soul, and the World*. Caspar was quite impressed with his mentor's botanical work and his philosophical principle that everything that happens must have an adequate reason for doing so, for it would be absurd for something to come from nothing. In 1759 Caspar Wolff produced a dissertation, "Theory of Generation," which became a landmark in the history of embryology. Yet, his work brought him little recognition during his lifetime. Denied a position at the Medical College of Berlin, Wolff eventually emigrated to St. Petersburg where he became Academician for Anatomy and Physiology.

Because his theoretical framework was inconsistent with prevailing ideas, Wolff was ignored and misunderstood by his contemporaries. The leading preformationists, Haller, Bonnet, and Spallanzani, saw life and science from a profoundly religious orientation. Consciously or subconsciously, they wanted to preserve and enhance evidence for the biblical account of creation, but scientifically they wanted to place generation within a generally mechanistic physiology. Wolff, in contrast, believed that studies of generation could only be purely descriptive, since the actual mechanism of development was impossible to determine.

"Theory of Generation" was a combination of observations on plants and the purely philosophical assumption that development was necessarily epigenetic. Apparently Wolff was unaware of the strong attachment of others to preformationism. Indeed, Wolff sent his thesis to von Haller. When Haller rejected it on religious grounds, Wolff retorted that a scientist must search only for truth and could not prejudge material on religious rather than scientific grounds. Yet Wolff's own theoretical preconceptions led him to support and seek out observations favorable to a purely descriptive epigenetic theory of development.

Wolff was one of the first biologists to enunciate "naturphilosophie" ("nature philosophy"). According to this mode of thought, the world was constructed in conformance with a pre-established harmony. Any apparent

changes were due to the inner nature of things because development took
place according to predetermined laws. Understanding life meant attaining full
descriptive knowledge of the gradual development of different living forms.
Life could not be explained in mechanical or physicochemical terms, according to nature philosophy.

As naming plants and animals was the primary task of science for Linnaeus,
for Wolff, description of vital phenomena was the task of biology. He seems
to have been the first naturalist to make the study of development and differentiation of plants and animals his special concern. It was in the development of the embryo of plant or animal that nature philosophy found its most
typical expression. As Wolff said: "We may conclude that the organs of the
body have not always existed, but have been formed successively; no matter
how this formation has been brought about. I do not say it has been brought
about by a fortuitous combination of particles, by a kind of fermentation,
through mechanical causes or through the activity of the soul, *but only that it
has been brought about*." [13]

The famous essay on generation begins with a series of definitions. "Generation" is defined as the formation of a body by the creation of its parts.
Because the actual mechanism of generation is unknowable, a vague descriptive
term, the "vis essentialis," or "inner force," is introduced to suggest what
brings about change. Much of the research that Wolff carried out utilized the
chick embryo, which had served proponents of preformation so well—but he
concentrated particularly on the metamorphosis of plants. Basic to all of
Wolff's conclusions is the assumption that plants and animals are essentially
the same in terms of their undifferentiated materials. This emphasis on the
original undifferentiated state was important to this theory, but beyond that,
may well have laid the groundwork for the fundamental generalization of cell
theory—the universality of the cell as the unit of life.

Metamorphosis in plants was appropriate for studies of development in
Wolff's mind in that he was not interested in fertilization or the ultimate
origin of the primary undifferentiated material, but only in the differentiation
pattern once initiated. Yet while the system suited his philosophy, we should
not overlook the fact that this enabled him to circumvent the technological
limitations of microscopy. Microscopy had not really improved since the time
of Leeuwenhoek; indeed, in many cases standards of workmanship had degenerated. Given the poor quality of microscopes and the lack of staining
techniques, plant materials were preferable since more details could be seen in
plants than in unstained animal tissues. Thus, much of what Wolff says about
embryogenesis is an extrapolation of what he had seen in developing plant
parts. Wolff argued that the rudiments of leaves and the parts of flowers are
both derived from essentially undifferentiated material. The premise underlying all of his work is that in animals just as in plants development proceeds
through the differentiation of some primordial undifferentiated material.

According to Wolff, the inner life force of the plant causes liquid to be drawn up from the soil. The liquid collects at the growing point of the stem and becomes a kind of thin jelly. As evaporation of the liquid occurs, small sacs or vessicles are formed. Further development results in a hollowing out of tracks in the jellied mass of vessicles. In this manner the ducts of the plant vascular system develop. These observations in the growing plant colored Wolff's interpretation of what happened in the developing chick embryo.

Wolff believed the chick was formed from a mass of little sacs which contained no bodily structure or parts. Eventually, the vascular system formed out of this material as it did in plants. Indeed, Wolff gave as his first proof of epigenesis in the chick his observation that the blood vessels of the chick blastoderm are not present at the onset of incubation even when alcohol was used to make the structures more opaque. His erroneous analogy between the blood vessels of the animal and those of the plant probably led him to concentrate on this aspect of the problem. To lend more credence to his plant-animal analogies, he concentrated on the development of intestines, blood vessels, and kidneys because these parts have a final structure which is tubular. He even used the term for leaf in his description of the way that the intestine and probably central nervous system are formed via the folding of homogeneous layers into tubes.

Many important concepts are foreshadowed in the work of Caspar Wolff. The entire concept of change is of course fundamental to his work, but also there are intimations of the structural similarity of plant and animal life, cell theory, and the description of the embryo in terms of layers out of which new structures develop. The little vesicles or sacs which Wolff described were probably cells. Certainly, he was not the first to see cells. Robert Hooke had described the cells found in cork in his *Micrographia* in 1665. Others had reported seeing little globules, although in many cases these seem to have been artifacts or aberrations produced by uncorrected lenses. Most important, Wolff recognized that at an early stage of development plants and animals alike consist of minute units which are not miniature versions of adult organs.

Although Wolff seems to have realized that his little sacs, or "Kügelchen," could change into other parts, his concept is not identical to modern cell theory. His "cells" were secondary structures in a mass of "jelly" rather than the primary unit of origin. Wolff did not introduce the concept of epigenesis, which as we have seen is a very ancient idea, but he provided a new way of looking at the problem of development. Although often abstract and speculative, Wolff felt constrained to provide observational proofs for his theory. The marvelous way in which his observations fit his theory is a direct result of his approach; he knew just what he was looking for.

Although Wolff could not convert his contemporaries to epigenesis, his work was neither totally unknown nor without influence. T. H. Huxley regarded Wolff as a man of genius so far ahead of his time his contemporaries had failed to understand him. Ultimately, preformation could only discourage research on the embryo, for if development were simply the expansion of a miniature individual, then knowledge of the more easily studied adult would be equivalent to knowledge of the earliest stage of development. There would be no compelling reason to study the minute, fragile precursors of the adult when it would be so much easier to study the completed product. Wolff's work showed that there really was something interesting to study—there was change as well as growth during development. To Wolff, the philosophical limitations of preformation rendered it useless as a guide to scientific research. "Those who adopt the system of predelineation," wrote Wolff, "do not explain the development of organic bodies but affirm that it does not occur."[14] Despite Wolff's total rejection of preformation, as in so many scientific controversies there were elements of truth on both sides of the picture which required a new theoretical framework to encompass the conflicts. The data themselves were valid; the theories were incomplete.

A change in the climate of opinion late in the eighteenth century made Wolff's philosophy more congenial to scientists. Under the influence of nature philosophy many scientists rejected the application of seventeenth century mechanism to living systems. Despite changes in philosophy, as a matter of scientific fact the debate between preformation and epigenesis was not satisfactorily resolved until the nineteenth century. Two fundamental changes were required: First, cell theory provided a framework for a more accurate understanding of sperm, egg, and developing embryo; second, scientists went beyond simplistic mechanism and nature philosophy to the realization that they need not treat living organisms as machines, nor give up all hope of ever explaining the mechanisms that governed living beings.

A continuation of the story of generation will take us into the nineteenth century where the problem is very much tied to the development of cell theory and the improvement of microscopic techniques.

Nature Philosophy or Romanticism

In the early nineteenth century "nature philosophy," or "Romanticism" (as the more florid form was called), permeated the thinking of many practitioners of the more "insecure" sciences. The Romantic emphasized emotion, diversity, the subjective, and favored a historical description of growth and change. The romantic philosophy affected biology, a young and self-conscious science, by stimulating interest in the forms found in nature. Indeed, the "nature philosophers" used the concept "nature" as something equivalent to "God." Interest was directed towards variation as a reflection of a limited number of "archetypes," or "ideal forms." The archetype was a kind of generalized form which was a reflection of nature's basic plan. Knowing the archetype was the key to understanding living organization. Thus, Romanticism generated an interest in comparative anatomy and morphology as well as the history of institutions, customs, races, and the individual organism. In terms of its "history," the organism could be understood through a detailed description of development and differentiation. Many biologists touched by Romanticism pursued embryological study; these included von Baer, Remak, Oken, Serres, Mickel, Pander, and Dollinger. Although philosophy provided the new impulse, it was through patient and meticulous microscopic research that Karl Ernst von Baer and others found proof that development proceeded from homogeneous to heterogeneous, from general to particular forms.

Karl Ernst von Baer (1792–1876)

Many of the scientists involved in the development of cell theory shared an interest in nature philosophy, sometimes in its most extreme forms, and some seem to have suffered from severe mental and emotional problems. Karl Ernst von Baer was certainly typical in this. Born in Russian Esthonia, to a family of impoverished German nobility, von Baer reluctantly entered medical school as a way of supporting himself and becoming a scientist. Although medical studies did not greatly appeal to him, he found it an improvement over his original goal of a military career. Unhappy as a practicing physician, he turned to the study of anatomy, embryology, physiology, and comparative anatomy. Finally, having exhausted his funds and pawned all his possessions, he walked to Berlin where he tried to continue his studies. After years of wandering and poverty he obtained a position at Königsberg, where he carried out most of his embryological work. Although he is justly famous as the founder of modern embryology, von Baer was a naturalist whose interests encompassed anthropology and ethnography. Work and honors did not bring him peace of mind. Once von Baer suffered from a depression so severe he could not work at all and remained shut up in his house for a whole year. After this he worked sporadically, but found that travel and scientific expeditions were necessary to his mental equilibrium. His excursions took him from Lapland to museums all over Europe and on inspection tours of the fisheries of the Russian Empire. When his sight and hearing were failing he retired, but continued some researches until his death at 84 years of age.

In 1827 von Baer published a milestone in the history of embryology: "On the Origin of the Mammalian and Human Ovum." This work contained the first accurate report of the mammalian ovum. When von Baer presented his discovery of the mammalian ovum at a meeting of Oken's Society of Natural Scientists and Physicians in 1828, his paper was received in polite but bored silence. Cell theory had not yet been formulated and neither von Baer nor his audience understood the implications of the ovum as a single cell: the unit of organization, reproduction, and development.

Surprisingly, the mammalian egg had not been discovered earlier, since at about 1/200 of an inch it is the largest cell in the body, while the sperm, barely $1/10^5$ of an inch across, had been discovered by the first wave of microscopists. Although Harvey's concept of generation was epitomized in the phrase "omnia ex ovo," he had seen only the eggs of birds, reptiles, amphibia, and fishes. These eggs are large enough to be visible to the naked eye and can easily be studied with only a simple hand lens. Ovism received support from Regnier de Graaf's (1641–1673) discovery in 1672 of bumps on the ovaries of rabbits, ewes, and women, which he assumed to be ova. Later these structures were called Graafian follicles. Attempts to trace the movement of these "eggs" to the uterus were frustrating. Haller thought that fluid in the Graafian follicle coagulated to form the ovum. In 1667 Nicolaus Steno

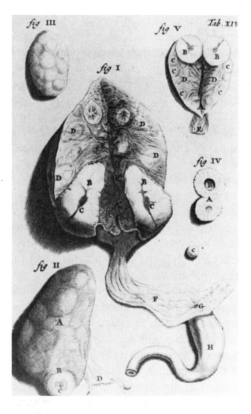

Illustrations from De Graaf's treatise on the ovaries of cows and sheep. [From J. F. Fulton and L. G. Wilson (1966). *Selected Readings in the History of Physiology.* **Courtesy of Charles C Thomas, Publisher, Springfield, Illinois.]**

(1638–1686) had described the female genital organs of dogfishes and the follicles in their ovaries. Steno argued that the "testis" of women should be regarded as analogous to the "ovary" of ovipara. Unfortunately, Harvey's search for the "egg" was not reconciled with Steno's approach to the "ovum."

During a comparative study of the ovaries of dogs and other animals, von Baer discovered the ovum in the Graafian follicle of a pregnant bitch. For several reasons this discovery was quite surprising. Anatomists expected the mammalian ovum to be transparent, but the ovum of the dog is actually yellow and opaque. Furthermore, von Baer did not anticipate finding ova in an animal which was already pregnant. In other studies, von Baer had seen minute objects in the oviducts of rabbits. Although he could not then find these same structures in the ovaries and knew they were too small to correspond to Graafian follicles, he assumed these "Körperchen" must be ova. After his accidental discovery of the tiny yellowish speck floating in the fluid of the follicle, he was able to see these objects with the naked eye and with his microscope in all mature follicles. After opening the follicle and examining the speck more closely, he determined that it was identical to the minute body he had found in the oviduct.

Von Baer's second great contribution was a generalization about the process of development which is known as the germ layer theory. Heinrich Christian Pander (1794–1865) had previously described the germ layers of the chick embryo, but it was von Baer who generalized the phenomenon. Purely a descriptive guide, rather than an attempt to deal with the mechanism of development, the theory simply states that similar organs are derived from comparable germ layers. Despite the great diversity of forms in vertebrate adults, von Baer had found a useful framework for guiding embryological studies of all manner of species.

Although von Baer followed Wolff's suggestion and originally postulated four germ layers, later work showed that his two middle layers were better considered as a single germ layer. In 1855, Robert Remak (1815–1865) refined the concept and renamed the three germ layers: ectoderm, or "outside skin," gives rise to skin and nervous system; mesoderm, or "middle skin," produces muscles, skeleton, and the excretory system; endoderm, or "inside skin," differentiates to form notochord, digestive system, and associated glands. Thus, each germ layer had a "specific histological future." This simplistic germ layer rule is often the only vestige that reluctant students carry with them from introductory embryology. But contradictions to the basic rules have often been found.

The third major contribution of von Baer to embryology was his discovery of the notochord in the chick embryo. Von Baer proved that this often transient endodermal structure is present in the embryos of all vertebrates. The notochord is a cellular rod that runs the length of the vertebrate embryo and eventually develops into the backbone. While the structure cannot be seen in adult vertebrates, except for certain fish, it serves as a very important tool for discovering the vertebrate nature of questionable organisms.

Another powerful descriptive generalization formulated by von Baer became widely known as the "biogenetic law" or the "law of corresponding stages." Essentially this law states that during development the embryo of a higher animal passes through stages which resemble stages in the development of lower animals. Actually, von Baer gave four descriptive propositions which are connected with this law. First, during development general characters appear before specialized ones. Second, the most general characteristics gradually develop towards the less general and then to the most specific. For example, limb buds become limbs and later differentiate into hands, wings, or flippers. Third, during development the embryo of a given species continuously diverges from those of other species. Fourth, the embryo of a higher species goes through stages which resemble the stages of development of lower animals.

Karl Ernst von Baer. [From William Bullock, *The History of Bacteriology.* **Courtesy of Oxford University Press, London.]**

Scientists less critical and cautious than von Baer transformed these inno-cuous general principles into the dogmatic and deceptive biogenetic law. It was Ernst Heinrich Haeckel (1834-1919) who formulated the "law" and the catchy but inaccurate motto "ontogeny recapitulates phylogeny." In the 1870s embryology became the captive of evolutionary zealots who confounded development of the individual with evolution of species. Some workers firmly believed that during development a higher organism went through stages that were identical to the forms of lower animals, so that development proceeded from egg to worm to fish to amphibian and on to embryos of increasing per-fection. As T. H. Huxley (1825-1895) put it, "In the womb, we climb the ladder of our family tree." While providing a persuasive but superficial sketch of development, the biogenetic law is not strictly true. Von Baer himself never accepted such an interpretation of his work. Charles Darwin (1809-1882), realizing the value of the biogenetic law to his evolutionary theory, called the field of embryology "second in importance to none in natural history." However, Darwin devoted only a small part of a chapter entitled "Mutual Affinities of Organic Beings" to embryology. Darwin increased the confusion over what von Baer actually said and did, since in early editions of *Origin* he quoted von Baer on the similarity of embryos, but incorrectly gave credit to Louis Agassiz (1807-1873). Ironically, both von Baer and Agassiz rejected Darwin's evolutionary hypothesis, while other scientists, under the spell of Haeckel, Huxley, and Darwin, tried to force the facts of ontogeny to fit their concept of phylogeny.[15]

Better research in embryology and the use of comparative methods were stimulated by von Baer's example. Having shown how complex and absorbing embryological development really was, he also provided the guidelines for others to follow. The next phase of embryological research—experimental embryology—seems at first sight quite different from the descriptive phase of nineteenth century work. Yet the fascination with change proved a thread of continuity in all embryological research.

To understand the next phase of research in generation, it is necessary to study the maturation of another great generalization of nineteenth century biology—cell theory (see Chapter 9). We can summarize progress in embry-ology in terms of changes in the level of resolution at which scientists were able to work. First studied as whole organisms, embryos were next viewed in terms of the germ layers. Following examination of the individual embryonic cells, scientists probed the components of cells, especially the nucleus, nucle-olus, and chromosomes. Today, the process continues at the molecular level as scientists examine the macromolecules of nucleus and cytoplasm.

Let us then turn to some of the late nineteenth century workers, who had the benefit of cell theory and at least intimations of biochemistry to stimulate their imagination as they approached the ancient question of the progressive development of the embryo.

Experimental Embryology

Wilhelm Roux (1850–1924)

Without doubt the patron saints of early experimental embryology were
Wilhelm Roux, Hans Driesch, and E. B. Wilson. Each tried to answer the
question of how factors intrinsic or extrinsic to an egg or its parts could
govern the development of the embryo or a part of the embryo. Wilhelm
Roux, like his mentor Ernst Haeckel, was a rabid evolutionist. Becoming convinced that descriptive and comparative studies of embryonic development
were inadequate, he demanded a new experimental approach and saw himself
as the founder of "developmental mechanics" or "causal analytical embryology." In celebration of the founding of a new discipline, the appropriate
nineteenth century rite was the formation of a new journal, in this case the
Archiv für Entwicklungsmechanik der Organismen (*Archive for Developmental
Mechanics of Organisms*). Roux exhorted his colleagues: "After sufficient
description, it is time to take the further step towards knowledge of the
processes that produce them." The protocol required resolving developmental
processes into simpler though still complex functional processes. In turn,
these functional processes were to be reduced to their components, so that
eventually the truly simple processes might be resolved into physicochemical
terms.[16] Not a modest man, Roux predicted that several centuries later
students would read his work with the same intensity of interest with which
he had studied Descartes.

Roux's work helped resolve the ancient conflict between preformation and
epigenesis. He defined development as "the production of perceptible manifoldness." This however involved two aspects: first, a true increase in manifoldness which could be called neoepigenesis; second, the transformation of
imperceptible manifoldness which could be called neopreformation. He believed that two mechanisms of differentiation were theoretically possible and
that experimentation could decide between them. Self-differentiation would
involve an independent or mosaic development of each part, whereas correla-
tive dependent differentiation would require the interaction of cells or groups
of cells. Actually, Roux favored the mosaic model since he tended to visualize the fertilized egg as similar to a complex "machine" and development as a
process that involved the distribution of parts of the machine to the appropriate daughter cells.

Peculiar environmental conditions, including changes in gravity, light, terrestrial magnetism, and electricity, generally allowed normal development,
which Roux interpreted as proof for self-differentiation. Obviously some extreme conditions did affect the embryo in a gross way, but Roux believed
there was no proof that such factors determined which part of the egg produced the eyes, or neural groove, or any specific part.

Confident that external forces could be neglected, Roux set out to study formative forces within the egg. The first question was whether all the parts of the egg must collaborate to cause normal development, or if the separate parts could develop independently. To answer this question, Roux destroyed one of the blastomeres of a frog at the two-cell stage. When one cell was injured with a cautery needle, the undamaged cell developed into a half-embryo. Such experiments encouraged Roux's belief that each cell normally develops independently of its neighbor and that total development is the sum of the separate differentiation of each part. While Roux continued to believe in mosaic development, the experimental work he helped to inspire soon destroyed his simplistic theory and made room for more productive concepts.

Recreating Roux's famous experiment with different conditions and species, other scientists obtained quite opposite results. In 1891 Oscar Hertwig separated cells of the sea urchin embryo. In these experiments both cells survived and each developed into a whole animal, thus rendering Roux's concept of mosaic development untenable.

Hans Adolf Eduard Driesch (1867–1941)

Hans Driesch, one of the first to follow Roux's protocol for experimental embryology, provided evidence that development was, contrary to Roux's theory, decidedly not mosaic. The embryo seemed to Driesch to develop epigenetically as a harmonious equipotential system. Because Driesch and Roux differed so much in philosophical viewpoint and technical capability, it is interesting to note how much of their background was similar—for both had studied with Ernst Haeckel. Driesch's interests were very broad, encompassing mathematics, physics, and philosophy. Restless by nature, Driesch travelled widely, giving lectures in philosophy and science in the Far East and the United States. Even as a doctoral candidate, Driesch had questioned the wisdom of his mentors. Indeed his work presented a direct challenge to the eminent August Weismann as well as to Haeckel and Roux. Eventually relations between Driesch and Haeckel deteriorated to the point where Haeckel advised his former student to take some time off and spend it in a mental hospital. Although Driesch never took Haeckel's advice, he did finally abandon biology for philosophy.

Several important differences in experimental conditions led to correction of Roux's dogmatic extrapolation of results obtained with the injured but unseparated frog egg. First, Driesch used echinoderm eggs instead of frog eggs. Secondly, he used different methods to separate the cells. As he discovered in 1892, sea urchin embryos at the two-cell stage could be separated into individual cells merely by shaking. The separated cells formed ciliated blastulae, gastrulated, and developed to the pluteus stage: normal in configuration, but half normal size. With improved techniques, Driesch was able to extend his results to the four, eight, and even later cell stages. He could also displace the cells of a sea urchin embryo into atypical relationships by shaking the embryos. Development into normal plutei indicated that the ectoderm and endoderm were interchangeable. As characteristic of good experiments, this work quickly suggested many new problems. Driesch felt he had answered the primary question concerning the "potency" of the two first blastomeres. The first two nuclei did not each contain half the determinants of the whole fertilized egg, but both contained all the information needed for full development.

Obviously Roux's concept of mosaic development was wrong. Driesch was not particularly happy about disproving a nice hypothesis. "That this is a particularly pleasing result one could scarcely contend;" he wrote, "it seems almost a step backward along a path considered already established." Driesch consoled himself that although this could be considered a step backward in theoretical conceptions, it had to be considered as progress since "the establishment of facts . . . always constitutes progress."[17]

Driesch concluded that since a new organism can be regenerated from parts of the embryo it cannot be regarded as a complex machine, because the parts of a machine cannot reproduce the whole, being necessarily simpler than the original machine. He therefore saw the developing egg as a harmonious equipotential system. This implies that all the parts are uniform. The process of development is harmonious because the parts normally work together to form one individual even though each could form an independent individual. In contrast to Roux, Driesch emphasized the epigenetic nature of early development. Driesch compared the presumptive significance of an embryonic part—which is what it usually forms—and its prospective potency—what it might form in a variety of situations. Experiments had demonstrated to him that the cell's prospective potency was much greater than the presumptive significance. The fate of a given cell was, therefore, a function of its relative position in the whole. This refers to function in the mathematical sense, "as in a system of coordinates in analytical geometry."[18]

Instead of finding that more experimental work rewarded him with a clearer picture of the problem, Driesch was eventually driven to abandon hope of ever finding a mechanistic explanation for development. There was no way he could picture a machine which when divided could develop into two whole new machines, identical to the original. When only vitalism could answer his questions, Driesch turned to Aristotle for inspiration and revived that venerable old *deus ex machina,* the *entelechy*—an internal perfecting principle.

Advances in Experimental Embryology

Following the resolution of ancient controversies through the work of Roux, Driesch, and other pioneers of experimental embryology, more modern hypotheses were propounded. Probably most significant was the competition between Hans Spemann and C. M. Child. Child and his school conceived a quantitative model of development in terms of the "axial gradient," while Spemann and co-workers advocated the "organizer concept." The organizer concept applies to rather specific stages of development in vertebrate embryos and explained regional differences on a qualitative basis.

Hans Spemann (1869–1941)

In many ways Spemann's discovery of the organizer and his concept of induction represent the culmination of the impulse generated by Roux, Driesch, and Wilson. By sharpening the focus of embryological experimentation, Spemann tried to discover the precise moment when a particular embryonic structure became irrevocably determined in its path of differentiation. His work was to make the idea of "determination" a dominant theme in development research for some time. For Spemann, embryonic development was the study of the "physiology of development" and like all natural phenomena could only be treated "from the standpoint of strict causation" by the scientist.[19] In his work on experimental embryology two specific methods were systematically exploited: isolation and transplantation.

Without doubt, Spemann was a master of the art of microdissection. For almost a quarter of a century he carried out a series of experiments on the developing newt's egg which gradually led to the discovery of the "organizer" effect. Constriction experiments in which the blastomeres at the two-cell stage were separated at the first cleavage furrow by use of a loop of blond baby hair convinced Spemann that even in the two-celled stage the dorsal half of the embryo differs qualitatively from the ventral half. The dorsal half was able to form an embryo, but the ventral half lacked some quality necessary for correct development. Further studies involved transplanting a definite portion from one embryo into a definite area of another, using embryos of different species so that color differences would indicate the identity of the transplated areas during development. Spemann had noted that dorsal lip grafted to a host embryo tended to lead to the formation of a new embryo. Still, the question remained as to whether development was due to self-differentiation and/or to induction.

The work of Spemann's student, Hilde Mangold, made it possible to finally answer this question. The dorsal lip of one species was grafted onto a host of another species which differed in the amount of dark pigment. The crucial experiment was a demonstration that both the dorsal lip and the host embryo participated in the formation of the secondary embryo. These experiments had a profound effect on the study of development and won for Spemann the Nobel prize in Physiology and Medicine in 1935. The ancient doctrine of epigenesis was given a new and more meaningful expression since Spemann's work showed that one step in development is a necessary condition for the next and led to the addition of parts which were not there before.

During the "golden days" of induction theory, after Spemann's dramatic experiments, there was hope that some "magic molecules" would be found with organizing and inducing powers. As no such molecule was discovered, the idea lost much of its appeal. Interest in organizers gradually dissipated, but organizer theory did help to stimulate a chemical approach to embryology. With further progress in development, genetics, and molecular biology, it became apparent that development proceeds in accordance with instructions somehow mapped out in the genetic material. However, the exact chain of events by which instructions encoded in nucleic acid finally cause a cell or group of cells to differentiate in time and space is still obscure. Indeed many pioneers in gene theory and embryology saw the two fields as separate and distinct rather than as different aspects of the same problem. Modern studies of cytogenetics were encouraged by E. B. Wilson's attempt to explicate the relationships among the data of embryology, heredity, and cytology in his monograph *The Cell in Heredity and Development*.

The molecular explanation has proven more successful in genetics than in embryology. Exploiting microorganisms to such spectacular effects, molecular geneticists have lavished their attention on creatures which face a problem quite the opposite of that with which the embryo must contend. Bacteria must retain their identity in spite of a changing environment, while the embryo must transform cells carrying identical genetic information into an integrated, harmonious population of cells expressing very specific portions of the common heritage.

To follow the problem of embryology further into the present century would require another book in itself. Today the study of development and differentiation at the molecular level has replaced the older themes of generation. Yet the essential core of such studies remains the wonderful coordination of dynamic process in time and space. Although macromolecules, nucleic acids, repressors, inducers, organelles, and membranes have displaced the cell as the object of primary interest, the cell remains a natural unit of structure and function, the link between generations, and the unit through which the egg becomes embryo and adult. To see how biologists first came to this realization, we will trace the history of cell theory in the next chapter.

Notes

1. E. Gaskings (1967). *Investigations into Generation 1651-1828*. Baltimore: Johns Hopkins Press, p. 7.

2. H. B. Adelmann (1942). *The Embryological Treaties of Hieronymus Fabricius of Aquapendente*. Ithaca: Cornell University Press, p. 37.

3. Adelmann, p. 64.

4. Adelmann, p. 221.

5. W. Harvey (1965). *The Works of William Harvey, M.D.* Translated and with "A Life of the Author" by Robert Willis, Sydenham Society, London, 1847. New York: Johnson Reprint Corp., p. 116–167.

6. J. Needham (1934). *A History of Embryology*. Cambridge: Cambridge University Press, p. 48.

7. Needham, p. 191.

8. E. Rádl (1930). *The History of Biological Theories*. Translated by E. J. Hatfield. Oxford: Oxford University Press, p. 263.

9. I. Sandler (1973). The Re-examination of Spallanzani's Interpretation of the Role of the Spermatic Animalcules in Fertilization. *Journal of the History of Biology 6:* 193–223, p. 195–196.

10. F. N. L. Poynter (1968). Hunter, Spallanzani, and the History of Artificial Insemination, in L. G. Stevenson and R. P. Multhauf, eds., *Medicine, Science, and Culture*. Baltimore: Johns Hopkins Press, pp. 97–113 (pp. 99–100).

11. Sandler, pp. 214–215.

12. Sandler, pp. 221–222.

13. Gaskings, p. 97.

14. J. Oppenheimer (1967). *Essays in the History of Embryology*. Cambridge: MIT Press, p. 133.

15. Oppenheimer, "Embryos and Evolution" and "An Embryological Enigma in the *Origin of Species*."

16. Oppenheimer, pp. 66–72; B. H. Willier and J. Oppenheimer, eds. (1974). *Foundations of Experimental Embryology*. New York: Hafner, pp. 3–37.

17. Willier and Oppenheimer, pp. 48–49.

18. Openheimer, p. 74.

19. H. Spemann (1938). *Embryonic Development and Induction*. New Haven: Yale University Press, p. 4.

References

Adelmann, H. B. (1966). *Marcello Malpighi and the Evolution of Embryology*. 5 vols. Ithaca: Cornell University Press.

Adelmann, H. B. (1942). *The Embryological Treatises of Hieronymus Fabricus of Aquapendente*. Ithaca: Cornell University Press.

Aristotle. *De Generatione Animalium*. Translated by A. L. Peck (1943). London: W. Heinemann.

Aristotle. *Historia Animalium.* Translated by A. L. Peck (1965). Cambridge: Harvard University Press.

Aromatari, J. of. (1694). *Epistola de Generatione Plantarum es Seminibus.* (1625). Translated in *Philosophical Transactions.* Vol. XVIII. London: Royal Society of London.

Balinsky, B. I. (1953). *An Introduction to Embryology.* Johannesburg: Saunders.

Bowler, P. J. (1973). Bonnet and Buffon: Theories of Generation and the Problem of Species. *Journal of the History of Biology 63:* 259–281.

Bulloch, W. (1938). *History of Bacteriology.* London: Oxford University Press.

Burton, J. (1973). Virgin Birth in Vertegrates. *New Scientist 59:* 334–335.

Castellani, C. (1973). Spermatozoan Biology from Leeuwenhoek to Spallanzani. *Journal of the History of Biology 6:* 37–68.

Cole, F. J. (1944). *History of Comparative Anatomy, from Aristotle to the Eighteenth Century.* London: MacMillan.

Cole, F. J. (1930). *Early Theories of Sexual Generation.* Oxford: Clarendon.

Driesch, H. (1908). *Science and Philosophy of the Organism.* London: Black.

Foster, M. (1901). *Lectures on the History of Physiology during the Sixteenth, Seventeenth, and Eighteenth Centuries.* Cambridge: Cambridge University Press.

Franklin, K. J. (1933). *De Venarum Ostiolis 1603, of Hieronymus Fabricus, of Aquapendente (1533?-1619).* Springfield, Ill: Thomas.

Gaskings, E. (1967). *Investigations into Generation 1651-1828.* Baltimore: Johns Hopkins Press.

Goldschmidt, R. R. (1956). *Portraits from Memory.* Seattle: University of Washington Press.

Harvey, W. (1965). *The Works of William Harvey, M.D.* Translated with *A Life of the Author* by Robert Willis, Syndenham Society, London. 1847. New York: Johnson Reprint.

Keele, K. D. (1965). *William Harvey: The Man, the Physician and the Scientist.* London: Nelson.

Keynes, G. (1966). *The Life of William Harvey.* Oxford: Clarendon.

Lillie, F. R. (1916). The History of the Fertilization Problem. *Science 43:* 39–53.

Meyer, A. W. (1936). *An Analysis of William Harvey's Generation of Animals.* Stanford: Stanford University Press.

Meyer, A. W. (1939). *The Rise of Embryology.* Stanford: Stanford University Press.

Needham, J., ed. (1970). *The Chemistry of Life.* Cambridge: Cambridge University Press.

Oppenheimer, J. (1967). *Essays in the History of Embryology and Biology.* Cambridge: MIT Press.

Power, D'A. (1897). *William Harvey.* New York: Longmans, Green.

Poynter, F. N. L. (1968). Hunter, Spallanzani, and the History of Artificial Insemination. In *Medicine, Science and Culture.* Edited by Lloyd G. Stevenson and Robert P. Multhauf. Baltimore: Johns Hopkins University Press.

Punnett, R. C. (1928). Ovists vs. Animalculists. *American Naturalist 62:* 481–507.

Radl, E. (1930). *The History of Biological Theories.* Translated by E. J. Hatfield. London: Oxford University Press.

Sandler, I. (1973). The Re-examination of Spallanzani's Interpretation of the Role of the Spermatic Animalcules in Fertilization. *Journal of the History of Biology 6:* 193–223.

Singer, C. (1925). *The Evolution of Anatomy.* London: K. Paul, Trench, Trubner.

Spallanzani, L. (1789). *Tracts on the Natural History of Animals and Vegetables.* Translated by John Graham Dolzell (1803). Edinburgh: Creech and Constable.

Spemann, H. (1938). *Embryonic Development and Induction.* New Haven: Yale University Press.

Stevenson, L. G. (1953). *Nobel Prizewinners in Medicine and Physiology, 1901–50.* New York: Schuman.

Tortonese, E. (1948). Lazzaro Spallanzani Founder of Experimental Physiology. *Endeavor 8:* 92–96.

Willier, B. H., and Oppenheimer, J. M., eds. (1964). *Foundations of Experimental Embryology.* Englewood Cliffs, N.J.: Prentice-Hall.

Wilson, E. B. (1925). *The Cell in Development and Heredity.* New York: MacMillan.

9

CELL THEORY

The Fabric of the Body

The formal statement of "cell theory" is invariably associated with Matthias Jakob Schleiden and Theodor Schwann, but as we shall see the development of cell theory was actually the work of many scientists. Unlike evolutionary science and the germ theory of disease, this nineteenth century discipline seems to have grown and gained support rather smoothly by the accumulation of new data through increasingly sophisticated microscopic observations. Except for a few small pockets of resistance, the theory was granted quite rapid acceptance. Apparently, no "revolution" was needed for scientists to see the fabric of the body in terms of cells and cell products once that doctrine had been clearly enunciated.

The cell theory is a fundamental part of modern biology. It is prerequisite to and implicit in our concepts of the structure of the body, the mechanism of inheritance, fertilization, development and differentiation, the unity of life from simple to complex organisms, and evolutionary theory. To say that the cell is the fundamental unit of life is to provide a tremendously powerful generalization to unify the study of life. This fundamental unity is not necessarily self-evident, since even to the untrained eye it is evident that all cells are not identical in character. Schleiden and Schwann were pioneers whose great insight was to recognize the common cellular basis of living forms. Even though some aspects of cell theory as originally formulated proved to be incorrect, the concept of the cell as the basic unit of life, structure and function, reproduction, growth and differentiation remains essentially valid.

Although cell theory is often described as the direct result of microscopical progress, it is equally valid to look at it from the viewpoint of long-term progress in the study of anatomy. Cell theory is one possible answer to the ancient question, what is an organism? As we have seen ancient anatomists

214

speculated about the constitution of the human body—even going beyond the visible structures to minute components seen only with the mind's eye, such as the three-fold system of vessels of the Alexandrian anatomist, Erasistratus, and the similars, dissimilars, and fibers of Aristotle and Galen. Modern anatomy can be regarded as the epoch begun by Vesalius, who entitled his great work the *Fabric of the Human Body* to express the way he saw its organic constitution. In the eighteenth century, many anatomists concentrated on the structure and function of organs and systems of organs, but different levels of resolution seemed attainable and valuable to other investigators.

Tissue doctrine, elaborated by the great French anatomist Marie Francois Xavier Bichat, was an attempt to systematize the study of the construction of the body. Although the development of cell theory was to supersede this idea, it did serve a valuable role as a transitional theory which replaced previous vague and disconnected references to "tissues."

Marie Francois Xavier Bichat (1771–1802) and Tissue Theory

The indefatigable Bichat essentially lived in the anatomical theater and dissection room of the Hôtel Dieu. A prodigious worker, he accomplished at least 600 autopsies in the space of a year. This regimen obviously was not conducive to health and long life. In 1802, when barely 31, Bichat died of a "fever," probably tuberculosis, leaving his work incomplete.

During Bichat's years of study, French physicians were laboring to transform medicine into a science which, like other sciences, would adopt the method of philosophical "analysis." This would transform unwieldy observations of general and complex phenomena into precise and distinct categories. Attempting to go beyond humoral pathology, which was still quite influential, Philippe Pinel (1755–1826) contended that diseases must be understood by tracing them back to the organic lesions which were their sources. Because organs were composed of different elements, research must be directed towards revealing these constituents. Organs which manifested analogous traits in health or disease must share some common structural and/or functional components.

Failing to find this analogy at the organ level, Bichat conceived the idea that there might be some such analogy at a deeper level—a finer level of resolution. He set about to know the body in terms of organs which could be "decomposed" and "analyzed" into their "intimate structures," the tissues. The "tissue" to which Bichat compared his structural component of the body was a very special and luxurious kind of cloth. The tissues, Bichat presumed, were the fundamental structural and vital elements of the body. Having decomposed the body into 21 kinds of tissues (nervous, vascular, mucous, serous, connective, fibrous, etc.), Bichat grouped them into organs which in turn formed more complex systems (respiratory, nervous, digestive). The actions of tissues were explained in terms of "irritability," "sensibility," and "sympathy" (the effect organs exert on each other in sickness and health).

Bichat's objective was not merely to extend knowledge of descriptive anatomy but rather to provide a scientific language with which to describe pathological changes. Through an understanding of the specific sites of disease should emerge better therapeutic methods and the means of assessing the efficacy of such treatments.

A perverse effect of Bichat's work was that some scientists and physicians came to regard the tissue as the body's ultimate level of resolution. Particularly among reactionary French physicians, even after the cell theory had been well grounded for plants and animals, the tissue was still considered the natural unit of structure and function. One reason for this was Bichat's well-known, and at the time possibly well-deserved, contempt for microscopic work of all kinds. The microscope was denounced because, as Bichat wrote, "when one gazes into the darkness, everyone sees in his own way and is affected accordingly."[1]

Intimations of Cell Theory

Microscopes and Observations of Cells

Ever since microscopic investigations began in the seventeenth century, observations of "cells" had been reported. Yet handicapped by what has been called "imperfect vision, both optical and intellectual,"[2] scientists could not incorporate their observations of sperm or blood corpuscles into a general framework.

In 1665 the word *cell* and illustrations of cellular structures appeared in Robert Hooke's *Micrographia*. Studying sections of cork under his microscope, Hooke "could exceeding plainly perceive it to be all perforated and porous, much like a Honeycomb. . . ." There was very little substance in cork other than the walls around the pores or "cells" which were like little boxes.[3] Hooke's illustration of the underside of a leaf of the stinging nettle also clearly shows the outlines of the cell walls. Work on plant structure was furthered by Nehemiah Grew, who published over 100 engravings from microscopical drawings of plants. Both of these men used the same compound microscope of the Royal Society, an instrument which probably had a magnification of about 30 diameters. Nehemiah Grew spoke of the structures seen in plants as "cells" or "bladders." Younger parts of the plant had cells which appeared to be "juicy," with thin walls, and were close together. Grew called these parts "parenchymatous" (from the Greek "poured in beside," a term which had been used by Erasistratus. He first thought the parenchyma might be crystalline but then realized it must be more complex. The parenchyma was "a body very curiously organiz'd," which looked rather like the "froth of beer."[4] Similarly, Malpighi found plants to be made up of little bodies, or "utriculi," surrounded by a wall. At the same time, Leeuwenhoek was making even more acute observations at high magnifications of infusoria, bacteria, sperm, and blood corpuscles. During the eighteenth century there were no microscopists to compare with the pioneers of the past century, and wealthy amateurs tended to take over the field and compete for newer kinds of instruments.

As we have seen (Chapter 7), by the beginning of the nineteenth century better instruments were being made and the problems of chromatic aberration and spherical aberration were on their way to being solved. During the 1830s, the major centers of biological research were able to obtain the improved microscopes. An avalanche of new discoveries, unprecedented since the seventeenth century, rapidly followed. Plant anatomy was the first field to be rejuvenated, but embryology, where very high magnification was not essential, also progressed. Not all discoveries were dependent on improved microscopes. Robert Brown (1773–1858), aware of the problems with the compound microscope, used a simple microscope with a magnification of about 300 diameters. With this instrument, he carried out his studies of Brownian movement (the random thermal motion of small particles) and discovered the nucleus in the cells of flowering plants.

Cell theory had been clearly enunciated and was finding acceptance before the major wave of technical improvements in the microscope and sample preparation. The optical performance of a compound microscope is intimately related to the possibility of correct interpretation of the images it produces. A high degree of chromatic and spherical aberration due to uncorrected lenses and low numerical aperture was common in microscopes before 1840. A very small opening was used to allow only the central rays to form the image—but while this did not reduce aberrations very much, it did reduce the amount of light available quite effectively. Often, high magnifications ($\times 500$) were used but the images were confused and optical artifacts, such as diffraction haloes, were common. Imaginative microscopists may have found such artifacts as fascinating as the objects they were ostensibly investigating.

By the beginning of the nineteenth century, controversies about "globulist" theories of the structure of biological materials were raging. These disputes may often have been the outcome of diligent studies of optical artifacts. While some investigators may indeed have seen cells, in other cases these "globules" were just circles produced by optical interference. The globules of Milne-Edwards, one of the last reports on globulism (1823), seem to have been of this nature. He claimed that every organ of the body was made up of globules, almost all uniformly measuring 1/300 of a millimeter in diameter. Histological studies with improved lens systems revealed not a trace of Milne-Edwards' globules.

Recognition of the Cell—Seeing vs. Knowing

Even in the first quarter of the nineteenth century, a major impediment to
understanding the nature of the cell was too strict attention to the legacy of
Robert Hooke and Nehemiah Grew. Although these seventeenth century microscopists had examined the "juices" of plant tissues as well as the empty
boxes in cork, the connotation and thrust of the word *cell* evoked the idea of
walls—as in prisons—surrounding empty spaces. Although the fact that plants
contained walls enclosing some fluid seemed evident enough, there seemingly
were no comparable structures in the animal body. Certainly the corpuscles
which Leeuwenhoek, Malpighi, and Swammerdam had found in the blood were
unlike plant cells. The careless use of terms like "cellular tissue" and "globule"
further added to the intellectual confusion. ("Cellular tissue" was later renamed "areolar connective tissue.")

In the 1830s the picture changed dramatically as observations new and old
began to assume a coherent and meaningful pattern. By this time the nucleus
had been recognized in plant cells and in the early stages of animal development. Cells had been recognized in animal tissue and the nature of the "juice"
within the cell was being studied under its new name, *protoplasm*. The foundations of cell theory were built in this period, although a false notion of the
formation of new cells persisted until the work of Rudolf Virchow. The
giants in this field were men who could carry out their own investigations as
well as interpret and integrate the work of other less gifted scientists who
could not see the generalization to be drawn from their unsystematic work.
For after all, seeing cells is to the appreciation of cell theory, as seeing rocks
is to the science of geology.

Animal cells were so much more heterogeneous than plant cells that the
concept of unity was long delayed. Cells and nuclei of various forms were
given separate and distinct names. For example, Purkinje called the large
nucleus of the ovarian egg in the hen the "germinal vesicle." For a long time
there was resistence to the idea that the egg was a cell and the germinal
vesicle was its nucleus. Skin cells were a popular object of study since they
seemed to resemble plant cells in their structure and manner of growth. Because growth of animal tissues was generally believed to be directly related to
blood vessels (quite unlike plants where each part grew independently), the
idea of a "plant-like" growth in epithelium was a powerful analogy at the
time—though meaningless or ridiculous today.

Although less famous than Schleiden and Schwann, Charles Mirbel, Lorenz
Oken, Johanness Purkinje, and Johannes Müller made valuable contribution to
the development of cell theory.

Charles Brisseau-Mirbel (1776–1854)

Mirbel, an eminent French botanist, combined microscopic observation with speculation about the fundamental nature of plant structure. Convinced that cells were found in all parts of the plant, he tried to account for their origin through an idea that may be traced back to Empedocles. He suggested that cells were formed in a primitive fluid and compared the resulting cells and tissues to the network of membranes found in the foam of a fermenting liquid. Unfortunately, zoologists and botanists paid scant attention to each other at the time. Schleiden's better known work is similar in conception, but better grounded in observation. Both Mirbel and Schleiden share in the blame for the incorrect notion of the *de novo* formation of cells out of a fluid.

Schleiden called Mirbel's work "highly distinguished" while heaping scorn on other predecessors.

Lorenz Oken (1779–1851) and
Johann Wolfgang Von Goethe (1749–1832)

An early version of cell theory was created by Lorenz Oken, a German naturalist heavily influenced by "nature philosophy" of the most extreme sort. Nevertheless, Oken was a prolific and creative scientist when he restrained his imagination. While in retirement at a remote island in the North Sea, he distracted himself with the study of marine animals, much as Aristotle had so many centuries before. It was during this period that Oken received his inspiration about the structure and evolution of animal life. Along with his sometime friend and associate Johann Goethe, the poet, Oken stimulated many young scientists to search for the "archetype"; the generalized form which reflects nature's basic plan and is the key to understanding all living organisms.

Combining microscopic observations with purely metaphysical speculations, Oken postulated the existence of a primitive, undifferentiated mucus-like fluid called "urschleim." Spherical vesicles arose from this jelly-like material to produce "infusoria," the simplest living things. In "Faust," Goethe used this concept in describing the "primal sea jelly" out of which all life arises. According to Oken, complex organisms were actually aggregates of these simple living things, so that "all flesh may be resolved into infusorians." Every animal or plant is, therefore, a "colony" of infusoria which have given up their independence to subordinate their life to the organism as a whole.[5]

Certainly Oken's concept of the structure of complex organisms lacked rigorous experimental verification. Yet he may have stimulated more sober scientists to undertake the necessary investigations.

Johannes Evangelista Purkinje (1787-1869)

Concerned with many areas of physiology as well as histology, Purkinje was truly innovative in his teaching as well as his research methods. Because university officials were often reactionary and reluctant to meet his needs for space and equipment, much of this work was carried out in his own home laboratory, which became known as the "cradle of histology." At the institutes for histological research he founded, first at the University of Breslau and later at Prague, Purkinje and his students made important discoveries and improvements in techniques. In 1832 Purkinje persuaded the University of Breslau to purchase one of the new achromatic microscopes. Three years later he and his student, G. G. Valentin (1810-1883), described ciliary movement and the epithelium of various animals. Purkinje invented a knife which was a precursor of the microtome and could even section bones and teeth; developed the use of balsam sealed preparations; and adapted Louis J. M. Daguerre's (1789-1851) methods to produce the first photographs of microscopic material. To demonstrate the action of the heart valves, Purkinje developed the kinesiscope—a device which showed pictures on a moving drum.

Purkinje seems to have anticipated some aspect of cell theory before Schwann, although Schleiden had already enunciated the concept for plant tissues. The importance of Purkinje and Valentin's observations on animal tissues prior to the work of Schwann has generally been neglected. Indeed, in some respects Purkinje came closer to a true understanding of cell theory than Schleiden and Schwann. By 1837, Purkinje had described the body as composed of fluids (including blood, plasma, and lymph), fibers in loose connection (tendons), and "körner," or "granules," which were animal cells. From his descriptions of certain cells it seems he had also observed the nucleus. Between 1837 and 1838 he published his most famous observations pertaining to cell theory. He described nerve cells, their nuclei and dendrites, myelinated nerve fibers, and the large flask-shaped cells in the cerebellar cortex which are named for him.

Purkinje also helped to promote the idea of the cell as composed of "protoplasm" and seems to have been first to use that term in a scientific sense in 1839. Other scientists learned from Purkinje just how powerful a tool the new microscope could be in biological research.

Johannes Peter Müller (1801-1858)

A great physiologist and comparative anatomist in his own right, Müller deserves the honor due a founder of modern scientific medicine and the mentor of many famous students, including Jacob Henle, Robert Remak, Hermann von Helmholtz, Rudolf Virchow, and Theodor Schwann. Like many of the others connected with the development of cell theory, Müller had a colorful life and an unsettled mind. He was the son of a shoemaker who considered saddle making the proper station in life for his son. But Müller aspired to and succeeded splendidly in medical studies. In 1833 he assumed the Chair of Anatomy and Physiology and the Directorship of the Museum of Comparative Anatomy at the University of Berlin. Müller travelled widely to carry out zoological studies. Shipwrecked during one of his collecting expeditions, he nearly died. When the revolution of 1848 broke out, Müller was Rector of the University of Berlin and charged with controlling the violent disturbances among the students and staff. (Student demands included free education and an end to examinations.) The student rebellion was severely put down by the military, but Müller's physical and mental health broke down as a result of the continuing hostility and sabotage.

A man of broad interests, Müller taught human and comparative anatomy, embryology, physiology, and pathological anatomy at the University of Berlin. He made original contributions to each field during his rather short life. When he died (probably from overwork), three people had to be appointed to replace him. Influenced by the work of Purkinje, Müller became one of the first to use the microscope in pathological studies. In 1838 he published the first part of a work on the microstructure of malignant growths. The work was never completed. Among his contributions to histology were studies of various glands and the structure of cartilage, bone, and connective tissues.

The Formal Statement of Cell Theory

Matthias Jacob Schleiden (1804-1881)

It is one of the mysteries of intellectual history how a man as unsuccessful and unbalanced as Matthias Schleiden, a mental and emotional cripple, could have become an outstanding scientist. By all accounts he was one of the most bizarre scientific personalities we might ever meet. After studying law at Heidelberg, he practiced as a barrister in Hamburg, but he was such a failure that he attempted suicide. Even with a gun in his hand, aimed at his forehead, he was unsuccessful. After recovering from the self-inflicted but superficial wound, he decided to switch from law to natural science. In 1931 he began to study botany and medicine at the Universities of Göttingen and Berlin, earning doctorates in both medicine and philosophy. For many years he was a Professor of Botany at the University of Jena. Despite his great scientific reputation, he resigned after 12 years and embarked on a life of wandering around Germany. Like von Baer, he seemed to believe that travel was good for the nerves. Perhaps his wanderlust was the result of quarrels with the religious authorities about his natural science point of view.

From 1864 he lived as a private teacher and peripatetic physician. Among his strange acts was a visit to Wiesbaden to celebrate the fiftieth anniversary of his law degree although he had left the law 45 years before after a suicide attempt. One fortunate aspect of his travels was his historic meeting with Schwann in Berlin. As a person, Schleiden was characterized as arrogant, choleric, temperamental, and unable to view his own hypotheses critically. These traits sometimes led him into errors, which he stubbornly stuck to, but he also was a man of great ability who helped to revive and reform the science of botany with novel theories and innovative techniques.

Heaping invectives on pedantic systematic botany as practiced by disciples of Linnaeus, Schleiden redefined botany as an inductive science comprehending "the study of the laws and forms of the Vegetable Kingdom." Unfortunately, it had established but few facts, "few indications of natural laws, and no fundamental principles and ideas by which it might be developed." Only the chemistry and physiology of plants were truly significant, Schleiden argued. Mere knowledge of the systematic arrangement of plants he dismissed as either a "pastime" or "waste of time." Botany must deal with "actual existences—natural bodies" examined in all possible ways, including microscopic examination of parts invisible to the naked eye.[6]

In 1838 Schleiden published "Contributions to Phytogenesis" in Müller's *Archives for Anatomy and Physiology*. Taking the important but neglected work of Robert Brown on the nucleus as indicative of a special relationship to cell development, Schleiden focused his attention on this structure. Soon he came to regard the nucleus as a "universal elementary organ of vegetables" which he renamed the "cyloblast." For Schleiden, all plants of any complexity were aggregates of "fully individualized, independent, separate beings, namely the cells themselves." Within the plant, each cell led a double life: "an independent one, pertaining to its own development alone; and another incidental, in so far as it has become an integral part of a plant."[7] Thus, all aspects of plant physiology and comparative physiology were fundamentally manifestations of the vital activity of cells.

Because the cell was the ultimate unit of structure and function, the origin of the cell was a critical problem. Schleiden described several methods of cell formation in "Contributions to Phytogenesis" and later in *Principles of Botany*. Unfortunately he emphasized the erroneous theory of "free-cell formation" which likened cell growth to the simple physical process of crystallization. According to this theory, a nucleolus grew by accumulation of minute mucus granules out of a fluid (cytoblastema) containing sugar, dextrin, and mucus (protein). Continued accretion of mucus transformed part of the fluid into a relatively insoluble substance and formed the cytoblast (nucleus) around the nucleolus. When the cytoblast attained full size, the young cell began to develop as a delicate transparent vesicle on its surface. As the vesicle gradually expanded, the complete cell was formed by conversion of the jelly in the cell wall into cellulose. Plants could also grow by the formation of cells within cells. This process also required the presence of the cytoblastema. In this case the entire contents of a cell were divided into two or more parts, and a "tender gelatinous membrane" immediately formed around each part. However, wood seemed to form by the sudden consolidation of an organizable fluid into a uniquely formed tissue of cells.

Although Schleiden's theory of free-cell formation suggested the simple analogy of crystal growth, he rejected the idea that organized life forms could arise through spontaneous generation. Even the lowly algae, lichens, and fungi reproduced their own kind. Confining his own work and speculations to the plant world, Schleiden observed that many eminent men had struggled to establish an analogy between the animal and vegetable kingdoms. Failure had been generally acknowledged, although few understood that the precise reason was the impossibility of applying the idea of individual, as used in the animal kingdom, to the plant world. Only the very lowest orders of plants, consisting of single cells, could be called "individuals." Yet by clearly enunciating the idea that the plant is a community of cells Schwann provided the key principle unifying plant and animal life. Indeed, by communicating his unpublished results to Theodor Schwann in October, 1837, Schleiden brought about the extension of cell theory from the plant world to the animal kingdom.

Theodor Schwann (1810–1882)

Quite a contrast with the abrasive, heterodox Schleiden, Schwann seems to have been a timid, introspective, and excessively pious person. Schwann received his early education at the Jesuits' College in Cologne, and then studied medicine at the Universities of Bonn, Würzburg, and Berlin. He graduated in 1834 at Berlin, where he had become one of Müller's favorite pupils and co-workers. During the time he had Müller's encouragement, energy, and will power to keep him at work, Schwann made numerous contributions to histology, physiology, and microbiology in addition to his famous work on cell theory. While making histological preparations for Müller, Schwann discovered the sheath surrounding nerve fibers which has been named for him. In studying the process of digestion he discovered the ferment (enzyme) named pepsin. Schwann also studied the respiration of the chick embryo and its need for oxygen. A vigorous program of experiments on fermentation carried out by Schwann challenged the theory of spontaneous generation and suggested that living microorganisms are responsible for the chemical changes in putrefaction and fermentation. To counter the argument that "tortured" air had lost its "vital force," Schwann proved that a frog could live quite well in air that had been heat-sterilized.

So gentle he could not accept a professorship in a German university, Schwann virtually retreated into scientific exile. Violent quarrels were part of academic tradition, but Schwann could not endure the vitriolic criticism which emanated from the highest echelons of the German scientific establishment. Learning and teaching anatomy consumed the remainder of his life.

Schwann had observed nucleated units in his preparations of notochord— but until he talked with Schleiden, he apparently did not think very much about the meaning of his observation. He later described how he began thinking along the lines that led to his famous expression of cell theory:

> One day, when I was dining with M. Schleiden, this illustrious botanist pointed out to me the important role the nucleus plays in the development of plant cells. I at once recalled having seen a similar organ in the cells of the notochord, and in the same instant I grasped the extreme importance that my discovery would have if I succeeded in showing that this nucleus plays the same role in the cells of the notochord as does the nucleus of plants in the development of plant cells.[8]

When Schwann began his microscopic researches, some resemblances between plant cells and certain animal structures had been noted, but the great variety of forms found in the animal kingdom seemed more significant than any similarities. Even when animal cells, fibers, corpuscles, etc., were seen, they remained "merely natural history ideas." Without a unifying theory, studies of the mode of development of one kind of cell could not be extended to any other kind. Animals appeared to be more diversified than plants in both their internal and external forms. Even with improved microscopes, at 450-fold magnification, it was difficult to see animals cells, as they were generally quite transparent.

Like Schleiden, Schwann found the nucleus his key to elucidating the cellular nature of animal organization. Beneath the myriad forms which animal tissues had assumed, even so diverse as a "muscular fibre, a nerve-tube, an ovum, or a blood-corpuscle," all had in common the cell nucleus. Schwann's purpose as set forth in *Microscopical Researches into the Accordance in the Structure and Growth of Animals and Plants* (1839) was: "to prove the most intimate connexion of the two kingdoms of organic nature."[11] At this he succeeded magnificently, while also providing evidence that even though parts of an animal may seem quite distinct in a physiological sense, they all develop according to the same laws. Schwann's classic treatise, *Microscopical Researches,* consists of three sections. In the first, Schwann describes the structure and growth of the chorda dorsalis of the tadpole and cartilage from various sources. Section two provides evidence that cells are the basis of all animal tissue, no matter how specialized. In his third section, the theory of cells is expounded.

Close examination of the chorda dorsalis and cartilage indicated that "the most important phenomena of their structure and development accord with corresponding processes in plants." Schwann's main conclusion in the first section of *Microscopical Researches* was that certain animal tissues did indeed originate from cells which were in all respects analogous to the cells of plants. "We have now thrown down a grand barrier of separation between the animal and vegetable Kingdoms," wrote Schwann, "diversity of structure."[9] Animal tissues, like plant tissues, contained cells, cell membranes, cell contents, nuclei, and nucleoli. The generation of cells within cells and the formation of cells around the nucleus seemed to take place in cartilage just as described by Schleiden in vegetable cells. Having provided proof for the analogy between plant and animal tissue in a particularly favorable system, Schwann undertook the ambitious project of proving that most or all animal tissues developed from cells.

Proof of the cellular nature of other tissues was difficult because of the minuteness of some cells and the delicate nature of the cell membrane. Magnification of 400–500 diameters did not compensate for the problem caused by the similar power of refraction possessed by both the animal cell wall and cell contents. Because of these technical problems and the diversity of forms, investigators had described cells as mere globules or granules. Here, Schwann introduced a criterion by which true cells could be distinguished—the presence or absence of the nucleus provided the most important and abundant proof of the existence of a cell. Frequently it was necessary to trace a tissue back to an earlier stage of development to observe its cellular origins. The primary cells of a tissue were frequently formed within previously existing parent cells, but this was not the only means of cell generation which Schwann described.

To prove that all tissues originate from or consist of cells, Schwann began with an examination of the ovum (or "vesicle of Baer") as the "common basis of all the subsequent tissues" and proceeded to a discussion of the "permanent tissues" of the animal body. Microscopical investigations revealed that the entire animal is composed of cells or cell products. As Schleiden had described the double life of plant cells, so too did Schwann see animal cells; the independent life of each cell was subordinated to the functioning of the organism as a whole.

The most scientific classification of tissues, Schwann argued, must be founded on the "degree of development at which the cells must arrive, in order to form a tissue." Two major modifications of cell life were defined. For *independent* cells, the cell wall remained clearly distinguishable from neighboring structures. *Coalesced* cells were those in which the walls blended, partially or entirely, with neighboring cells, or intercellular substances to form a homogeneous substance.[10] As a preliminary sketch, Schwann devised five classes of tissue based on the organization of their cells. Class I consisted of isolated, independent cells such as blood and lymph cells. In Class II were independent cells united into continuous tissues, such as horny tissues and the crystalline lens. Tissues where the cell walls had coalesced, such as cartilage, bone, and teeth, were placed in Class III. Fiber cells made up Class IV. Finally, Class V consisted of tissues where both cell walls and cell cavities had coalesced, i.e., muscle, nerve, and capillary vessels.

Although Schwann described some instances in which cells were generated from parent cells, unfortunately he also accepted the concept of cell formation from a structureless fluid, or cytoblastema, by a process analogous to crystal growth. Schwann emphasized the point that cell growth was only *figuratively* similar to crystallization, but he did seem to find the metaphor of crystallization out of the mother-liquor very powerful and satisfying. While pointing out similarities and differences between cell and crystal growth, Schwann urged other scientists to pursue this analogy as an important avenue of research. Although Schwann wrote that "everything alive has a cellular origin," he did not mean that cells come only from pre-existing cells in the modern sense. He and Schleiden liked the analogy to crystal growth as a way out of the overly metaphysical theories of the nature philosophers, but their concept of the cytoblastema and the origin of cells from this material is rather similar to Oken's concept of the "urschleim" (primeval mucus) from which primitive cells arose. Cells originated from the undifferentiated cytoblastema but went on to differentiate in the direction dictated by the "blind laws of necessity." Cells produced all the body's most differentiated tissues, including hooves, feathers, lenses, cartilage, bones, teeth, muscles, nerves. No matter how unique and noncellular bodily parts might appear, if traced backwards in terms of embryonic development, all the most complex and specialized tissues and parts of the animal were found to be derived from cells. Thus, Schwann's "unitary theory of development" served as a link for developmental phenomena and structural studies of the adult.

In the third section of *Microscopical Researches,* Schwann summarized all his researches and explicitly elaborated the very powerful generalization known as "cell theory." His first major proposition states:

> There is one universal principle of development for the elementary parts of organisms, however different, and that this principle is the formation of cells.

Generation of cells is described in his second proposition:

> A structureless substance is present in the first instance, which lies either around or in the interior of cells already existing; and cells are formed in it in accordance with certain laws, which cells become developed in various ways into the elementary parts of organisms.[11]

Schwann's theoretical section, "Theory of the Cells," is of particular interest in the history of biochemistry. Here he attempted to deal with the basis of cellular phenomena—the fundamental powers of an organized body—and whether these be "vital" or "mechanical." Actions carried out altogether blindly in accord with the laws of necessity, Schwann wrote, may *seem* to be adaptive to some higher purpose. Thus, organized bodies which seem to possess powers not found in inorganic nature might even so be acting in terms of physical and chemical laws. Understanding organized bodies in a scientific sense resolves itself into a question of studying the fundamental powers of the individual cells.

Cell phenomena fall into two natural groups. First, *plastic* phenomena are those related to the combination of molecules to form a cell. Second, chemical changes in the component particles of the cell itself, or in the surrounding cytoblastema, Schwann called *metabolic* phenomena. He coined the word "metabolic" from the Greek to describe "that which is liable to occasion or to suffer change." Schwann recognized metabolic power as a universal property of cells and, therefore, of life. He suggested that "animal heat" is derived from cell metabolism.

Fermentation was used as the "best known illustration of the operation of the cells." Schwann argued that the "fermentation-granules" were fungi and that fungi are basically cells like all other cells. While Schwann was quite correct in his theory of intracellular fermentation and broader view of the cell as the unit of metabolism, this concept was accepted only after bitter controversy between the eminent German chemist, Justus von Liebig, and Louis Pasteur. Liebig and his associate, Friedrich Wöhler, published an anonymous satire of Schwann's theory. They portrayed the dispersal of yeast in solution giving rise to eggs which rapidly hatched into animals in the shape of alembics. These creatures threw themselves upon the sugar in solution, greedily devour-it. Having digested their meal, the creatures vulgarly belched forth carbon dioxide and excreted alcohol. By temperament, Schwann was unsuited to such disputes. He was destroyed psychologically and left the field to be defended and conquered by Pasteur—a man who was described as never one to put up his sword until all enemies had been conquered or killed. (Pasteur's part of the story will be told in Chapter 10.)

Corrections of Cell Theory

While Schwann and Schleiden had provided a powerful new framework for understanding the structure, development, and functions of plants and animals, their theory was still deficient in several important respects. Their concept of free-cell formation and the notion of the cytoblastema were the major defects

in an otherwise handsome ediface. It took the work of many other scientists
to repair this framework and fully develop cell theory.

In 1839 Purkinje used the term "protoplasm" to describe the cell substance. The word "protoplast" had been used by theologians for Adam, the
"first formed." For Purkinje the term covered whatever was first produced in
the development of the individual plant or animal cell. Grew's word "cam-
bium" had enjoyed a similar usage among botanists, without reference to
cellular structure. Karl Nägeli (1817-1891) and Hugo von Mohl (1805-1872)
further characterized the contents of cells. Using many different species,
Nägeli made microscopic studies of cell formation during the reproductive
process and at growing points in plants. He found that the lower Algae were
favorable subjects for studies of cell division and for watching the movement
and behavior of the cell substance. Although Nägeli, co-editor with Schleiden
of a short-lived journal, at first defended Schleiden's theories, his comparative
studies of the production of cells in various groups of plants eventually convinced him that "free cell-formation" was incorrect. New cells did not arise
by precipitation around the nucleus, but arose from the division of pre-existing
whole cells. However, Nägeli was unsuccessful in determining the origin of
the nucleus and his views were not totally consistent.

At about the same time, Hugo von Mohl was studying the cells of higher
plants. In studies of plant tissue preserved in alcohol, where the contents of
the cells had contracted away from the wall, he discovered what he called the
"primordial utricle." He later found that the same effect could be produced
in fresh material with strong acids. When treated with iodine, this cellular
material turned yellow, indicating the presence of nitrogenous material. Apparently independently of Purkinje, von Mohl used the word "protoplasm" to
describe the part of the plant cell within the cell membrane. In a series of influential articles, von Mohl made the word "protoplasm" a general part of the
biological vocabulary. He concluded that nothing like Schleiden's universal
theory of free cell formation was ever observed in nature. Showing little
deference towards the founder of cell theory, he even suggested that
"Schleiden has never once observed the division of cells."

"Protoplasm" was introduced to the general public at Edinburgh in November 1868 in an address by T. H. Huxley (1825-1895) called "The Physical
Basis of Life." When uttered by Huxley, the word was thoroughly divorced
from its previous religious associations. "All vital action," Huxley announced,
"may be said to be the result of the molecular forces of the protoplasm which
display it." As "uhrschleim" had appeared in Goethe's "Faust," so "protoplasm" appeared on stage in Gilbert and Sullivan's "Mikado." Pooh Bah says
proudly: "I can trace my ancestry back to a protoplasmal primordial atomic
globule." [12]

The confusion over the origin of cells hindered the interpretation of various observations. Studies of early stages of development displayed a great lag between observation and interpretation. The cleavage of eggs had been seen in the seventeenth century by Swammerdam and by Spallanzani at the end of the eighteenth century. Yet when Prévost and Dumas saw the segmentation of frog eggs in 1824, they still could not understand the real meaning of the phenomenon. In 1854, Martin Barry (1802-1855) published illustrations of cleavage of the rabbit egg, which clearly showed the nuclei. But Barry believed that cells of later stages were derived directly from nuclei.

Max Schultze (1825-1874), professor of anatomy at Bonn, attempted the synthesis of ideas about protoplasm, protozoa, and the egg-cell. In 1861 he inelegantly, but succinctly defined the cell as a lump of nucleated protoplasm. According to Schultze, protoplasm was the physical basis of life. In all forms of life studied, plants and animals, higher and lower forms, it provided unity of structure and function.

Albert Kolliker (1817-1905), who studied the development of the eggs of the cuttlefish, was probably the first to lay great stress on the division of the nucleus during the process of segmentation of the ovum. He applied Schwann's theory to embryonic development in a great number of animal species and tissues. Franz Leydig (1848) and Robert Remak (1852) finally clarified the behavior of the nucleus during cell division. Remak (1815-1865) carefully described the equal division of embryonic blood corpuscles in the developing chick. He believed that in normal growth cells increase in number by the division of one cell into two new cells, but he still believed that some other process might be involved in pathological conditions. Schleiden's old error continued to be a real possibility to many scientists—until Rudolf Virchow convinced them that every cell is the product of a pre-existing cell.

Rudolf Ludwig Carl Virchow (1821-1902) and Cellular Pathology

It was Rudolf Virchow who formulated cell theory in its modern form and incorporated it into pathology as the foundation stone of scientific medicine. Virchow was a man of many talents and accomplishments, a prominent member of the intellectual and social movements of the nineteenth century. Like many other famous German scientists of his generation, Virchow was a student of Johannes Müller. By 1845 Virchow had published one of the first systematic studies of "white blood" (leukemia). Between 1847-1848 Virchow was assigned to investigate an outbreak of "typhus" in an industrial district in Silesia. His report was very critical of the government which he blamed for the terrible social and sanitary conditions which he believed were the true causes of the epidemics rampant in that district. This report led to summary dismissal from his position. Following the revolution of 1848, his reputation

for radicalism and sympathy for the opposition made it expedient for him to
leave Berlin. Fortunately, he obtained a position at the University of Würzburg, the first Chair of Pathological Anatomy in Germany. He spent seven
very productive years there, until the chair at Berlin was vacant in 1856 and
Virchow was offered that position. He accepted on condition that he be
given an institute of pathology for his research. This request was satisfied and
Virchow remained at Berlin until his death. In 1847, Virchow and Benno
Reinhardt (1819–1852) founded the *Archiv fur pathologische Anatomie
(Archives for Pathological Anatomy)*. Editing the journal gave Virchow
another outlet for his liberal ideals, as he judged an article according to
whether it contained an important contribution to knowledge. Thus, he included articles on comparative anatomy and physiology, anthropology, ori-
ental languages, medieval translations from Greek and Arabic, as well as the
expected papers on the histology of tumors and infectious diseases.

In Berlin, Virchow participated in political as well as scientific life. In
1850 he was elected to the Berlin City Council and in 1861 to the Prussian
Diet where he opposed the policies of Bismarck. He remained on the City

**Rudolf Virchow, founder of cellular pathology. [© 1961, Parke, Davis and
Company.]**

Council his whole life and initiated many social, sanitary, and medical reforms in the city. During the wars of 1866 and 1870, Virchow was responsible for the development of the first hospital trains and for military hospitals. He served as a member of the Reichstag from 1880 to 1893. Among Virchow's many interests was physical anthropology. In 1869 he founded the Berlin Society of Anthropology, Ethnology and Prehistory, serving as President until his death. He also worked for the founding of the Berlin Ethnological Museum and Folklore Museum.

In the first volume of the *Archives for Pathological Anatomy* (1847), Virchow reviewed the prevailing ideas on the organization and growth of tissues. Reflecting the influence of Schleiden and Schwann he described the origin of cells from the differentiation of a "formless blastema," a fluid exuded from vessels. At the time, studies of inflammation seemed to support this general theory. (White corpuscles from the blood enter a wounded area in great numbers and become macrophages.) Yet even as a student making observations of the healing of the cornea, Virchow had seen phenomena inconsistent with prevailing views of cell growth and began making intensive microscopic studies of pathological processes. He reached the same conclusion that Robert Remak had come to from embryological studies. He and Remak rejected free cell formation and spontaneous generation which he believed it resembled. By 1854 Virchow was convinced that "there is no life but through direct succession."

In 1855 Virchow published a paper on "Cellular Pathology" in his *Archives,* which included the famous "*omnis cellula e cellula*" (all cells come from pre-existing cells). He complained that the microscopic method was still new to medicine and was not deep in the thinking of the older physicians, while others overvalued the instrument and used it uncritically. His experiences might have been similar to those of a student attending Henry Acland's courses in microscopical anatomy at Oxford in the 1840s. Some of the senior men would attempt to look at the microscopic preparations which illustrated the lectures. A certain Dr. Kidd, on examining a delicate morphological preparation, proclaimed "first, that he did not believe in it, and, secondly, that if it were true he did not think God meant us to know it."[13]

Virchow's views were expressed in a series of lectures given in Berlin under the general title "Cellular Pathology." Virchow began with a review of plant structure and corrected earlier facile comparisons of plant and animal cells. He then analyzed the structure of the tissues of every organ in the human body in health and disease. Virchow vigorously opposed the ancient idea of "general disease," so inextricably linked with humoral pathology. Instead, he demanded that medical and scientific inquiry should focus on the question, "Where in the body's cells is the disease?" Virchow argued that there is no essential difference between normal and pathological states. All disease is simply modified life. Thus, study of pathology as a science was to be inextricably linked to the study of physiology to determine the disturbances that take place in a pathological state. For example, extensive studies of cancer convinced Virchow that cancer cells differ from normal cells primarily in their behavior rather than in structure.

For Virchow, the cells were "the last constant link in the great chain of mutually subordinated formations that form tissues, organs, systems, the individual." In 1858 Virchow published his lectures and research in a book, *Cellular Pathology*. While others had advanced some of the same ideas he espoused, his vigorous campaign for the critical use of the microscope in medicine, the revised cell theory, and the recognition of the cell as the locus of disease was more organized and influential than any other statements of these concepts.[14]

Refinements of Cell Theory

In his influential treatise, *The Cell in Development and Heredity,* E. B. Wilson outlined three stages in the development of cell theory after its enunciation by Schleiden and Schwann. Between 1840 and 1870, scientists labored at the foundations of the theory—marking out the fundamental outlines and principles of genetic continuity. The second period, from 1870 to 1900, witnessed the maturation of cytology and cellular embryology, as well as the development of new concepts of the physical basis of heredity and the mechanism of development. Wilson wrote from the perspective of a scientist in the third stage, after the rediscovery of Mendel's laws, a time of "modern and more searching inquiries into the mechanism of sex and heredity."

Scientists variously focused their attention on cells and parts of cells, from nucleus to cell walls, to protoplasm, back to the nucleus, to chromosomes, and currently to DNA. Cell theory had served nineteenth century science as a link between plant and animal, ovum and adult, health and disease. Yet new studies of physiology, which both Schleiden and Schwann had called for, were slow in coming, despite the heroic efforts of Claude Bernard, Louis Pasteur, and others. Still, Pasteur's studies of fermentation and spontaneous generation did provide unequivocal support for Schwann's prescient but largely unsubstantiated metabolic theory of the cell. Studies of the fine structure of the cell required improvements in microscopes and, more particularly, improved methods of preparing and staining samples for study.

Henle, Schwann, and others usually prepared their materials by teasing out or squashing fresh material into a thin layer and studying this directly under the microscope. The structure of relatively uniform tissues might be adequately studied this way, but the interrelationships of different types of cells in complex organs would be completely confused.

Fixatives were first used as preservatives for gross specimens. Robert Boyle (1627-1691) had suggested many preservatives, but it is uncertain whether he ever tested any of these possibilities. Microscopists used various chemicals, such as chromic acid and its salts, primarily as hardening agents. Tissues would be soaked until they became rigid enough for free-hand slicing. Studies of the nervous system were aided by use of such agents. Alcohol or acetic acid or mixtures of the two were commonly used. Canada balsam was used for mounting after treatment with turpentine. Max Schultze (1825–1874) discovered that osmic acid fixation could preserve fine, life-like cellular details.

Staining methods were generally quite crude. Carmine was used in the 1850s, but in the next decade hematoxylin and the first aniline dyes were available. Differential staining was not developed until the 1870s. By this time Walter Flemming (1843–1915) and others were refining techniques of fixation and staining to see details within the cell and nucleus. In conjunction with the oil-immersion lenses, these techniques were the foundation of modern cytology. Techniques of embedding and preparing sections also were refined in the last decades of the nineteenth century.

By the end of the nineteenth century the minute bodies which appeared during cell division had been investigated. Eduard Strasburger (1844–1912), professor of botany at Bonn, helped to unify the field with his monumental *Cell-Formation and Cell-Division,* editions of which appeared from 1875 on. His descriptions of the complex processes taking place in the division of plant cells were quite clear. Walter Flemming emerged as the outstanding figure in the study of cell division in animals. His work was published in 1882 as *Cell Substance, Nucleus, and Cell Division.* Flemming is best known for his description of the chromosomes in the late 1870s. However the term was not used until 1888 by H. W. G. Waldeyer (1836–1921) who was the first to use hematoxylin as a histological stain. Flemming used the term "chromatin" for the nuclear substance and gave the name "mitosis" to cell division. He described the longitudinal splitting of the chromosomes in 1880. At the same time, Strasburger was able to study the division of living staminal hair cells of *Tradescantia.* Flemming did similar work on thin slices of embryonic amphibian cartilage. From these studies emerged an appreciation of the stages of cell division as a sequence in time which was basically the same in plant and animal cells.

The period from 1875 marks the time when the scattered and diverse observations of cell division and nuclear inclusions were finally organized. Cell theory came to include two more generalizations: cells in animals and plants are formed by equal division of existing cells and division of the nucleus precedes division of the cell.

Another exciting field developed at the end of the nineteenth century and seemed to call into question the universality claimed for cell theory. Microbes seemed to be quite different from cells as generally understood in plants and animals. The component which signified the cellular nature of otherwise

nondescript "vesicles" or "globules"—the nucleus—was not even present in
bacteria. Eventually cell theory was able to accommodate the finding that
not all living organisms or tissue components contain cells in the classic sense
of a "blob of protoplasm containing a nucleus." On the other hand, the fine
structure of the cells of higher organisms is of bewildering complexity. Lewis
Thomas captured that intricacy in a portrait of the eukaryotic cell as an "eco-
system more complex than Jamaica Bay," harboring "little animals" like
mitochondria, with their own DNA and RNA, along with various centrioles,
basal bodies, and perhaps other "more obscure tiny beings at work."[15]

Notes

1. S. Bradbury (1967). *The Evolution of the Microscope.* Translated from the
   French by author. New York: Pergamon, p. 183.

2. A. Hughes (1959). *A History of Cytology.* New York: Abelard-Schuman.

3. R. Hooke (1665). *Micrographia.* Reprint 1961. New York: Dover, p. 112.

4. N. Grew (1682). *The Anatomy of Plants.* Reprint 1965. New York:
   Johnson Reprint. See introduction by C. Zirkle, p. xiii.

5. T. S. Hall (1969). *Ideas of Life and Matter.* Chicago: Chicago Univ. Press,
   II, pp. 121–304 ("Tissue, cell and molecule, 1800–1860").

6. M. Schleiden (1849). *Principles of Scientific Botany.* Reprint 1968. New
   York: Johnson Reprint, pp. 1–2.

7. T. Schwann (1847). *Microscopical Researches.* Translated by H. Smith.
   Reprint 1969. New York: Kraus Reprint. Includes M. J. Schleiden, "Contributions to Phytogenesis (1838)," pp. 231–232.

8. Hughes, p. 38.

9. Schwann, p. 34.

10. Schwann, p. 65.

11. Schwann, p. 165.

12. Hughes, p. 50.

13. Hughes, p. 12.

14. E. Ackerknecht (1953). *Rudolf Virchow.* Madison: University of Wisconsin
    Press.

15. L. Thomas (1974). *The Lives of a Cell.* New York: Viking, p. 4.

References

Ackerknecht, Erwin (1953). *Rudolf Virchow, Doctor, Statesman, Anthropolo-
    gist.* Madison: University of Wisconsin Press.

Baker, J. (1948). The Cell Theory: A Restatement, History, and Critique. Part 1. *Journal of Micro. Soc. 89:* 103.

Baker, J. (1949). The Cell Theory: A Restatement, History, and Critique. Part 2. *Journal of Micro. Soc. 90:* 87.

Bradbury, Savile, ed. (1967). *Historical Aspects of Microscopy: Papers Read at a one-day Conference held by the Royal Microscopical Society at Oxford, 18 March 1966.* Cambridge: Heffer, for the Royal Microscopical Society.

Conklin, F. G. (1939). Predecessors of Schleiden and Schwann. *Amer. Nat. 73:* 538–546.

Espinasse, M. (1956). *Robert Hooke.* Berkeley: University of California Press.

Gerould, J. H. (1922). The dawn of the cell theory. *Sci. Monthly 14:* 268–277.

Goldschmidt, R. B. (1956). *Portraits from Memory.* Seattle: University of Washington Press.

Haymaker, W., ed. (1970). *Founders of Neurology: One hundred and forty-six Biographical Sketches by eighty-eight authors.* 2nd ed. Springfield, Ill.: Thomas.

Hertwig, O. (1895). *The Cell.* London: Swan Sonnenschein.

Hooke, R. (1665). *Micrographia.* Reprinted 1961. New York: Dover.

Hughes, A. (1959). *A History of Cytology.* New York: Abelard-Schuman.

Lamarck, J. B. (1802). *Zoological Philosophy, An Exposition with Regard to the Natural History of Animals.* Translated by H. Eliot (1963). New York: Hafner Publishing Co.

Moulton, F. R., ed. (1940). *The Cell and Protoplasm.* Publ. of A.A.A.S. No. 14. Washington, D.C.: Science Press.

Nordenskiold, E. (1935). *The History of Biology: A Survey.* New York: Tudor.

Oppenheimer, J., and Willier, B. H. (1964). *Foundations of Experimental Embryology.* Englewood Cliffs, N.J.: Prentice-Hall.

Robinson, V. (1929). *Pathfinders in Medicine.* New York: Medical Life.

Rudnick, D., ed. (1958). *Cell, Organism and Milieu.* Society for Study of Development and Growth. New York: Ronald.

Sachs, J. von. (1909). *A History of Botany.* (English trans.) Oxford: Clarendon.

Schleiden, M. J. (1849). *Principles of Scientific Botany; or, Botany as an Inductive Science.* Translated by E. Lankester with a new introduction by J. Lorch. Reprinted 1968. *No. 40: The Sources of Science.* New York: Johnson Reprint.

Schwann, T. (1847). *Microscopical Researches into the Accordance in the Structure and Growth of Animals and Plants.* Reprinted 1969. Translated by Henry Smith. New York: Kraus Reprint.

Sigerist, H. E. (1933). *The Great Doctors; A Biographical History of Medicine.* Translated by E. and C. Paul (2nd Ed. 1958). Garden City: Doubleday.

Suner, A. P. (1955). *Classics of Biology.* New York: Philosophical Library.

Tempkin, O. (1949). Metaphors of Human Biology. In *Science and Civilization.* Edited by R. C. Stauffer. Madison: University of Wisconsin Press.

Thomas, L. (1974). *The Lives of a Cell, Notes of a Biology Watcher.* New York: Viking.

Wilson, E. B. (1925). *The Cell in Development and Heredity.* New York: Macmillan.

10

MICROBIOLOGY AND BIOGENESIS

Studies of Microorganisms

Nineteenth century scientists had several compelling reasons to study the long neglected "infusoria." First, these tiny creatures are of interest in themselves as the simplest organisms. While few serious scientists entertained the idea that these entities were "complete" animals with minute organ systems, certainly bacteria, protozoa, and molds displayed enough diversity of structure and life cycle to satisfy the curiosity of any microscopist worth his newly improved lenses. Second, contending that metabolism was an intracellular process, Schwann used fermentation as the "best known illustration of the operation of the cells, and the simplest representation of the process which is repeated in each cell of the living body."[1] Clearly, Schwann anticipated the use of microbes as representatives of all living beings for the study of vital functions—metabolism, growth, and reproduction. In a more specialized sense, particular microorganisms could be studied in relation to fermentations, putrefaction, and chemical transformations of great diversity.

Opponents and proponents of evolutionary theory also seized opportunities to turn studies of microbes into props to support their particular version of "creation." Clues gleaned from the examination of the simplest cells as units of life might illuminate the great question which Darwin himself avoided. This was the problem which T. H. Huxley termed "biogenesis," the origin of life. Some scientists believed that if microorganisms were the product of spontaneous generation they might fill in the gap between living and nonliving things. Of course, not every advocate of spontaneous generation accepted the theory of the transmutation of species.

Microorganisms could also be studied from the medical point of view. Infections and diseases could be analyzed as phenomena analogous to fermentation and putrefaction, processes already suspected of being caused by microorganisms.

These possible avenues of approach are not necessarily separate. As we have seen, scientists in the past were much less likely to specialize. While limited in many ways, scientists did not let disciplinary boundaries limit the scope of their research. This is especially obvious in the work of the chemist Louis Pasteur. While not interested in microbiology as a prop for evolutionary theory, Pasteur's research encompassed almost every other aspect of the subject: microorganisms as ferments, as agents of disease, and as living organisms that arise only from parents like themselves and not through spontaneous generation.

Although microbiology is closely linked to progress in microscopy, Leeuwenhoek's discovery of infusoria and assorted "wretched beasties" remained a mere "curiosity" until the second half of the nineteenth century. Only eccentrics continued to search for these "incredibly small" animals. It seems strange that no role was found for these minute creatures although they could be found in such great numbers. Leeuwenhoek suggested that "all the people living in our United Netherlands are not as many as the living animals that I carry in my own mouth this very day." While it is impossible to ignore Leeuwenhoek's accomplishments, his perverse insistence on secrecy seriously impeded further progress in bacteriology. Failing to duplicate his results, other microscopists became more than a little skeptical. This did not trouble Leeuwenhoek, who wrote to Boerhaave in 1717, "I know I am in the right."[2] In time, Leeuwenhoek was completely vindicated.

Various accounts of infusoria appeared in the eighteenth century. Noteworthy among these was the work of Louis Joblot (1645-1723), probably the first to confirm the existence of some of Leeuwenhoek's larger "animalcules." Although chiefly remembered for his opposition to spontaneous generation, Joblot also published (1718) a treatise on the construction of microscopes and his observations of animalcules. With its 12 engraved plates, the work was a first in the history of microbiology.

Linnaeus (1707-1778), the great arbiter of the names, classification, and (almost) the existence of all species of things, was quite skeptical of microscopic studies. Believing that an orderly arrangement of all species was the supreme achievement of a naturalist, Linnaeus found the creatures discovered by Leeuwenhoek, Joblot, and others a nuisance and an embarrassment. All such creatures were tossed into the all-purpose miscellaneous category— "Vermes"—in a class called "Chaos."

Although "living ferments" have been used for thousands of years to produce beer, wine, and bread, scientists did not begin to appreciate the true nature of fermentation until the 1830s. Unfortunately, fermentation was a problem long associated with the occult "science" of alchemy. The term "ferment" was used quite loosely to refer to an active substance which could transform passive (fermentable) substances into its own nature. After man had enjoyed the products of alcoholic fermentation for millenia, alchemists had studied the phenomenon for centuries, and more than 150 years after Leeuwenhoek described yeast, three investigators almost simultaneously realized that yeast is a living organism and the cause of alcoholic fermentation. Working with improved microscopes, Theordor Schwann, Charles Cagniard-Latour, and Friedrich Traugott Kützing observed the globular nature of yeast and its reproduction by budding (1836–1837).

Despite the lucidity of Schwann's arguments and the ingenious nature of his experiments and observations, Justus von Liebig (1803–1873) ridiculed the concept of cellular fermentation. To Liebig, Schwann and Latour were guilty of injecting occult vital phenomena into a matter that he preferred to regard as a purely chemical process. Convincing proof of the biological nature of fermentation and the subversion of the doctrine of spontaneous generation were finally provided by Louis Pasteur. Since Pasteur illuminated virtually every branch of bacteriology through his genius, there is no better way to look at the history of the subject than through the life and work of Louis Pasteur. Such a survey will take us from chemistry to fermentation, from spontaneous generation to the germ theory of disease.

Louis Pasteur (1822–1895)

Although his family was quite poor, Louis Pasteur's father, a tanner and ex-soldier of Napoleon's army, hoped that his son would become a scholar and professor. The boy was bright and a talented artist, so a professorship in either the arts or sciences was a possibility. Many of Pasteur's portraits of family and friends were considered quite good, but at 19 years of age he gave up painting and devoted himself strictly to science. Not all of his teachers considered Pasteur an apt pupil. Indeed, in chemistry he was rated as "mediocre." In the competitive examination for the École Normale Supérieure of Paris, Pasteur ranked only sixteenth. Although admission to the school was an honor in itself, Pasteur refused to enter the school until he could prepare himself better. In 1843 he competed again and advanced to fifth place, but his first attempt at student life in Paris ended in homesickness so acute that he had to return to his family.

The most important thing that Pasteur learned as a student was the applicability of the experimental method to a broad range of problems. While he never regarded science as the answer to all questions, he never hesitated to apply the experimental method to problems in biology and medicine, areas in which he had no training. Devoted above all to "pure science," eager to pursue the implications of his preliminary experiments, ready to see grand generalizations linking the stereochemistry of crystals to the "diseases" of

wine and beer and to the mechanism of infection, Pasteur introduced new
methods of food preservation, aided agriculture, and saved the wine, beer,
vinegar, and silk industries of France. An outsider to medicine, he advanced
the possibility of the prevention and treatment of disease more than any other
single individual.

Pasteur's first triumph was his study of crystal structure and stereoisomer-
ism. These studies seem quite remote from the biological and medical researches that made his name a household word, but seem to have given him
the impetus, insight, and inspiration for his approach to all later researches.
Jean Baptiste Biot (1774–1862) had demonstrated that even in solution some
organic chemicals such as sugars and tartaric acid rotate the plane of polarized
light. Another chemist, Eilhard Mitscherlich (1794–1863), found that in addition to the common large crystals of tartaric acid there were smaller crystals
which he called "paratartaric acid" or "racemic acid." Although both types of
crystals had the same chemical properties, racemic acid was inactive to polarized light. Pasteur seized upon this apparent inconsistency and devised experimental means of answering the question, "How could chemicals be the same and
yet different?" Pasteur soon prepared and crystallized 19 different salts of the
tartrates and paratartrates. Examining them under the microscope, Pasteur
found he could mechanically separate out two kinds of crystals—left- and right-
handed crystals which had equal and opposite polarizing properties. These experiments were repeated in Biot's laboratory using Biot's own chemicals and
polarimeter. Having proved his case, the young Pasteur won the lifelong admiration and fatherly devotion of the rather cantankerous old chemist.

From Crystals to Ferments

How does one get from the purely theoretical question of molecular dissymmetry to the chemistry of life? In this case it was the result of a "chance"
discovery that took place in Pasteur's laboratory in 1857. It was common
enough in warm weather for molds to grow on tartrate solutions. Most investigators would note this and throw away the spoiled preparations in disgust. But Pasteur wondered whether the effect of molds would be the same
towards the two forms of tartaric acid. Discovering that the molds used only
the D form, Pasteur developed a simple and ingenious method of separating
stereoisomers by using living agents. More important, it provided a key to
the problem in Pasteur's keen mind: Molecular dissymmetry was a criterion
for life processes which distinguishes them from chemical processes. This
association between dissymmetry and life led Pasteur into studies of fermentation and the role of microorganisms in the balance of nature.

Pasteur saw molecular dissymmetry in every aspect of the universe—living
and nonliving. "The universe is a dissymmetrical whole" wrote Pasteur. "I

am inclined to think that life, as manifested to us, must be a function of the dissymmetry of the universe and of the consequences it produces. The universe is dissymmetrical . . . . Life is dominated by dissymmetrical actions. I can even foresee that all living species are primordially, in their structure, in their external forms, functions of cosmic dissymmetry."[3]

In 1849, Pasteur was given a position at the University of Strasburg. His lectures in chemistry attracted many industrialists from the surrounding area. Although Louis wrote to his father that he had no intention of marrying soon and would like one of his sisters to come and look after him, only two weeks after his arrival Pasteur was asking the Rector of the Academy of Strasburg for permission to court his daughter Marie. He carefully explained to M. Laurent that he was completely without fortune, except for "good health, some courage," and his position at the University. Pasteur, who was then 26, modestly wrote to his future mother-in-law: "There is nothing in me to attract a young girl's fancy. By my recollections tell me that those who have known me very well have loved me very much." The proposal was soon accepted, and Louis and Marie Pasteur embarked on a 50-year marriage of devotion to each other and to science.[4]

While at the University of Strasburg, Pasteur continued his experiments on crystals, all the while drawing conclusions of the broadest possible implications—sometimes very speculative, but always very ingenious. In one experiment he broke off a piece of an octahedral crystal and replaced it in its mother liquor. The crystal grew larger in every direction by deposition of new particles, but the mutilated part grew even more actively. In a few hours the crystal was back to its original shape. Pasteur suggested that the healing of wounds might be compared to the physical phenomenon of crystal growth.

Having won recognition and honor for his chemical researches, Pasteur was in 1854 appointed Professor of Chemistry and Dean of Sciences at the University of Lille in northern France. The "mission" with which he was charged included meeting the needs of the industrial interests of the area. Pasteur tried to make education relevant to the needs of the district by encouraging more laboratory experience and creating a new diploma for those who wished to enter an industrial career at the level of foreman or overseer. But he also emphasized that the theoretical was as important in education as the practical. "Without theory, practice is but routine born of habit," Pasteur declared. "Theory alone can bring forth and develop the spirit of invention." When asked the use of a purely scientific discovery he liked to quote Franklin's answer: "What is the use of a new-born child?"

Although pure chemistry was his favorite research interest, Pasteur did not disdain investigations of the practical problems of commercial developments in France. Soon he became directly involved in problems of alcoholic fermentation, a major industry in the region around Lille. Because of his inquisitive mind and breadth of vision, he was able to apply principles learned in experimental chemistry to the problems of fermentation. Such researches were to continue for over 20 years, during which time the role of microorganisms in causing various desirable and undesirable fermentations was established.

Pasteur often said that "in the fields of observation, chance favors only the mind which is prepared."[5] He proved this was true for himself when a M. Bigo came to Pasteur for advice on his problems with the manufacture of alcohol from beetroot. Visiting the factory every day and studying samples of the fermentation juices with his microscope, Pasteur discovered that the globules of yeast were round when fermentation was healthy, became lengthened when the process began to alter, and became quite long when fermentation became lactic. M. Bigo was able to apply these observations to the workings of his factory with good success.

Like Schwann and others, Pasteur found various microorganisms in fermenting juices. But he also noticed optically active products of fermentation and, from the clue obtained with his studies of crystals, he formed the hypothesis that fermentation was a process carried out by "living ferments." His ideas were quite the opposite of the accepted concept of fermentation expounded by the most renowned chemists of the period. Berzelius, Liebig, and Wöhler believed that fermentation was a purely chemical process and that microorganisms were the product rather than the cause of fermentation. After analyzing many examples of the fermentation process, Pasteur suggested that all fermentations are caused by microorganisms and that each living ferment is specific for a particular kind of fermentation. Changes in environment, temperature, pH, composition of the medium, and poisons affect different ferments in particular ways. For example, yeast produce alcohol best under acidic conditions, whereas lactobacillus favored a neutral pH.

Pasteur continued his studies of fermentations and the related problems of spontaneous generation and diseases for the rest of his life. He worked on so many questions simultaneously that it is difficult to discuss them all in a logical manner. His mind was constantly full of ideas for new projects and stimulated by interactions between ongoing research problems.

In 1857 Pasteur left Lille and joined the École Normale in Paris as Assistant Director in charge of administration and direction of scientific studies with responsibility for "the surveillance of the economic and hygienic management, the care of general discipline, intercourse with the families of the pupils and the literary or scientific establishments frequented by them." Facilities and funds for research were practically nonexistent. Pasteur used a tiny attic under the roof of the École Normale as a laboratory and struggled to continue his research. Nevertheless, Pasteur devoted himself to the school and students, worrying over details of catering, the weight of meat per pupil, the ventilation of classrooms, and all factors pertaining to the health of the students.

In 1860 the Academy of Sciences awarded the Prize for Experimental Physiology to Pasteur. Claude Bernard wrote the report and emphasized the great physiological interest of Pasteur's studies on alcoholic, lactic, and tartaric acid fermentation. At the same time, Pasteur was studying the problem of spontaneous generation and reporting his preliminary experiments to the Academy. His old friend Biot advised Pasteur to put aside the question—it was a quagmire to be avoided. "You will never find your way out," warned Biot. "I shall try," Pasteur answered, although he knew that attempting to prove a universal negative is neither a scientifically nor a philosphically rewarding proposition.

Spontaneous Generation—Historical Aspects

Biot was giving Pasteur what he considered wise advice. The question of spontaneous generation was an ancient and controversial one. But Pasteur realized that microbiology and medicine could only progress when the idea of spontaneous generation was totally vanquished. Belief in the spontaneous generation of life was almost universal and unquestioned from the earliest times up to the seventeenth century. As suggested by the term "Mother Earth," many peoples have seen the whole world as a "living organism." Thus, it was not surprising if living creatures arose out of seemingly inanimate substrates. The lowest creatures, parasites, "vermin" of all sorts which often appear suddenly from no known parents, seemed to be the result of some kind of transmutation of lifeless materials into organized beings.

Aristotle lent his great prestige to the idea of spontaneous generation, even cataloging the particular species that would be generated from various substances. According to Aristotle, heat was necessary for generation by any mechanism: sexual, asexual, or spontaneous. While higher creatures reproduce by virtue of "animal heat," the lower forms arose from slime and mud in conjunction with rain, air, and the heat of the sun. A combination of morning dew with slime or manure would produce fireflies, worms, bees, or wasp larvae. From rotting corpses and excreta, tapeworms formed, while slime produced crabs, fish, frogs, and salamanders. Moist soil gave rise to mice. Joannes Baptista van Helmont (1579-1644) produced a recipe for mice from bran and old rags stuffed in a bottle, and left to develop in a dark closet. Even Newton (1642-1726) contributed new myths by suggesting that plants arise from the attenuated tails of comets.

Francesco Redi (1626-1698) initiated the experimental attack on the question of spontaneous generation. Redi was a court physician and member of the Academy of Experiments of Florence, Italy. Redi applied the experimental method with great zeal, conducting carefully controlled experiments. Various kinds of substrates, raw or cooked, were tested, including flesh from

Francesco Redi. [From W. Bullough (1960). *The History of Bacteriology.*
Courtesy of Oxford University Press.]

oxen, deer, buffalo, lion, tiger, duck, lamb, kid, rabbit, goose, chicken, swal-
low, swordfish, tuna, eel, sole, and snake. Redi noted the way different flies
behaved when attracted to these substances and suggested that maggots might
develop from the objects deposited on the meat by adult flies. He followed
the development of larvae from the eggs and noted that different kinds of
pupae gave rise to different species of flies. These studies were published in
1668 as *Experiments on the Generation of Insects.* While these experiments
did not altogether destroy the concept of spontaneous generation, they did
cause the transfer of the field of battle from the generation of macroscopic
creatures to the small new world of infusoria and animalcules. Their dis-
coveror, Antony van Leeuwenhoek, believed that his little beasties were living
organisms that arose from parents like themselves.

While Swammerdam and the preformationists rejected spontaneous generation as a contradiction of their encapsulation theory, other scientists revived
the idea of the great mathematician and philosopher Gottfried Wilhelm Liebniz
(1646-1716) of "monads" or living molecules. Most notable in this tradition
were Georges Buffon (1707-1788) and his friend, the English microscopist,
John Turbevill Needham (1713-1781). These two combined their not modest
talents to disprove the work of Louis Joblot (1645-1723). To prove that
infusoria are not spontaneously generated, Joblot boiled his medium and
divided it into two parts. One half was sealed off and the other left uncovered.
The open flask was soon teeming with life, but the sealed vessel was free of
infusoria. To prove that the medium was still susceptible to putrefaction,

Joblot exposed it to the air and showed that infusoria soon were actively growing. Thus, Joblot concluded that something from the air had to enter the medium to produce microorganisms. When Needham repeated Joblot's experiments, he found abundant "proof" for the theory of "living molecules" and spontaneous generation. In Needham's hands whether the flasks were open or closed, the water boiled or not boiled, infusions placed in hot ashes or not, all vessels "swarm'd with Life, and microscopical Animals of most Dimensions." Thus Needham concluded "there is a vegetative Force in every microscopical Point of Matter, and every visible Filament of which the whole animal or vegetable Texture consists . . ." When animals or plants died they slowly decomposed to one common principle which Needham viewed as "a kind of universal *Semen*." From this "Source of all" the "Atoms may return again, and ascent to a new Life."[6]

Needham's views were well known since he published them in the Philosophical *Transactions of the Royal Society of London* (1748). The claims of Needham and Buffon did not stand unchallenged for long. Lazzaro Spallanzani soon exposed the fallacious assumptions and ill-contrived experiments generally associated with advocates of spontaneous generation. As usual, Spallanzani's experiments were direct, simple, and logical. By heating a series of flasks for different lengths of time he determined that different microbes had characteristic susceptibility to heat. Whereas some "superior animalcula" were destroyed by slight heating (probably protozoa), other very minute entities survived in liquids boiled for almost an hour. Further experiments convinced Spallanzani that the animalculae entered the medium from the air. Although he believed there was a "vast variety of animalcular eggs, scattered in the air, and falling everywhere," Spallanzani cautiously concluded that the air "either conveys the germs to the infusions, or assists the expansion of those already there."[7]

Although Spallanzani's experiments directly answered many of the questions raised by advocates of spontaneous generation and proved the importance of sterilizing conditions, his critics claimed that he had tortured the all important "vital force" out of the organic matter by his cruel treatment of his media. The vital force by definition was capricious and unstable, rendering it impossible to expect reproducibility in experiments involving organic matter. A more serious objection was posed in 1810 by the French chemist Joseph Louis Gay-Lussac (1778–1850), who showed that the sterile vessels lacked oxygen and concluded that oxygen was necessary for fermentation and putrefaction. Fortunately for the food industry, Nicolas Appert (1750–1841), a French chef, ignored the theoretical aspects of the dispute and applied Spallanzani's techniques to the preservation of food. He placed foods in clean bottles, corked them tightly and raised them to the boiling point of water. His techniques were described in a book published in 1810 which heralds the beginning of the canning industry.

Ironically, during the nineteenth century, increasing knowledge of the process of fermentation and advances in chemistry led to further support for the theory of spontaneous generation. Although experiments against spontaneous generation also became more sophisticated, proponents of that doctrine still challenged the universality of negative experiments. Theodor Schwann repeated many of Spallanzani's experiments but added the refinement of heating the air as well as the medium. To prove that the "vital principle" had not been tortured out of the heated air, he proved that a frog could live quite happily when supplied with such air. Yet there were certain technical problems with Schwann's work and his results were not always consistent or reproducible. Other attempts to purify the air of microorganisms were conducted by Franz Schulze (1815–1873), who passed air through concentrated potassium hydroxide or sulfuric acid. Although his experimental results seem quite convincing, attempts to repeat them gave equivocal results and critics could still claim that the apparatus tortured the vital principle.

Heinrich Schröder (1810–1885) and Theodor von Dusch (1824–1890) introduced a new wrinkle into the debate by filtering the air entering the flask of putrefiable medium through a long tube of cotton-wool. While this treatment should have been gentle enough to satisfy the opposition, it was not uniformly successful. Indeed some test substrates invariably spoiled. Because any single case of apparent spontaneous generation could allow the proponents of the theory to maintain that it only occurred under special, favorable, perhaps even sympathetic conditions, the opponents were always on the defensive.

Pasteur vs. Pouchet

One of the most vigorous defenders of spontaneous generation was Félix Archimède Pouchet (1800–1872), Director of the Natural History Museum in Rouen, a member of many learned societies, and quite well known as a botanist and zoologist. In 1858 he began to present a series of papers to the Academy of Sciences of Paris, all purporting to show proof for spontaneous generation or "heterogenesis." In 1859 he published *Heterogenesis,* a massive treatise (about 700 pages) devoted to his philosophy and experiments. His writings amply justify Tyndall's characterization of Pouchet as "ardent, laborious, learned, full not only of scientific but of metaphysical fervour."

Since Pouchet believed it was self-evident that nature employed spontaneous generation as one of the means for the reproduction of living things, his experiments were designed not to determine *whether* heterogenesis took place, but only to discover the *methods* by which this takes place. Pouchet's assumptions were very similar to those of Buffon and Needham. He did not allow for the generation of life *de novo* from accidental aggregations of molecules or in solutions of mineral substances. Heterogenesis required the

Louis Pasteur holding a swan necked flask. [© 1962, Parke, Davis and Company.]

existence of a vital force coming from pre-existing living matter, and could only occur in solutions of organic matter which, originally formed in living beings, retained some vital properties. According to Pouchet, the factors which promoted heterogenesis were organic matter, water, air, and the proper temperature. Any alteration in these prerequisites could affect the kind of organisms produced. Thus, each factor should be eliminated in turn in a series of experiments to test its effect on heterogenesis. In Pouchet's hands, all the standard experiments on spontaneous generation gave positive results. Just as some gardeners seem to be blessed with a "green thumb," Pouchet apparently had a "vitalist thumb." Like his main opponent, Louis Pasteur, Pouchet was a vigorous and enthusiastic experimenter. Eventually, Pasteur forced his adversary to admit that the existence of germs in the air was the critical issue.

While Claude Bernard, Milne-Edwards, Dumas and others criticized Pouchet's work, it was Louis Pasteur alone who concentrated on the experimental basis of the debate. His work on fermentation had shown him that "ferments" are actually living organisms, and he believed that the air was the source of these germs. Rigorous experimentation, Pasteur argued, forced a reasonable person to the ineluctable conclusion that "in the present state of science . . . spontaneous generation is a chimera." Under present circumstances spontaneous generation of microorganisms does not occur. All previous "evidences" purporting to support the contrary proposition were the result of experimental artifacts and careless technique. Clearly Pasteur's experiments did not deal with the question of the ultimate origin of life, but proved that microbes do not arise *de novo* in sterile medium under conditions prevailing today.

In some 20 years of experimenting on the nature of ferments and battling the irrepressible heterogenesists, Pasteur studied the origin of ferments and germs in the air, their distribution in nature, and their care and feeding. His first priority was to demonstrate the germ-carrying capacity of the air. If the microorganisms associated with fermentation were brought to their substrate by the air, it should be possible to find them there. This was demonstrated by sucking air through cotton filters and trapping the dust particles. After washing the cotton in a mixture of alcohol and ether, the dust was collected and examined under the microscope. In addition to inorganic matter, various microbes were found. The actual number and kind varied with temperature, moisture, movement of air, and other environmental factors. Pasteur conducted numerous experiments to prove that the number of germs in air was quite variable, depending on season and locality. Hospitals were quite high in germ content while mountain air was low. Both Pasteur and Pouchet chased up and down mountains and climbed glaciers to test different kinds of air. Pasteur even considered a balloon ascent to get higher than Pouchet. Pouchet examined dust from many places, including the Natural History Museum, the roof of the Cathedral of Rouen, the banks of the Nile, the tomb of Rameses II, and the central chamber of the Great Pyramid of Gizeh. Needless to say, his observations were always quite different from Pasteur's.

One of the simplest and most convincing of Pasteur's experiments used the "swan neck flask." If liquids were sterilized in flasks with peculiar long necks drawn out and bent under a flame, the medium would remain sterile even though ordinary air could enter the flask. The germ-laden dust particles were trapped in the curve of the neck, and the medium remained sterile. But if the medium were tipped to slosh through the bend, or if the neck were broken off and dust allowed to enter, the medium soon seethed with microbial life.

Experimenting with various media, Pasteur proved that even the most easily decomposable fluids, such as milk, blood, or urine, could remain sterile if proper precautions were taken. He also showed that microbes could be grown on a simple defined medium, and that certain microbes grew only in the absence of oxygen. Admitting that even he could make mistakes, Pasteur argued that microbes remained true to their original type and were not simply transmuted into different species. He wrote:

> At a time when ideas on the transformations of species are so readily adopted, perhaps because they dispense with rigorous experimentation it is somewhat interesting to consider that in the course of my researches . . . I once had occasion to believe in the transformation of one organism into another . . . and that this time I was in error; I had not avoided the cause of illusion which my confined confidence in the theory of germs had so often led me to discover in the observation of others.[8]

These years of experimentation on the germ-laden dusts of the air affected some of Pasteur's daily behavior in a peculiar manner. He never used a plate or a glass without examining it and wiping it carefully. According to his biographer, Vallery-Radot, "no microscopic speck of dust escaped his short-sighted eyes. Whether at home or with strangers he invariably went through this preliminary exercise, in spite of the anxious astonishment of his hostess. . . ."

Through the sheer bulk of his successful experimentation and the pungency of his vituperations against his opponents, Pasteur effectively destroyed the reign of heterogenesis as a fashionable cause. In 1862, Pasteur was awarded a prize for this work by the Academy of Sciences, but Pouchet arose to challenge his conclusions. A commission was appointed to settle the debate.

Felix-Archimede Pouchet. [**From W. Bullough (1960).** *The History of Bacteriology.* **Courtesy of Oxford University Press.**]

Before the final confrontation between Pasteur and Pouchet, Pasteur discussed the question of spontaneous generation at a special lecture held at the Sorbonne (April 7, 1864). He told his audience of scientists and assorted celebrities that man was now thinking about great problems—"the unity or multiplicity of human races; the creation of man 1000 years or 1000 centuries ago, the fixity of species, or the slow and progressive transformation of one species into another; the eternity of matter; the idea of a God unnecessary." The subject of his lecture was equally momentous, but was uniquely capable of being resolved by logical and conscientious experimentation: "Can living beings come into the world without having been preceded by beings similar to them?" He showed his audience an infusion of organic matter, "as limpid as distilled water, and extremely alterable." If exposed to air the infusion would soon be full of microbial life. But a properly sterilized flask of the same material would remain clear. Showing the audience such a flask he dramatically intoned:

I have taken my drop of water from the immensity of creation, and I have taken it full of the elements appropriate to the development of inferior beings. And I wait, I watch, I question it, begging it to recommence for me the beautiful spectacle of the first creation. But it is dumb, dumb since these experiments were begun several years ago; it is dumb because I have kept it from the only thing man cannot produce, from the germs which float in the air, from Life, for Life is a germ and a germ is Life. Never will the doctrine of spontaneous generation recover from the mortal blow of this simple experiment.[9]

On June 22, 1864 a showdown between Pasteur and Pouchet was scheduled in Chevreul's laboratory in the Natural History Museum at the Jardin des Plantes. Pasteur prepared the requested 60 infusions for the commission, but Pouchet refused to carry out similar experiments and withdrew from the contest. Thus, Pasteur won that round by default but was often angered on hearing echoes of the old debate even in the 1870s. When Pasteur learned that Claude Bernard had expressed the opinion that living ferments might not be necessary for fermentation, Pasteur undertook a major experiment to disprove this. He had some hothouses built to seal off parts of his vineyard in August before yeasts become associated with the grapes. When harvested in October, these grapes did not ferment unless yeast was added. These experiments demonstrate Pasteur's gift for flair and showmanship as much as anything else.

The last major champion of spontaneous generation was Henry Charlton Bastian (1837–1915), author of a 1000-page tome modestly entitled *The Beginnings of Life*. "Dr. Bastian," according to Pasteur's student Duclaux, "had

some tenacity, a fertile mind, and the love, if not the gift, of the experimental method." For 40 years Bastian carried out his campaign for spontaneous
generation through numerous papers and six more books. Bastian out-did
other heterogenesists in his claims for "archebiosis"—the production of life *de
novo* from inanimate matter, even the simplest solutions. Bastian's extravagant
claims and persistence inspired Pasteur and his followers to even greater
heights of experimental rigor. One new insight was the great heat resistance
of certain germs.

In 1877 Pasteur wrote to Bastian:

> Do you know why I desire so much to fight and conquer you? It is be-
> cause you are one of the principal adepts of a medical doctrine which I
> believe to be fatal to progress in the art of healing—the doctrine of the
> spontaneity of all diseases. That is an error which, I repeat it, is harm-
> ful to medical progress. From the prophylactic as well as from the
> therapeutic point of view, the fate of the physician and surgeon depends
> upon the adoption of the one or the other of these two doctrines.[10]

He might have added, even more so the fate of the patient!

John Tyndall (1820–1893)

Further experimental refinements were introduced by John Tyndall, an English
physicist who had studied the optical properties of particulate matter—the
light-scattering or "Tyndall effect." Inspired by Pasteur's work, he aired his
views on the germs in the air at a lecture to the Royal Institution in London
(Jan. 20, 1870) and through the pages of *The Times*. Bastian viciously attacked Tyndall, warning him to leave the problem to biologists and physicians.
Tyndall's long series of experiments on sterilization methods resulted in a
book entitled *Essays on the Floating Matter of the Air in Relation to Putrefaction and Infection* (1881). Tyndall invented a chamber in which light
scattering by dust particles was used as a measure of the optical activity of
the air. A layer of glycerine was used to trap and hold dust particles and
render the air "optically empty." Test tubes were filled with the appropriate
infusions through a funnel and the tubes were lowered into a boiling water
bath. Among the fluids tested were urine, infusions of sole, cod, turbot,
herring, hare, rabbit, pheasant, grouse, mutton, beef, liver, and turnips. In
further tests of the thermoresistance of bacteria, Tyndall developed the method of fractional sterilization by discontinuous heating, known today as tyndal-
lization. He proved that active bacteria were easily destroyed by heat, but
that they might also exist in a heat-resistant spore form. After destroying the
active bacteria in the first step of sterilization, Tyndall allowed the infusion to

Tyndall's culture box. [**From W. Bullough (1960).** *The History of Bacteriology.*
Courtesy of Oxford University Press.]

cool and then heated it again. He found that discontinuous boiling for one
minute in five separate steps could sterilize an infusion which resisted one
hour of continuous boiling. These experiments, and Tyndall's elegant exposi-
tions, helped to destroy the idea of spontaneous generation and to establish
the germ theory of disease. Tyndall was awarded an honorary Doctor of
Medicine by the University of Tübingen.

Tyndall wrote to Pasteur that he had decided to render "a service to
Science, at the same time as justice to yourself" by taking up the battle with
Bastian who was sowing "confusion and uncertainty" among the English and
American public. In expressing his admiration for Pasteur's work, Tyndall
wrote:

> For the first time in the history of Science we have the right to cherish
> the sure and certain hope that, as regards epidemic diseases, medicine
> will soon be delivered from quackery and placed on a real scientific
> basis. When that day arrives, Humanity, in my opinion, will know how

to recognize that it is to you that will be due the largest share of her gratitude.[11]

It is quite remarkable how the physicist and the chemist gave such remarkable contributions to a largely reluctant and reactionary medical profession. Joseph Lister (1827-1869), founder of antiseptic surgery, was one of the few medical men to appreciate and apply Pasteur's work.

In 1870 T. H. Huxley coined the word "biogenesis" to express the idea that life always comes from pre-existing life. As the true nature of microorganisms was revealed it became increasingly obvious that they could not be simple blobs of organic material such as would arise from the putrefaction of dead organisms. Even the smallest bacteria are too complex to simply precipitate out of solution like crystals. The discovery of bacteriophage and other viruses stimulated a brief revival of the idea, but even viruses, while very small, are quite complex and have definite mechanisms for replication. Yet many still regarded spontaneous generation as a necessity in the sense that if life did not always exist on earth it must have been "spontaneously generated" at some point.

According to many scientists thinking seriously about the origin of life, somehow in the "primordial chicken soup" a process like spontaneous generation must have occurred. In 1924 A. I. Oparin, a Soviet scientist, proposed an explanation for the chemical evolution of life from the organic soup of the ancient seas when the earth was young and very different from today. Sidney Fox, Harold Urey, and Stanley Miller have investigated this possibility experimentally. Oparin stated that "the origin of life is not the result of some 'happy chance' . . . but a necessary stage in the evolution of matter."[12] (A major question in the exploration of the planet Mars has been the question of whether life also arose there.)

The Germ Theory of Disease

Historical Aspects

While Pasteur was certainly not the first to believe that epidemic diseases could be caused and transmitted by "germs," his great work provided proof for a scientifically grounded theory that was to revolutionize medicine. His work was carried out with clarity and simplicity, scientific rigor and a flair for the dramatic which reached and inspired many who could not really be touched by purely scientific arguments. Pasteur replaced ancient, vague, and confused notions of contagion with a scientific theory.

Primitive ideas about contagion actually are not directly related to disease at all. "Contagion" referred to the general notion of transfer through contact.

Just as heat and cold were transferred to neighboring bodies, so were putrefaction, uncleanliness, corruption, even disease. Epidemics were probably rare
in small bands of primitive peoples, but would have been terrifying events
once population density increased sufficiently to produce them. Many peoples believed diseases were sent by the gods as punishment for their sins, but
the Babylonians seem to have grasped the association of some diseases with
insect carriers. While some philosophers, poets, and architects expressed interest in the possibility that disease was acquired by contact with the sick person or contaminated articles, the Hippocratic writers had little interest in such
theories. The Hippocratic physician rejected the notion of contagion and its
primitive association with sympathetic magic. In seeking natural causes for all
phenomena, scientists gradually replaced supernatural or religious explanations
of disease with natural, though peculiar, causes such as comets, eclipses, floods,
earthquakes, or major astrological disturbances which charged the air with
"miasmata" or pollution.

Surprisingly, the germ theory of disease was elaborated by a physician as
early as 1546. Having observed the epidemics of syphilis, plague, typhus, and
foot-and-mouth disease which attacked Italy in the sixteenth century,
Hieronymus Fracastorius (ca. 1478-1553) asked himself, "What is contagion?"
His answer appeared in *Contagion, Contagious Diseases and their Treatment.*
(In 1530 Fracastorius had published a classic study of syphilis, *Syphilis, or the
French Disease,* which gave the disease its common name.) Beginning with an
analysis of the usage of the term contagion, Fracastorius proceeded to what
he considered a more scientific definition. Contagion, he stated, is "a certain
precisely similar corruption which develops in the substance of a combination,
passes from one thing to another, and is originally caused by infection of the
imperceptible particles." Three fundamentally different types of contagion
were possible. The first kind infected by direct contact only. The second
kind infected by contact and also by means of *fomites* which he defined as
clothes, linens, utensils, etc., that carried the original germs of contagion to
infect another victim. In the third category was contagion transmitted not
only by contact and fomites, but also capable of infecting at a distance.
Pestilent fevers, tuberculosis, certain eye diseases, and smallpox seemed to fall
into the third category.

Fracastorius compared the first class of diseases to the putrefaction or infection which occurs in fruits, spreading from grape to grape or apple to apple.
He noted that diseases might be quite specific, affecting only trees, or be peculiar to men, oxen, horses, etc. Some diseases infected only children, men
or women, or the aged. Some diseases were confined to a particular part of
the body while others invaded every corner. He then asked whether all infection is "a sort of putrefaction and also whether putrefaction is not itself infection. . .?"[13] The association of putrefaction, fermentation, and infectious
disease was to become a rewarding avenue of research.

Unfortunately, although Fracastorius retained an honored place in the medical literature, eventually his theories were forgotten. His works became classics—that is, physicians and scientists knew their titles, but no one read them. Nineteenth century scientists had to rediscover much of what Fracastorius had already known.

The invention of the microscope and Leeuwenhoek's discovery of minute animals led to a renewed interest in the old idea that invisible animals caused disease. Such ideas had been expressed by ancient Roman writers. In his encyclopedic writings, Marcus Terentius Varro (117–27 B.C.) discussed hygienic regulations for selecting building sites for houses. Swampy places were dangerous because small animals might live there. So minute as to be invisible, these animals "enter the body through the mouth and nostrils and cause grave disorders."

It may seem perverse to us that the merits of germ theory were so long ignored by the medical profession. Perhaps it would seem less a case of simple wrong-headed obtuseness if we recall that theories should be judged by their fruitfulness. As Hippocrates said, in the craft of medicine the proof of a theory should be its success in practice. In general, the idea of contagion through tiny "seeds" of disease did not prove very useful before the second half of the nineteenth century. (Why quite absurd and harmful theories so tenaciously maintained their place in the minds of the medical profession is another, and separate question.) The contagion theory led to attempts to stop epidemics by quarantine, isolation and disinfection, but these methods were not consistently successful. Indeed epidemiological measures against yellow fever, typhus, and cholera based on contagion theory were quite ineffective. It seemed more likely that these diseases spread by poisoned air which could not be contained by seclusion of the sick. Many very prominent scientists and physicians rejected the contagion concept in favor of the miasmata theory.

In 1840 Jacob Friedrich Gustav Henle (1809–1885), a prominent German pathologist, physiologist, and anatomist, revived the contagion theory and published his examination of the relationships among "miasmatic, contagious, and miasmatic-contagious diseases." He believed that malaria was a purely miasmatic disease, but smallpox, measles, scarlet fever, typhus, influenza, dysentery, cholera, plague, and puerperal fever were miasmatic-contagious. Diseases such as syphilis, mange, foot-and-mouth disease, and rabies were acquired only from contagion. Clearly separating the concept of "disease" from the concept of "parasite," he defined the material basis of contagion as organic and living a separate existence "which is related to the diseased body in the way that a parasitic organism is related to it." The contagion was not the disease but the inducer or cause of disease.[14]

Accepting the work of Cagniard Latour and Schwann on fermentation, Henle compared the action of the contagion to fermentation. Schwann and

Henle had been friends during their association with Müller. Henle's ideas stimulated the search for the unseen bodies of the parasitic organisms of disease. Quite prescient in his grasp of the difficulties confronting germ theory, Henle realized that finding some agent associated with a disease did not prove it had a causal role. The agent must be isolated and cultured—free from the diseased individual. Indeed, Henle came close to outlining what are usually called "Koch's postulates." A major problem in rigorously establishing the truth of germ theory was methodological. Recalling the alchemists' dream of transmutation, microbes seemingly changed their characteristics overnight. Obtaining pure cultures was a difficult and tedious procedure, almost impossible in the hands of any but the most meticulous experimentalists. Working out proper sterilization and culture procedures required a prior understanding and commitment to germ theory. Given the primitive methods used until rather late in the nineteenth century, it was no wonder new animalcules seemed to appear spontaneously in the tempting broths set out for them, as well as in the raw wounds of surgical patients.

Pasteur and the Germ Theory of Disease

Pasteur keenly understood the relationship between the acceptance of spontaneous generation and the retardation of medicine and surgery. Advocates of spontaneous generation like Bastian and Pouchet had to be utterly vanquished or their doctrine would be a perpetual source of conflict between scientists supporting the germ theory of disease and conservative physicians and surgeons. Confused by the claims of the heterogenesists, physicians tended to reject the whole germ theory or dismiss it as a laboratory curiosity, unrelated to clinical medicine.

In a totally unexpected sequence the study of crystals had equipped Pasteur for studies of ferments and the "diseases" of wine, beer, and vinegar. Soon he was called on to rescue the silk industry of France from a mysterious disease causing great despair and hardship among the sericulturists of France. Although Pasteur had never before seen a silkworm and knew nothing about the creatures, he set out in his usual energetic fashion to learn all he could about this new disease. Between 1865 and 1870 he "sacrificed" himself to a long, complicated study of the problem. More than a simple "germ" was involved—two different disease organisms as well as nutritional and environmental effects had to be disentangled. By unravelling the causes of the silkworm diseases, Pasteur was able to suggest measures the industry could take to rid itself of the problem. Soon the Pasteur method of careful microscopic examination was allowing the distribution of pure "seed" (as the eggs of the silkworm were called) throughout Europe and Japan. Obviously, even in the lowly silkworm, "disease" involves complex interactions among host, germ, and environment.

From silkworm diseases, Pasteur progressed to the riddle of disease in higher animals, and finally, to his triumph over rabies in man. The progress of his work was hardly interrupted by a series of tragedies that saddened his personal life—the deaths of his young daughters, his son's war experiences, even a stroke that left him partially paralyzed for the rest of his life.

On October 19, 1868 Pasteur noted a peculiar tingling sensation on his left side. The first symptom was followed by complete loss of speech and a paralysis which came and went. It was soon apparent that he had suffered a cerebral hemorrhage which gradually robbed him of the power of movement on the entire left side. His physicians prescribed the application of 16 leeches behind the ear—which caused the blood to flow abundantly. Throughout the experience his mind remained clear and active and he amazed his doctors with his desire to talk about science. He told a friend: "I am sorry to die; I wanted to do much more for my country." During the week when death seemed inevitable he dictated a memorandum on an ingenious test for detecting diseased silkworm eggs at a very early stage. By the end of January he had returned to his research on the silkworm problem. A fall on a stone floor retarded his recovery, but not his work.

Vaccine for Anthrax and Chicken Cholera (1877-1879)

From 1877, after 20 years of research on microbes, Pasteur studied the diseases of higher animals and man. Despite his excellent accomplishments, he faced great hostility from the conservative medical and veterinary professions. He was regarded as an outsider and a revolutionary. Even if his work was accepted as a "laboratory exercise" in "pure science," certainly, experienced medical men argued, it was of no account in the "real" world. Even as an obscure physician in Germany named Robert Koch, of whom more later, was providing a rigorous demonstration of the etiology of anthrax, Louis Pasteur set about saving French cattle, sheep, and farmers from the "curse" of charbon.

Agriculture in many provinces of France was being ruined by a mysterious disease variously called charbon, splenic fever, or anthrax. In the province of Beauce about 20 of every 100 sheep succumbed to the disease, while in Auvergne some 10-15 percent would die. Some farms seemed especially cursed by the scourge and were known as "charbon farms." When the disease struck, sheep would lag behind the others with drooping heads, quivering and gasping until paralysis set in. Bloody discharges would follow with death occurring sometimes just hours after the onset of the first symptoms. The dead animal would rapidly become distended. Autopsies revealed black, thick, viscid blood, and a black and liquid spleen. The disease struck one area in Russia with such virulence that 56,000 cattle died between 1867 and 1870 along with many horses, sheep, oxen, and 528 people.

Several microscopists noted the presence of rod-shaped bacteria in the blood of anthrax victims. Rather large for a bacillus, the causal agent of anthrax appears in the bloodstream in such large numbers it can easily be found by direct examination of blood. In 1850 Casimir Joseph Davaine (1812-1882) and P. F. O. Rayer (1793-1867) identified "filiform" bodies in the blood of sheep which had died of the disease. After reading Pasteur's paper on butyric ferment, Davaine began to think his observations might have been significant. In 1863 Davaine examined blood from sheep which had died of anthrax and inoculated some rabbits with the blood. If the filiform bodies were not present, anthracic blood was not infectious. Pursuing the relationship between bacteria and disease, Davaine identified the same rod-shaped bacteria in the "malignant pustule" characteristic of the severe localized human form of anthrax. In other experiments, Davaine diluted anthracic blood with water. After allowing the bacteria to settle for 24 hours, he injected guinea pigs with either the clear supernatant liquid or the lower—bacteria containing—layer. Only the lower layer transmitted the disease. Despite Davaine's research and his vigorous support for the germ theory of disease, there were many loopholes in his work which critics eagerly exaggerated. Furthermore, other scientists failed to find filiform bodies in rabbits after the injection of anthracic blood.

Even after Koch's classic work on the etiology of anthrax the battle continued. In 1877 Paul Bert (1833-1886) announced that oxygen could destroy the anthrax bacillus, but the blood would still cause the disease. Therefore, he argued that the bacillus was not the cause of anthrax. These controversies drew Pasteur into the question. Carrying out some 100 transfers he purified the bacteria in a medium of either broth or urine. Only a living organism which had multiplied through all those transfers could be the virulent agent. The dilution had been so great that no inanimate poison from the original sample could remain. Indeed the germs multiplied so well in culture that as Pasteur described the process it "ends by filling the whole liquid with such a thickness of bacteridia that, to the naked eye, it seems that carded cotton has been mixed with the broth."[15] As in Davaine's experiment, Pasteur demonstrated that if the bacteria were filtered out or allowed to settle, the remaining clear liquid was harmless.

Acknowledging the work of Davaine and Koch, Pasteur clarified some of the conflicts and ambiguities which had remained. Some confusion arose because experimenters used long-dead carcasses. Therefore, their guinea pigs were killed by the germs of putrefaction rather than anthrax. Since the "septic vibrio" was an anaerobe, it was lost in cultures exposed to air. Thus Pasteur convinced Paul Bert that he had been mistaken in the interpretation of his experiments.

Anthrax and Chickens

An opponent of Pasteur's theories had disputed his statement that hens did not take anthrax. Yet Dr. Colin failed to provide Pasteur with a hen that suffered from the disease. His excuses had a hollow ring—such as the claim that a dog had eaten his experimental chicken. Having wrung a confession from his very discomfitted adversary, Pasteur demonstrated to the Academy of Medicine how a hen could be subjected to anthrax. Pasteur hypothesized that hens did not take anthrax because their body temperature was several degrees above that of animals which did. He tested this by immersing an inoculated hen in a water bath. This hen soon took the disease and died. As his control, Pasteur presented two living hens. One had been placed in a similar bath but was not inoculated while another had gotten twice the inoculum but no bath. In another experiment a hen was inoculated, kept in a bath until diseased, and then removed, dried, and warmed. This hen recovered completely.

Despite these ingenious experiments, physicians and veterinarians still denied their relevance to medicine. They argued that diseases came from within the animal and created septic poisons which might then turn into germs.

Chicken Cholera and the Beginnings of Scientific Immunology

In addition to his studies of anthrax, Pasteur took up the problem of chicken cholera. The disease is unrelated to human cholera, except in the virulence of the infection and the rapidity of death. After staggering about lethargically or standing motionless, hens would fall dead the next day. Other chickens picking at foods soiled with the excreta of the dead animals were soon similarly affected. Sometimes 90 percent of a flock would die. Several veterinary surgeons had noted "granulations" or tiny specks in the corpses of such animals. After several attempts Pasteur found that the microbe could be cultured in a medium made of chicken gristle. A small drop of fresh culture would quickly kill a chicken, but the microbe only caused a small sore or abscess in guinea pigs.

One of the most celebrated of chance observations arose from Pasteur's studies of this humble disease. When some chickens were inoculated with an old culture, they barely sickened and then recovered completely. When inoculated with a fresh, virulent culture these hens were quite resistant to the disease. Soon it was determined that the virulence of the microbe could be attenuated by storing the culture for variable intervals until all virulence was lost. Nonvirulent cultures could be propagated in that condition. Pasteur could then give his experimental chickens an acquired immunity to chicken cholera. As Pasteur recognized, the use of vaccine for chicken cholera was analogous to Edward Jenner's use of cowpox vaccination to provide immunity to smallpox.

Vaccine vs. Anthrax

Having successfully developed a method of preventing chicken cholera, Pasteur set about doing the same for anthrax. Using techniques similar to those developed in the study of chicken cholera, Pasteur and his associates soon were able to tame the anthrax bacillus. This was done by cultivating the bacteria in chicken broth at a temperature of 42-43°C where spores could not develop and selecting nonvirulent strains.

In an example of his gift for the dramatic and his readiness to meet the opposition in open battle, Pasteur participated in a great public demonstration of his anthrax vaccine. Just one month before Pasteur's announcement of the anthrax vaccine, the editor of the *Veterinary Press* issued a sarcastic challenge: "Will you have some microbe? There is some everywhere. Microbiolatry is the fashion . . . henceforth the germ theory must have precedence of pure clinics; the Microbe alone is true, and Pasteur is its prophet."[16] Several months later the same editor began a campaign for subscriptions to a test, or more rightly, a contest, of Pasteur's vaccine. Some of Pasteur's friends and associates were a bit nervous about the conditions of the trial, but Pasteur assured them, "What has succeeded in the laboratory on fourteen sheep will succeed just as well at Melun on fifty." The Melun Agricultural Society contributed 60 sheep to the test. Although his experiments on cows were still in the preliminary stage, Pasteur agreed to include cows in the trial. The experiments began on May 5, 1881 before a great crowd including many skeptical physicians and veterinarians. Setting aside 24 sheep, 4 cows, and 1 goat as control animals, Pasteur inoculated 24 sheep, 6 cows, and 1 goat with five drops each of an attenuated strain of anthrax bacillus. On May 17, Pasteur's associates, Chamberland and Roux, reinoculated the test animals with a somewhat more virulent culture.

(Pasteur and his associates were not idle while waiting for the results of the great anthrax show, but were hard at work towards conquest of another disease—one of the most dreaded afflictions at the time—hydrophobia or rabies.)

Although Pasteur did not know it, one of his adversaries suggested that Pasteur would attempt to cheat in the final phase of the experiment by using an unmixed culture where the top would be harmless liquid and the bottom full of active bacteria. On May 31, when all animals were to be inoculated with a highly virulent strain, the "master of ceremonies" took it upon himself to violently shake the virulent culture and requested that a triple dose be given. Other veterinary surgeons requested that the virulent culture be injected alternatively into vaccinated and unvaccinated animals, rather than doing each group separately. Calmly, Pasteur went along with all these picayune demands. Pasteur suffered agonies of anxiety over the slightest signs of illness among the vacinnated animals, but soon he was joyously reporting to his son and daughter that the control animals were dying while the vaccinated animals were all well.

Early in June the contest was over and Pasteur was secure in a "stunning success." Many skeptics came to Pasteur as "converted and repentant sinners."

Some even volunteered for anthrax vaccination themselves. "Joy reigns in the laboratory and the house," wrote Pasteur.[17]

Pasteur's Critics

The impact of Pasteur's demonstration was so great that by 1894 3.4 million sheep and 440,000 cattle had been vaccinated against anthrax. Of these only about one percent of the former and 0.3 percent of the latter died of the disease. Still there were many who questioned the value of Pasteur's work. Not all critics were ignorant opponents of germ theory. Robert Koch used the *Record of the Works of the German Sanitary Office* to publicize his dispute with Pasteur. Koch charged that Pasteur was incapable of cultivating pure strains of microbes, ridiculed his discovery of the role of earthworms in propagating anthrax, his experiments on anthrax in chickens, and even the preventive value of vaccination.

Assisted by Chamberland and Roux, Pasteur began to study conditions in French hospitals. His associates noted the extreme revulsion that attacked Pasteur when visiting hospitals and observing autopsies. Although he often went home ill from such experiences, he would return the next day hoping to make some new contribution to science. He became particularly interested in puerperal fever, a disease that was killing many new mothers in the lying-in hospitals of Paris. Soon Pasteur had found a microbe in puerperal infections. When one of his pompous but ignorant colleagues was holding forth on the causes of epidemics in maternity hospitals, Pasteur could not restrain a challenge to established medical "wisdom." Although he was "merely" a chemist, he announced, "It is the nursing and medical staff who carry the microbe from an infected woman to a healthy one."[18] Unlike his predecessors in the battle against child-bed fever, Phillip Ignaz Semmelweiss (1818-1865) and Oliver Wendell Holmes (1809-1894), Pasteur knew very well what transmitted the infection, for he had seen the microbe in blood taken from victims of the disease.

Sometimes the disputes between Pasteur and his reactionary colleagues provoked explosions of temper. At one session of the Academy of Medicine, Dr. Jules Guerin jumped from his seat and rushed to attack Pasteur. A fight was barely avoided and the meeting quickly adjourned. The next day Guerin sent his seconds to demand Pasteur assuage his honor with a duel. Showing more restraint than usual, Pasteur offered an apology to end the quarrel. It is difficult to imagine the outcome of the duel so narrowly averted—Pasteur had never fully recovered from his stroke and Guerin was an octogenarian.

Rabies

Although not a common disease, rabies exerted a powerful grip on the popular imagination. Pasteur himself said he remembered the howls of wolves in the countryside and the screams of a man cauterized after the bite of a rabid animal. Generally no matter what the treatment, the disease was fatal and gruesome in its progress. Pasteur began his work on a five-year-old victim of rabies in the Sainte Eugenie hospital of Paris. Collecting saliva four hours after the child's death, he injected it into two rabbits. They died of a disease which could be transmitted to other rabbits. Here Pasteur was at first misled by finding a peculiar microbe shaped like a figure 8 in a kind of capsule. Soon he realized that the same organism appeared in the saliva of normal individuals and could not be causally connected with rabies. (The microbe actually is a pneumococcus which can be transmitted to rabbits.) No visible agent could be found associated with rabies.

Pasteur and his associates found that the long incubation period which followed subcutaneous inoculation (or dog bite) could be shortened by inoculation directly under the dura mater of dogs or rabbits. In 1884 Pasteur described his studies of the mechanism of action of rabies and his method of protective inoculation for dogs. Finally in 1885 came the grand announcement that dogs could be protected after being bitten by a rabid dog. Pasteur was able to prepare a rabies vaccine by suspending spinal cords of rabbits which had died from rabies in dry, sterile air. After about two weeks the material became virtually harmless. By proceeding from harmless to more virulent material in a series of inoculations, dogs could be protected even after being bitten by rabid animals.

On July 6, 1885, 9-year-old Joseph Meister was brought to Pasteur after an attack by a rabid dog left him bitten on hand, legs, and thighs. After doctors pronounced his case to be hopeless, Pasteur began the first human test of rabies vaccine—60 hours after the bites had been inflicted. From material attenuated for 14 days, Pasteur proceeded to inject successively more virulent material. The last inoculation on July 16 contained material removed from a dead rabbit only one day before. Meister recovered completely and later became Concierge at the Pasteur Institute. Pasteur's second case, Jean Baptiste Jupille, a 15-year-old shepherd boy, was also successfully treated. After these first successes, people from all over Europe were sent to Pasteur, the only man in the world who could save them from rabies. By October 1886 about 2500 people had been treated. Pasteur estimated that the failure rate had been about 1 in 170. By 1935 some 51,057 people had been inoculated at the Pasteur Institute, Paris. The mortality was about 0.29 percent or 151 deaths.

Even after Pasteur's great victory over rabies, he still had detractors who found all his work worthless and a subject for ridicule. There were anti-Pasteurian lecture series and an International Society of Antivaccinators. On the humorous side, in 1887 Paul Boullier asked the question, "Do you know what is the new treatment of M. Pasteur?" "It consists in making from sixteen to *forty inoculations* depending on the gravity of the bite," came the answer. "Is that called *inoculate*? I call that *tattoo*."[19]

Pasteur sacrificed much in terms of personal wealth and comfort for the love of his country and his humanitarian ideals. Although he took out patents for the process of pasteurization, and could have made a fortune from industrial applications of his work, he released the patents for the general good and did not make any money from them. When he was offered a chair of chemistry at the University of Pisa, with very attractive conditions, he turned down the offer to remain and encourage French science. Deeply troubled by the French defeat in the Franco-Prussian War, he blamed the neglect of science for the loss and attempted to raise the level of French science and industry.

Pasteur's work on rabies brought him international acclaim. In 1888 the Pasteur Institute was dedicated to him. Unfortunately, by then Pasteur was so ill and weak he could not speak himself at the ceremonies. The speech Pasteur's son read to the assembled dignitaries read in part:

> . . . Two contrary laws seem to be wrestling with each other nowadays: the one, a law of blood and of death, ever imagining new means of destruction and forcing nations to be constantly ready for the battlefield—the other, a law of peace, work and health, ever evolving new means for delivering man from the scourges which beset him.[20]

To the workings of the second law, Pasteur had dedicated his life.

Robert Koch (1843–1910)

Very different from Pasteur in training, temperament, and methodology was Robert Koch, the German physician who most precisely formulated the principles and techniques of bacteriology. Koch was born in a small village in the Harz Mountains to hard-working, diligent, and highly fertile parents. Robert was the third of 13 children. Although Robert was only a mediocre student, he displayed great enthusiasm for nature studies, avidly collecting plants, insects, and minerals. Even though he took care of the family chickens, cow, pig, horse, rabbits, and guinea pigs, he had time for his hobbies, photography, chess, and music.

His teachers at the gymnasium ranked him from satisfactory to very good, but the family saw promise enough to send him to the University at Göttingen, where he first studied mathematics, but transferred to the medical program. Although bacteriology was not then taught at the University, Koch later said, "I owe a debt of gratitude to my former teachers . . . who awakened my interest in research." Even as a student, Koch demonstrated a talent for research, winning the university prize of 30 ducats for a paper entitled "Ganglion Cells in the Nerves of the Uterus." The money was used to finance a trip to a scientific meeting of the Society of German Naturalists and

Physicians. He learned about the hardships involved in research when he served as an "experimental animal." This required eating such delicacies as 20 to 30 grams of potassium malate, a pound of asparagus, and a half-pound of butter per day for five days. After graduating eximia cum laude in 1866, Koch spent 12 years in virtual exile from the scientific world.

Koch debated with himself over his future career. A desire for travel suggested work as a ship's doctor or military surgeon, but circumstances eventually forced him into a career as a rural physician. While visiting his family, he became engaged to Emmy Fraatz after a courtship described as "not overly emotional." During Koch's temporary assistantship at the General Hospital of Hamburg the city was struck by a cholera epidemic. A visitor described seeing Koch studying some "intestinal specimens while his breakfast 'mush' was sitting on a nearby table."[21]

The next position the young physician obtained was as house physician in an asylum for idiot children at Langenhagen. He was also allowed a private practice and was soon making enough money to marry Emmy. Unfortunately, in 1868 his job was eliminated by an administrative reshuffling and Koch had to earn his living in progressively more and more isolated areas. His interest in research continued, despite the burden of his private practice. A small laboratory was set up next to his consultation office and furnished with an incubator, sink, darkroom, and work bench. Koch used aniline dyes to stain

Robert Koch. [**From W. Bullough (1960).** *The History of Bacteriology.* **Courtesy of Oxford University Press.**]

his preparations, a Zeiss oil immersion microscope, and the illuminating apparatus invented by Abbe. The combination, in his skilled hands, allowed him to get clear images of bacteria and distinguish what he called the "structure picture" from the "color picture."

Studies of Anthrax

Although Koch was isolated from the scientific community, he read the works of Pasteur, Lister, and others and appreciated their import for medicine. Thus, he was well aware that anthrax was one of the diseases then serving as a focus for proponents and opponents of germ theory. Fortunately for Koch, the disease was raging among the animals in his district. Since the disease attacked farmers as well as cattle and sheep, such studies were within Koch's administrative responsibilities.

Various investigators at the time were finding microbes in diseases humans and animals but, since they could not cultivate pure strains, germ theory remained a controversial hypothesis. Koch realized that the germ theory of disease was not only obstructed by the obstinacy of conservative physicians and scientists, but it was also hurt by ardent and careless workers whose speculations exceeded their experimentation. When he was sure his own results provided unequivocal evidence that a particular bacterium caused anthrax, Koch was ready to publish. To reach the scientific community, Koch applied to Ferdinand Cohn for advice and criticism. Cohn (1828–1898), a botanist, was the first prominent German scientist to take an interest in microbiological research. Testing the sensitivity of microbes to different means of sterilization, Cohn had discovered a new species, *Bacillus subtilis,* which formed extremely heat-resistant spores.

In 1876, Koch wrote to Ferdinand Cohn and requested his advice and criticism before submitting his work for publication. Since he could not preserve his experimental results, he offered to demonstrate the critical experiments over a period of several days at the Institute of Plant Physiology. Receiving a favorable reply, Koch brought his microscope, animals, and equipment to the University at Breslau. Cohn and his associates were very impressed by Koch's demonstration. Later Cohn remembered: "Within the very first hour I recognized that he was the unsurpassed master of scientific research." Koch's landmark paper "The Etiology of Anthrax Based on the Developmental Cycle of *Bacillus anthracis*" was soon published in Cohn's journal, *Contributions to Plant Biology.*[22]

Because bacteriology was such a rudimentary science, Koch's apparatus, instruments, methods of inoculation and observation were all improvised and innovative. Early attempts to infect mice on the ear or middle of the tail were unreliable because the uncooperative rodents sometimes succeeded in removing the inoculum by licking or rubbing. Inoculation into an incision at the base of the tail or back proved more suitable. When using fresh anthracic material, Koch was able to produce anthrax quite reliably in his experimental animals. Even after twenty passages through mice the characteristic symptoms always appeared; the spleen was enlarged and numerous little rods were found in spleen and blood.

The life cycle of *Bacillus anthracis* proved quite complex, but such knowledge immediately threw light on the mystery of the disease in nature. When grown in a suitable medium in the presence of air, the bacilli formed long, unbranched threads and many spores. This transition could be followed under the microscope. The discovery of spore formation finally explained the mystery of the persistence of anthrax in "cursed" pastures. Anthrax spores were very resistant to harsh conditions and could survive in the soil for years.

Koch improvised many simple, ingenious, and elegant techniques to prove that it was specifically *Bacillus anthracis,* and not some poison in the blood, which was the agent of the disease. Bits of spleen were cultured in the aqueous humor of a cow. Observations conducted every 10-20 minutes showed the remarkable growth of the bacilli into networks of filaments. Continuous observation of the free end of a filament revealed the transformation of the "strongest and most luxuriantly growing filaments" into "strongly refractile, oval-shaped spores." Once the spores had become separate, the preparations would remain unchanged for weeks. The spores could be transformed into the original active form under specific conditions of temperature, nutrient medium, and access to air. Although fully convinced that experiments with small rodents elucidated the life cycle of *Bacillus anthracis* and the etiology of the disease, Koch scrupulously noted the difference between the natural disease of ruminants and the laboratory version. While several days usually elapsed before the death of large animals, infected laboratory mice died within 24-30 hours. Furthermore, the effects of *Bacillus anthracis* were found to be quite distinct in various species.

Knowledge of the life history of *Bacillus anthracis* made it quite unlikely that the disease was naturally transmitted by active bacilli, yet this could occur in people who slaughtered, butchered, or skinned infected animals. In animals, the disease could be transmitted by dried bacilli which might remain viable for about five weeks, but spores were the most general means of disseminating the disease. Spores, Koch reported, were "incredibly resistant, surviving year-long dryness, moisture, or alternate wetting and drying." One corpse of a creature that had died of anthrax could serve as a source of innumerable spores. Koch proved that the bacillus itself and not some poison in the blood or tissue of infected animals was the cause of the disease by performing serial transfers through beef aqueous humor. After 10-20 transfers, a pure bacterial culture would kill a mouse just as quickly as blood taken directly from a diseased sheep. In his enthusiasm for completeness, Koch tested his cultures on mice, guinea pigs, rabbits, and sheep. In all cases, the pure laboratory culture transmitted anthrax.

Having satisfactorily elucidated the role of *Bacillus anthracis* in the etiology of the disease, Koch proposed various prophylatic measures to prevent further injury in animals and men. The first step was the proper disposal of cadavers. Sheep buried in shallow wet graves served as excellent foci for the development and dissemination of spores. Since his research proved that spore formation required air, moisture, and a temperature above 15°C, at least one of the requirements had to be eliminated. Since the average soil temperature at a depth of 1–10 meters in Germany was below 15°C, burial far from farms in deep dry trenches would render the cadavers harmless.

The recommendations are of interest in showing how a combination of practical and theoretical knowledge could serve agriculture and medicine. Surveys of the disease pattern in sheep, cattle, and horses suggested that sheep were the normal reservoir of infection. Simply separating other animals from sheep interrupted the chain of infection. Koch then argued that typhoid fever and cholera were similar to anthrax in mode of dissemination. The problem with investigating these two diseases was lack of an experimental animal. He concluded: "Despite these obstacles we should not be deterred from proceeding as far as available methods can carry us. One should first investigate the problems with attainable solutions. With the knowledge thus gained, we can proceed to the next attainable objectives. Diseases such as diptheria, which can be transmitted to animals, appear immediately amenable to successful investigation. With a knowledge of comparative etiology of infectious diseases we can learn to hold at bay the epidemic diseases of man."[23] Apparently even as an obscure country doctor studying anthrax among sheep and cattle, Koch had conceived a magnificent program to eliminate epidemic diseases of man.

Wound Infection

Even after his contemporaries acknowledged Koch as a master of scientific method, he was still obliged to support his family with a time-consuming private practice and his job as district medical officer. Still working in his improvised home laboratory, Koch turned to the study of "traumatic infective diseases." By this time, scientists generally accepted the idea that traumatic fever, purulent infection, putrid infection, septicemia, and pyemia were essentially of the same nature, although disputes continued as to whether the microorganisms found in these infections were the cause or result of disease. Part of the problem was the difficulty of determining whether microbes were present in the blood of healthy animals. With improved illumination and suitable stains Koch was able to demonstrate that *"bacteria do not occur in the blood nor in the tissues of the healthy living body either of man or of the lower animals."*[24]

When "putrid materials" were injected into mice, various septic diseases were produced, but the fundamental cause was always bacterial infection. Koch found that the clinical picture "varied with the nature and quantity of the putrid fluid. Blood and meat infusion putrefied for a long time are less injurious than those putrefied for a few days." Koch was able to perform a series of 17 passages with edema fluid or blood from infected mice. Although the blood of septicemic mice was quite virulent, his preliminary experiments did not reveal the causal microbe; more refined techniques were needed to reveal it.

Having observed the subtleties of gangrene in mice, spreading abscess, pyemia, septicemia, and erysipelas in rabbits, Koch concluded that the criteria for associating the diseases with the microbes had been adequately met. The research had obvious clinical importance because these experimental diseases "resembled human traumatic infectious diseases with regard to origin from putrefying substances, and character at post mortem examination."

Each distinct disease in Koch's study was invariably associated with a definite form of microorganism. For Koch the most telling evidence for the applicability of germ theory to traumatic infective diseases was "the differences which exist between pathogenic bacteria and the constancy of their characters." While many distinguished scientists believed bacterial forms were easily interconvertible, Koch maintained that "the differences between these bacteria are as great as could be expected between particles which border on the invisible." Above all Koch emphasized the importance of using pure cultures. The appearance of new forms should not be regarded as proof of transformation of type, but more likely a break in proper technique which allowed the introduction of contaminants. Before rigorous methods for bacteriology were worked out, even Koch denied the possibility of obtaining pure cultures of the smallest, nondescript bacteria. However, "pure" cultures could be cultivated "in the animal body." Before Koch and others developed new methods for the study of bacteria there was truly "no better cultivation apparatus than the animal body itself."[25]

Only one year after the publication of his paper on traumatic infective diseases, Koch had developed most of the methods needed for the cultivation of separate species of bacteria outside the animal body. Having accomplished so much in rural isolation, Koch was finally appointed to a position commensurate with his genius for research. In 1880 Koch joined the staff of the Imperial Health Office in Berlin, where he was soon directing many brilliant young associates.

The Methodology of Modern Bacteriology

Obtaining pure cultures, the most needed improvement in methodology, became Koch's obsession. "I do not believe it is too much to say," he wrote, "that the most important point in all studies on infectious diseases is the use of pure cultures." A new way of obtaining pure cultures was based on the use of solid media. Slices of potato were first used to prepare pure cultures. Experience soon proved that some pathogenic bacteria refused to grow on them, but Koch realized that the principle of solid medium could be adapted to other conditions.

After many unsuccessful experiments, Koch concluded it would be impossible to construct "a universal medium . . . equally suitable for all microorganisms."[26] Therefore, he searched for a means of converting the usual nutrient broths to a form that was firm and rigid. Medium solidified by the addition of gelatin was useful for cultivating many different organisms. Koch found it advantageous to imitate the shape of potato slices with his nutrient gelatin. However, gelatin has disadvantages; it melts at body temperature and can be liquified at room temperature by some bacteria. An ancient Far Eastern culinary technique soon came to the rescue. Frau Hesse, the wife of one of Koch's associates, suggested the use of agar-agar which her mother used in making jellies. A Dutch friend who had been to Java had brought the use of agar back to New Jersey. Nutrient agar remains solid up to 45°C and is inert to bacterial digestion.

In 1881 during the International Medical Congress in London, Koch demonstrated his methods at Lister's laboratory. Pasteur said to Koch, "C'est un

Robert Koch's laboratory. [**From W. Bullough (1960).** *The History of Bacteriology.* **Courtesy of Oxford University Press.**]

grand progres, Monsieur."[27] This was one of the few friendly remarks that
ever passed between the Frenchman and the German.

Koch and Tuberculosis

There was another side to Robert Koch, indefatigable researcher; this obstinate
and secretive nature was revealed quite clearly in his work on tuberculosis and
the fiasco into which his "tuberculin cure" degenerated. Returning from his
triumphant reception at the International Medical Congress of 1881, he began
to study tuberculosis. If Pasteur had taken rabies, one of the most dramatic
but rare threats to human life, as his greatest challenge, Koch chose tuberculosis, a much more commonplace, indeed almost ubiquitous, scourge, for his
greatest work.

Working feverishly, Koch was ready to present his preliminary report to the
Physiological Society in Berlin on March 24, 1882. As one observer described
the event, Koch spread out 200 preparations "like a cold buffet" before his
audience. The room was filled to capacity, but the grand arbiter of German
scientific medicine was not there. Rudolf Virchow remained sceptical of the
"juvenile work" of the bacteriological school of research. Because of Virchow's
hostility, Koch's paper was not presented to the Berlin Medical Society. The
preliminary report was such a masterpiece of scientific research that Paul
Ehrlich recalled, "All who were present were deeply moved and that evening
has remained my greatest experience in science."[28]

Tuberculosis was clearly a terrible menace in the nineteenth century, as
Koch reminded his audience:

> If the number of victims which a disease claims is the measure of its
> significance, then all diseases, even the most dreaded infectious diseases,
> such as bubonic plague, Asiatic cholera, etc., must rank far behind tu-
> berculosis. Statistics teach that one-seventh of all human beings die of
> tuberculosis, and that, if one considers only the productive middle-age
> groups, tuberculosis carries away one-third and often more of these.[29]

For many reasons the tubercle bacillus was particularly difficult to observe
and grow in pure cultures. Special fixation and staining methods were needed
before the "fine rod-like forms" could be distinguished in the tuberculous
tissues. The organism is only a tenth the size of the anthrax bacillus and is
covered by a waxy layer which made staining very difficult. Great patience
and a firm conviction that tuberculosis was caused by bacteria were needed in
these studies because the organism grew very slowly. Although tuberculosis
could be found in the cow, horse, pig, goat, sheep, chicken, monkey, rabbit,

and guinea pig, not all strains of the bacillus are pathogenic for each species. Fortunately, the human infection could be transferred to a convenient laboratory animal, the guinea pig.

Having discovered the agent of the human disease, Koch considered the problem of the propagation of the bacillus. Fortunately, it did not form spores as easily as the anthrax bacillus. The tubercle bacillus could only grow in specially constituted medium at temperatures greater than 30°C, but even then several weeks were needed for good growth on solid media. Therefore, these microbes were "true parasites, dependent upon the animal body for survival" rather than "facultative parasites, like the anthrax bacillus." The pulmonary form of tuberculosis was the most common and also the "most efficient source of dissemination." Koch found that patients with pulmonary tuberculosis "discharge large amounts of sputum containing great numbers of . . . tubercle bacilli. . . . The infectious germs are thus disseminated on floors, lint, etc. . . . a considerable number remain viable and virulent for many days, or even months."[30] This was especially true in poorly ventilated, dirty, dusty tenement rooms where the cleansing rays of sunlight rarely penetrated. Tubercular domestic animals were regarded as of minor etiologic importance, since they do not discharge sputum. Although the milk of infected animals could be a source of infection, Koch underestimated this danger.

Koch's Postulates

During his work on tuberculosis, Koch formalized the criteria needed to prove unequivocally that a particular microbial agent caused a specific disease. "Koch's postulates" were first made explicit in 1884 in "The Etiology of Tuberculosis." First it was necessary to find the microbe invariably associated with the pathological condition. A thorough study of the relationship of the microbe to the history of the disease might be sufficient proof of the causal relationship. However, skeptics might argue that even if the microorganism was associated with the disease, it was not the cause of disease. Rigorous proof of a causal relationship required complete separation of the parasite from the diseased animal and all possible products of the disease. After the alleged disease-producing organism had been grown as a pure laboratory culture, it should be introduced into healthy animals. If the disease was induced "with all its characteristic symptoms and properties," the most skeptical scientist would have to be persuaded that the microbe was truly the cause of the disease.[31]

Although these procedures should be carried out for an unequivocal proof, it is not possible to do so in all cases. To prove that a microbe causes human disease would require unethical experimentation on human beings.

The Tuberculin Tragedy

In 1890 Koch announced that he had succeeded in developing a therapeutic agent for tuberculosis. Germany was host that year to the Tenth International Medical Congress. Although Koch may have been somewhat reluctant to commit himself at the time, he was pressured by Von Gossler, the Minister of Culture, to make the announcement. Koch claimed to have found a substance that could inhibit the growth of the tubercle bacillus in cultures and in animals. He hoped that the agent would be useful in humans. Three months later, having conducted tests in humans, Koch published a "Progress Report on a Therapeutic Agent for Tuberculosis." The agent was described as a "stable, clear, brown liquid," but its nature was kept secret. Human beings were so sensitive to the agent that the guinea pig could tolerate 1500-fold the reactive dose for man on the basis of body weight.

The mysterious agent discriminated between normal and tuberculous patients. While a healthy person had almost no reaction to a small dose, a tuberculous patient experienced severe reactions, including vomiting, fever, and chills. Koch claimed that the reactions disappeared after 15 hours. The agent was, as Koch predicted, "an indispensable diagnostic aid." Early stages of tuberculosis could be discovered when sputum and physical examinations were still negative. Koch's claim that tuberculin would be more valuable in therapy than diagnosis proved tragically misleading. Patients from all over the world soon flocked to Berlin for the new "wonder drug," but sometimes died from their reactions to tuberculin. Public opinion rapidly turned against Koch when it became obvious that his agent was useless as therapy, but two years after his death the Institute for Infectious Diseases was renamed the Robert Koch Institute in honor of his discovery of the tubercle bacillus.

While his work on tuberculosis was his greatest accomplishment, Koch and his students fought many other diseases, including cholera, malaria, rinderpest, and plague. His methods were exploited successfully in the search for the agents of typhus, leprosy, ray fungus, erysipelas, diptheria, tetanus, pneumonia, cerebrospinal meningitis, dysentery, relapsing fever, and other diseases.

Progress in Microbiology

Immunology: Cellular vs. Humoral Explanations

On the foundations built by Pasteur, Koch, and others, microbiology exploded into numerous subdisciplines. Progressive physicians and researchers were distinguished by their appreciation of the role of microbes in disease and the implications for therapy. A new generation of scientists was obsessed with microbe hunting and the race to be first to isolate and characterize the agents of disease. Further research elucidated the modes by which infection was transmitted—air, water, contaminated foods, human carriers, and insect vectors. Bacteria themselves were not the only danger. Sometimes, as in diptheria, specific toxins released by the microbes were more directly involved in the disease process.

It was obvious that the human body was equipped with powerful weapons to fight microbial invaders. Study of defense mechanisms was as valuable as microbe hunting. Because he was not a physician himself, Pasteur was more interested in prevention of disease than in therapeutics. His success in developing vaccines formed the foundation of the science of immunology. By the end of the nineteenth century, two schools of thought concerning the mechanism of immunity were competing for supremacy. According to the humoral theory, immunity was dependent on the induction of certain factors in the blood and fluids of the body. Other investigators contended that serum factors were insignificant compared to the role of certain cells as soldiers in the body's defense against microbial invaders. A leading advocate of the theory of the cellular basis of immunity was Elie Metchnikoff, discoverer of "phagocytosis."

Elie Metchnikoff (1845–1916)

Honored for his discovery of phagocytes ("eating cells") and the process of phagocytosis, the Russian zoologist Elie Metchnikoff was almost alone as a champion of the cellular mechanism of defense against infection. While antibodies and antitoxins were creating new triumphs in therapeutics, Metchnikoff pursued his idiosyncratic investigations of phagocytic cells. Indebted almost as much to Darwin and Wallace as to Virchow, Pasteur, and Koch, Metchnikoff began his studies of inflammation with simple unicellular organisms and then confidently pursued the mechanism of response to injury and microbial invasion step by step up the evolutionary ladder.

Although he was not the first to observe the presence of bacteria in blood cells, it is indisputable that he was the first to recognize that intracellular digestion of microbes could have an important place in natural immunity to disease. Koch had seen anthrax bacilli inside whole blood cells, but he concluded that bacteria must penetrate the white blood cells and reproduce there. According to Metchnikoff's studies, it was more likely that the blood cells had "eaten" the parasites. Furthermore, contrary to prevailing opinion, Metchnikoff argued that the inflammatory response was not a purely passive and deleterious reaction to injury. The heat, swelling, and pain of the inflammatory reaction were part of a beneficial and protective, but imperfect, process.

In 1888 the Russian zoologist joined the Pasteur Institute, bringing with him an unusual combination of experience and interests ranging from

Elie Metchnikoff. [From W. Bullough (1960). *The History of Bacteriology.*
Courtesy of Oxford University Press.]

embryogenesis and the evolution of digestive functions, to inflammation, immunity, and senescence.

Studies of a disease of Daphnia (the common water flea), caused by a
yeast, led Metchnikoff to conclude that "the blood corpuscles have the role
of protecting the organism from infectious materials." This simple system was
advantageous since Daphnia are small, simple, and transparent, and the fungus
which attacked them is rather large and easily observed in unstained preparations. Indeed, the patient Metchnikoff would observe one organism for many
hours—watching the "battle between two living beings—the fungus and the
phagocytes." In the phagocyte, Metchnikoff saw "the bearers of nature's
healing power" and "a new support for certain basic ideas of cellular
pathology."[32]

Although this work helped lead the way to a more modern appreciation of
the role of cellular and humoral factors in immunity, nineteenth century
workers failed to see this essential compromise. In 1890 Robert Koch proclaimed the triumph of the humoral theory over the phagocytic school. Just
six years later, Lord Lister believed that the humoral school had been
vanquished by advocates of the cellular basis of immunity. By 1908 both
schools had achieved independent but undeniable advances and a truce of
sorts. In that year Metchnikoff and Paul Ehrlich, a leading advocate of the hu-

moral theory, shared the Nobel Prize in Physiology and Medicine—awarded in recognition of their work in immunity. Despite the preconceptions of scientists, it seemed reasonable that the body in its wisdom might actually use both mechanisms and that humoral substances would be derived from special cells.

Serum Therapy

The discovery of antitoxin to diptheria and tetanus by Emil von Behring (1854–1917) and Shibasaburo Kitasato (1852–1931) in 1890 overshadowed the cellular theory for some time. Blood from animals injected with increasing but nonlethal doses of toxins had the power of neutralizing the toxin. Quickly, von Behring and Kitasato transformed their discovery from a laboratory curiosity into a powerful new therapy. The first human being treated with the diptheria antitoxin was a child in a Berlin clinic on Christmas night, 1891. Because diptheria was at the time a major killer of children, "serum therapy" aroused international interest. In 1901 Behring was awarded the first Nobel Prize for Physiology and Medicine.

While von Behring's first human tests generated great hope that serum therapy would prove a generally effective means of combating all infectious diseases, production methods for diptheria antitoxin were not at first entirely satisfactory. Not until Paul Ehrlich devised quantitative methods for measur-

Emil von Behring. [From W. Bullough (1960). *The History of Bacteriology.* Courtesy of Oxford University Press.]

ing the antitoxin level of serum and methods for producing diptheria antitoxin in large amounts was the antitoxin to become a practical weapon in the therapeutic arsenal. Ehrlich, the founder of chemotherapy, made numerous contributions to pharmacology, toxicology, histology, and immunology.

Paul Ehrlich (1854-1915)

In this survey of germ theory, Ehrlich's work nicely reflects its progress. The first major milestone along the way was the study of syphilis and the theory of contagion by Fracastorius. Ehrlich's multifaceted career included innovative theories of immunity and the synthesis of a drug, a "magic bullet," specific for syphilis.

When only eight years old, Ehrlich proved his interest in therapeutics by presenting the local chemist with his own prescription for cough medicines. Ehrlich studied medicine at the Universities of Breslau, Strasbourg, Freiburg-im-Breisgau, finally graduating from Leipzig in 1878. Karl Weigert (1845-1904), his cousin and a pioneer in histology, served as a model for the young Ehrlich. His doctoral thesis, "Contributions to the Theory and Practice of Histological Staining," reflects this influence and foreshadows many of Ehrlich's later ideas. Becoming a physician required passing state exams. Ehrlich's attitude toward such formalities and ordeals was well known. During a tour of the laboratories, when Robert Koch first visited Breslau, Ehrlich was introduced as a promising scientist "very good at staining, but he will *never* pass his examinations."[33] Ehrlich did indeed survive the ordeal of exams and became Senior House Physician at the Charité Hospital in Berlin. Here he worked with the physician R. T. von Frerichs (1819-1885), who recognized Ehrlich's talent for research and allowed him to continue his studies of staining. Conditions changed abruptly on the death of von Frerichs. His replacement, Prof. Gerhardt, believed that Ehrlich should be restricted to the traditional clinical work.

After Koch demonstrated his discovery of the tuberculosis bacillus by a tedious staining method, Ehrlich immediately ran to the laboratory and began to experiment on better methods. His procedure took advantage of the acid-fast characteristics of the microbe. Methods currently in use are modifications of Ehrlich's original technique. Staining properties led Ehrlich to consider the special characteristics of the tubercle bacillus. Using his new, more sensitive method, Ehrlich diagnosed himself as a victim of tuberculosis and left the Charité, where he found conditions intolerable, to recuperate in Egypt. On his return to Germany he became a Reader in Medicine in the University of Berlin and joined Koch and Emil von Behring at the Institute for Infectious Diseases. Behring's diptheria antitoxin generated great hopes and bigger disappointments, until Ehrlich made the antitoxin a practical therapeutic agent.

Paul Ehrlich, founder of chemotherapy, and his associate, Dr. Sahachiro Hata. [© 1963, Parke, Davis and Company.]

The association with Behring was to prove a very bitter one for Ehrlich because of Behring's financial machinations. According to Martha Marquardt, Ehrlich's secretary, Ehrlich expected that the contract between the inventors and the pharmaceutical factory would give him some 30–40,000 marks per year for at least 14 years. He claimed that Behring talked him into giving up his share of the profits in exchange for the directorship of a State research institute. Behring kept his percentage, became a very rich man himself, and did nothing for Ehrlich.

Despite his disappointment with Behring, the antitoxin success did lead the Prussian Minister of State, Friedrich Althoff, to found a new institute for serology and serum testing. In 1896 Ehrlich began his career as director. Although he had only a very small laboratory, and a very, very small budget, he and his associates accomplished a prodigious amount of work. Between 1877 and 1914, Ehrlich published 232 papers and books. Ehrlich often said, "As long as I have a water tap, a flame and some blotting paper, I can work in a barn."[34] He might have added cigars and mineral water to the list, for he practically lived on these two items. His thought processes were so dependent

on smoking that he always carried an extra box of cigars under his arm. As
director, Ehrlich managed his assistants and associates with a firm hand, every
day giving each person a series of "blocks," index cards in various colors, on
which he gave instructions for the day's experiments. He literally exploded
at the assistant who dared to ignore the instructions on the card. Ehrlich won
many honors for his pioneering research in pharmacology, toxicology, and
therapeutics. He virtually founded chemotherapy with his drug "salvarsan"
or "606" for syphilis, but the last year of his life was embittered by controversies over his cure for syphilis.

As a student Ehrlich had conducted systematic studies of the aniline dyes
which led him to discover the different types of white blood cells and to
categorize the leukemias by the prevailing cell type. During a study of lead
poisoning Ehrlich noticed that the same organs which were most affected by
the poison in the living being were the ones which most avidly accumulated
lead when tissues were stained with lead solutions. Thus he early conceived
the importance of specific interactions between particular chemicals and tissue
components.

Staining methods for bacteria were very difficult and unsatisfactory when
Ehrlich introduced the methylene blue stain in 1881. Koch used this in many
of his studies and with this dye was able to identify the tubercle bacillus.
Vital staining techniques evolved from Ehrlich's work on the oxygen affinities
of various tissues. All these triumphs in histology may be regarded as the
working out of the ideas expressed in his doctoral thesis. In his thesis,
Ehrlich suggested a direct link between the growth of the chemical industry
and progress in histology. "Theory follows practice in every field," he wrote.[35]

Discovery of Aniline Dyes

Aniline dyes were discovered in 1856 when William Henry Perkin (1848–
1907), a student at the Royal College of Chemistry, attempted to synthesize
quinine from toluidine. Since little was known about these chemicals beyond
their elemental analysis, the project did not seem too unreasonable at the time.
Perkin obtained only a brown sticky mess. Repeating the experiment with
aniline, he produced another mess. When cleaning the reaction vessel with alcohol, Perkin was struck by the deep purple color of the resulting solution.
Against the advice of his distinguished professor, August Wilhelm Hofman
(1818–1892), Perkin abandoned his schooling and went into the dye business
with his father and brother. British dyers, used to natural dyes, resisted the
synthetic "mauve" for some time. The new dye became popular with French
fashion houses before the British learned to accept them, but thereafter chemists were kept busy meeting the demand for gaudy new colors. Working
empirically, chemists produced Bismark Brown, Imperial Blue, Quinoline Blue,

and Hofman Violet. Perkin was able to retire from business at 36 and devote the rest of his life to chemical research and music. Three of his sons became distinguished chemists.

Ehrlich's Magic Bullets

What primarily interested Ehrlich was the possibility of developing selective stains for different kinds of cells, components of cells, and microbes. He later extended this concept to the basis of a scientific, experimental pharmacology with a new goal—the design of "magic bullets." That is, he believed scientists could synthesize compounds which would specifically interact with microorganisms without damage to normal cells.

Pursuing the study of serum-mediated immunity, he discovered in 1891 that the injection of poisons such as ricin led to the production of antibodies. In the same year Ehrlich clarified the latent period in the development of active immunity. The next year he showed the difference between active and passive immunity. These studies led to his famous "side-chain" theory of immunity. Ehrlich believed that the highly specific action of antibodies was based on the presence of specific groups or "receptors" on bacterial cells and toxins to which antibodies became attached.[36] Whereas antibodies were nature's own "magic bullets," chemotherapy was an attempt to imitate nature by creating drugs which would seek out and destroy microbial invaders without harming the host. Although Ehrlich often spoke of the "four big G's" necessary for success—Geduld, Geschick, Geld, and Glück (patience, ability, money, and luck)—when congratulated on the success of Salvarsan, he replied, "For seven years of misfortune I had one moment of good luck." Many of Ehrlich's ideas were remarkably prescient and stimulated further advances in immunology.

Controversies initiated in the nineteenth century continue to echo through current research on the mechanism of immunity. Although the lacunae in our knowledge of this complex system remain large, progress has been significant. The complete elucidation of the structure of an antibody by Gerald M. Edelman, Nobel laureate of 1972, is a major milestone in the study of immunity. Such knowledge is more important than ever in an era when infectious diseases seem less menacing than autoimmune diseases, cancer, and the rejection of grafts or transplants.

Alexander Fleming (1881-1955)

A major advance in chemotherapy was the discovery by Alexander Fleming that the mold *Penicillium* produced a substance inimical to the growth of certain bacteria. Although Fleming published detailed studies of the

antibacterial action of *Penicillium,* he did not seem to anticipate its potential
in therapeutics. Perhaps the failure of lysozyme, his earlier discovery, to be
therapeutically useful had discouraged him. H. W. Florey (1898-1968) and
others eventually made it possible to use penicillin as a therapeutic agent.
Not until the sulfa drugs had been successfully exploited and World War II
made chemotherapy a high priority field was penicillin to become a significant
factor in medicine. Since that time, the search for substances elaborated by
one microorganism which kills other microbes has proved very fruitful.

Gerhard Domagk (1895-1964)

Like Paul Ehrlich, Gerhard Domagk systematically searched for chemical substances that could destroy bacteria without damaging the patient. In 1935 he
published a landmark paper, "A Contribution to the Chemotherapy of Bacterial Infections." Few effective drugs were then available for microbial infections and none of them were useful for coccal infections. Using the
methods developed by Ehrlich, Domagk tested many chemicals in infected
animals. He found that Protosil, a red dye with no *in vitro* antibacterial
activity, protected mice from streptococcal infections. Because the dye is
broken down in the body into the active agent sulfanilamide, it was not possible to predict the value of Protosil from *in vitro* tests. Eventually, further
chemical transformations established a whole family of sulfa drugs from sulfanilamide. Since streptococcal infections are common and often lethal, the
sulfa drugs provided a major advance in therapy. Domagk's own daughter,
accidentally infected in the laboratory, was one of those saved by his discovery.
    Although Domagk won the Nobel prize for medicine in 1939, the Nazis
would not allow him to accept it. He received the medal in 1947, but not
the prize money.

Viruses

Even before viruses were seen or their nature understood, folklore and science
had been successful in combating certain virus-caused diseases. Lady Mary
Wortley Montagu introduced the Turkish method of inoculation against smallpox to England in 1718 and Cotton Mather championed the method in the
colonies. In 1798 Edward Jenner (1749-1823) succeeded in providing protection against smallpox by the use of cowpox—a method called "vaccination,
which eventually saved millions from that dreaded disease. A more direct and
scientific attitude enabled Pasteur to develop a vaccine for rabies—a disease he
was sure was caused by a living microbe even though he could not see the
agent. It is hard to imagine the more stolid Koch creating a vaccine against
an unseen enemy.

One of Pasteur's associates, Charles Chamberland (1851-1908), introduced a method that eventually led to the discovery of the "filterable viruses." In 1884 Chamberland published a paper describing the use of a porous porcelain vase to prepare "Physiologically Pure Water." This method, which could separate even heavily contaminated liquids from all visible microorganisms, was used in Pasteur's laboratory to separate microbes from culture medium. Chamberland suggested using the method to get perfectly pure drinking water. It seems he had at least a trace of Pasteur's mania for eliminating germs from his diet. Another methodological contribution of Chamberland was the autoclave.

In 1892 Dimitri Ivanovski (1864-1920) read a paper to the Academy of Sciences of St. Petersburg on tobacco mosaic disease. Using the filter method of Chamberland, he found that the filtered extract produced the same disease as an unfiltered extract. He thought that the simplest explanation was the presence of a toxin in the filtered sap or that some bacteria had passed through the pores of the filter. No such breaks were found, nor did bacteria grow in the medium put through the filters. This work did not make a big impression, but other scientists were turning to similar studies and their discoveries made more of an impact.

In 1898 Friedrich Loeffler (1852-1915) and P. Frosch published their "Report of the Commission for Research on the Foot-and-Mouth Disease." Both authors had been associated with Koch before this study and Loeffler had discovered the causal agent of diptheria. Studies were performed on freshly removed vesicles from the mouth and udders of sick animals so that the contents of the lesions could be observed free from contamination with other microbes. Surprisingly, no bacteria were observed in culture media inoculated with fluid from the lesions. Unless accidentally contaminated, the culture medium remained free of bacteria. However, these apparently sterile fluids could be used to transmit the disease to experimental animals. Furthermore, filtered lymph, although free of bacteria, transmitted the disease just as effectively as unfiltered lymph. Experimentally infected animals were capable of transmitting the disease to other animals. This suggested that a living agent, rather than a toxin, must have been present in the filtrate. As Koch had said, the animal body is a very efficient incubator for the multiplication of living agents of disease. Having provided evidence that foot-and-mouth disease was caused by a filterable virus, Loeffler and Frosch suggested that other infectious diseases, such as smallpox, cowpox, and cattle plague, were also caused by very minute organisms.

At about the same time Martinus Willem Beijerinck (1851-1931) was coming close to an appreciation of the nature of filtrable viruses through his studies of tobacco mosaic disease. He seems to have been unaware of Ivanovski's work, which was less developed than his own anyway. Beijerinck concluded that a "contagium vivum fluidum" which could pass through the finest filter and reproduce only in living plant tissue was the cause of tobacco mosaic disease. Although he found it difficult to imagine the growth and reproduction of such a "soluble body," he made the ingenious suggestion that "the contagium must be incorporated into the living protoplasm of the cell in order to reproduce. . . ."[36] Based on reports in the botanical literature, Beijerinck thought many other plant diseases were caused by similar soluble germs.

Within a score of years, filterable viruses were suspected of being the cause of various plant, animal, and human diseases. In 1915 Frederick William Twort (1877-1950) suggested that even bacteria were victims of diseases caused by the invisible filterable virus. Many attempts to grow viruses in artificial media had proved futile, when he noticed that colonies of certain contaminating cocci growing on agar could not be subcultured, but after some time these colonies became glassy and transparent. If pure colonies of the micrococcus were touched by a tiny portion of material from the glassy colonies, they too became transparent. Twort was cautious in his interpretation of this phenomenon. Because of World War I, he could not continue these studies. His paper, "An Investigation on the Nature of Ultra-microscopic Viruses," had little impact on microbiology. Felix d'Herelle later initiated extensive research on bacterial virus—which were sometimes called Twort-d'Herelle particles. Twort later became obsessed with speculative work on the possibility that bacteria evolved from viruses which had developed from even more primitive forms.

Felix d'Herelle (1873-1949) independently recognized the existence of a bacterial virus. In 1917 he published his experiments on "An Invisible Microbe that is Antagonistic to the Dysentery Bacillus." Because the invisible microbe could not grow on laboratory media or heat-killed bacilli, but grew well in a suspension of washed bacilli in a simple salt solution, d'Herelle concluded that the anti-dysentery microbe was an "obligate bacteriophage" (bacteriophage = eater of bacteria). D'Herelle had obtained the invisible microbe from the stools of patients recovering from bacillary dysentery. When an active filtrate was added to a culture of Shiga bacilli, bacterial growth soon ceased. Bacterial death and lysis followed. A trace of the lysate produced the same effect on a fresh Shiga culture. More than 50 such transfers gave the same results, indicating that a living agent was responsible for lysis of the bacteria.

Speculating on the general implications of the phenomenon he had discovered, d'Herelle predicted that bacteriophages would be found for other pathenogenic bacteria. Although the natural parasitism of the invisible microbe seemed "strictly specific," d'Herelle believed that laboratory manipulations could force the bacteriophage to "acquire activity against various germs."[37] Hopes that bacteriophage could be trained as weapons in the war on bacteria were never realized. The bacteriophage was not the "microbe of immunity" d'Herelle predicted. Until pressed into service as the "experimental animal" of molecular genetics, bacterial viruses remained a laboratory curiosity.

In 1935 Wendell M. Stanley (1904-1971) reported that he had isolated a "crystalline protein" having the infectious properties of tobacco mosaic virus. This was an advantageous system for the first successful attempt to crystallize a virus because infected plants have large numbers of active virus particles. Standard purification techniques yielded the crystalline chemical which possessed certain attributes previously invariably associated with living organisms. Stanley concluded that tobacco mosaic virus was "an autocatalytic protein which, for the present, may be assumed to require the presence of living cells for multiplication."[38] Stanley won the Nobel Prize in Chemistry in 1946 for this provoking work which suggested that a chemical substance might be alive. Further work showed that the virus was not a pure protein but rather a combination of protein and nucleic acid, but the importance of nucleic acid was not appreciated at the time. (See Chapter 14.)

It was not until the development of the electron microscope that viruses were removed from the category of "invisible entities." In 1939 scientists first viewed the tobacco mosaic virus under the electron microscope.[39]

As L. Marton summarized the situation in 1941, the limits of the light microscope had been reached and were frustrating further progress in microbiology.[40] The electron microscope was a valuable new tool for bacteriological research. In the short time since the first working models had been developed, the range of microscopic observation had been extended about 50- to 100-fold. Since that time, the new tool has been well exploited in understanding bacteria and the more minute agents of disease—the virus and rickettsiae.

Notes

1. T. Schwann (1847). *Microscopical Researches into the Accordance in the Structure and Growth of Animals and Plants.* Reprint 1969. New York: Kraus Reprint, pp. 197–198.
2. C. Dobell (1960). *Antony van Leeuwenhoek and His "Little Animals."* New York: Dover, pp. 243, 76.
3. R. Vallery-Radot (1923). *The Life of Pasteur.* New York: Doubleday, p. 72.
4. Vallery-Radot, p. 49.
5. Vallery-Radot, p. 76.
6. T. Brock, ed. (1975). *Milestones in Microbiology.* Washington, D. C.: American Society for Microbiology, p. 12.

7. Brock, p. 15.

8. Vallery-Radot, p. 219.

9. Vallery-Radot, pp. 108–109.

10. Vallery-Radot, p. 256.

11. Vallery-Radot, p. 253.

12. A. I. Oparin (1965). *The Origin of Life.* 2nd Ed. Translated by S. Morgulis. New York: Dover.

13. Brock, pp. 69–74.

14. Brock, pp. 76–79.

15. Vallery-Radot, p. 260.

16. Vallery-Radot, pp. 313–314.

17. Vallery-Radot, p. 325.

18. Vallery-Radot, p. 291.

19. H. Lechevalier and M. Solotorovsky (1974). *Three Centuries of Microbiology.* New York: Dover, p. 61.

20. R. Dubos (1960). *Pasteur and Modern Science.* New York: Doubleday, p. 147.

21. Lechevalier and Solotorovsky, pp. 64–66.

22. Lechevalier and Solotorovsky, p. 69.

23. Lechevalier and Solotorovsky, p. 79.

24. Brock, pp. 96–97.

25. Brock, p. 100.

26. Brock, p. 104–105.

27. Lechevalier and Solotorovsky, p. 84.

28. Lechevalier and Solotorovsky, p. 84.

29. L. Clendening (1960). *Source Book of Medical History.* New York: Dover, pp. 392–393.

30. Lechevalier and Solotorovsky, p. 106.

31. Brock, pp. 116–117.

32. Brock, pp. 132–133.

33. M. Marquardt (1949). *Paul Ehrlich.* New York: Schuman, p. 15.

34. Lechevalier and Solotorovsky, p. 439.

35. F. Himmerlweit and M. Marquardt (1957). *The Collected Papers of Paul Ehrlich.* New York: Pergamon, I, p. 67.

36. J. Parascondola and R. Jasensky (1974). Origins of the Receptor Theory of Drug Action. *Bull. Hist. Med. 48:* 199–220; J. Parascondola (1971). Structure-Activity Relationships: the Early Mirage. *Pharmacy in History 13:* 3–10.

36. Brock, p. 156.
37. Brock, pp. 157–159.
38. Brock, p. 162.
39. T. F. Anderson (1966). "Electron Microscopy of Phages," in *Phage and the Origins of Molecular Biology*. J. Cairns, G. S. Stent, and J. D. Watson, eds. New York: Cold Spring Harbor Laboratory of Quantitative Biology, pp. 63–78.
40. L. Marton (1941). The Electron Microscope. A New Tool for Bacteriological Research, *J. Bacteriology 41:* 397–413.

References

Adelberg, E. A. (1960). *Papers on Bacterial Genetics.* Boston: Little, Brown.

Anderson, T. F. (1966). "Electron Microscopy of Phages," in *Phage and the Origins of Molecular Biology*. J. Cairns, G. S. Stent, and J. D. Watson, eds. New York: Cold Spring Harbor Laboratory of Quantitative Biology. pp. 63–78.

Barlow, C., and P. (1971). *Robert Koch.* Geneva: Heron.

Brock, T., ed. (1961). *Milestones in Microbiology.* Englewood Cliffs, N.J.: Prentice-Hall.

Bulloch, W. (1938). *The History of Bacteriology.* London: Oxford University Press.

Bulloch, W. (1930). *Systems of Bacteriology.* New York: Medical Research Council.

Casida, L. E., Jr. (1976). Leeuwenhoek's Observations of Bacteria. *Science 192:* 1348–1349.

Collard, P. (1976). *The Development of Microbiology.* Cambridge: Cambridge University Press.

Conant, J. B., ed. (1952). *Pasteur's Study on Fermentation.* Harvard Case Histories in Experimental Science. Cambridge, Mass.: Harvard University Press.

De Kruif, P. H. (1932). *Microbe Hunters.* New York: Harcourt, Brace.

De Kruif, P. H. (1950). *Men against Death.* New York: Harcourt, Brace.

Doetsch, R. N., ed. (1960). *Microbiology: Historical Contributions from 1776–1908.* New Brunswick, N.J.: Rutgers University Press.

Dubos, R. J. (1950). *Louis Pasteur: Free Lance of Science.* Boston: Little, Brown.

Dubos, R. (1960). *Pasteur and Modern Science.* Garden City, N. Y.: Doubleday Anchor.

Duclaux, E. (1920). *Pasteur, The History of a Mind.* Translated by E. F. Smith and F. Hedges. Philadelphia: Saunders.

Edelman, G. M. (1973). Antibody Structure and Molecular Immunology. *Science 180:* 830–840.

Eve, A. S., and Creasey, C. H. (1945). *Life and Work of John Tyndall.* London: MacMillan.

Gladston, I. (1954). Ehrlich, Biologist of Deep and Inspiring Vision. *Sci. Monthly 79:* 395–399.

Heidelberger, M. (1956). *Lectures in Immunochemistry.* New York: Academic.

Himmelweit, F., and Marquardt, M. (1957). *The Collected Papers of Paul Ehrlich.* New York: Pergamon.

Hitchens, A. P., and Leikind, M. C. (1939). The Introduction of Agar-Agar into Bacteriology. *J. Bacteriol. 37:* 485–493.

Koch, R. (1880). *Investigations into the Etiology of Traumatic Infective Diseases.* Translated by W. W. Cheyne. London: New Sydenham Society.

Lechevalier, H., and Solotorovsky, M. (1974). *Three Centuries of Microbiology.* New York: Dover.

Lechevalier, H. (1972). Dimitri Iosifovich Ivanovski (1864–1920). *Bacteriol. Rev. 36:* 135–145.

Lohnis, F. (1921). Studies upon the Life Cycles of the Bacteria, Part I, Review of the Literature–1838–1918. *Mem. Nat. Acad. Sci. 16,* Second Memoir, Washington, D.C.: Government Printing Office.

Maurois, A. (1959). *The Life of Sir Alexander Fleming.* Translated by G. Hopkins. London: Cape.

Marton, L. (1941). The Electron Microscope. A New Tool for Bacteriological Research. *J. Bacteriol. 41:* 397–413.

Marquardt, M. (1949). *Paul Ehrlich.* New York: Schuman.

Metchnikoff, E. (1905). *Immunity in Infective Diseases.* Translated by F. G. Binnie. London: Cambridge University Press.

Metchnikoff, E. (1968). *Lectures on the Comparative Pathology of Inflammation. Delivered at the Pasteur Institute in 1891.* New York: Dover.

Metchnikoff, O. (1921). *Life of Elie Metchnikoff.* London: Constable.

Muir, R. (1915–1916). Paul Ehrlich: Obituary. *J. Pathol. and Bacteriol. 20:* 350–360.

Oparin, A. I. (1938). *The Origin of Life.* New York: MacMillan.

Paget, S. (1914). *Pasteur and After Pasteur.* London: Black.

Parish, H. J. (1965). *A History of Immunization.* Edinburgh: Livingstone.

Parascondola, J., and R. Jasensky (1974). Origins of the Receptor Theory of Drug Action. *Bull. Hist. Med. 48:* 199–220.

Parascondola, J. (1971). Structure-Activity Relationships: The Early Mirage. *Pharmacy in History 13:* 3–10.

Robinson, R. (1956). The Perkin Family of Organic Chemists. *Endeavour 15:* 92.

Twort, F. W. (1915). An Investigation on the Nature of Ultra-microscopic Viruses. *Lancet 2:* 1241–1243.

Vallery-Radot, R. (1923). *The Life of Pasteur.* Translated by R. L. Devonshire. Garden City, N. Y.: Doubleday, Page.

Vandervliet, G. (1971). *Microbiology and the Spontaneous Generation Debate during the 1870's.* Kansas: Coronado.

Waksman, S. A. (1954). *My Life Among the Microbes.* New York: Simon and Schuster.

Waskman, S. A. (1947). *Microbial Antagonisms and Antibiotic Substances.* New York: Commonwealth Fund.

Walker, M. E. M. (1930). Pioneers of Public Health: *The Story of Some Benefactors of the Human Race.* Edinburgh: London, Oliver and Boyd.

Whetzel, H. H. (1918). *An Outline of the History of Phytopathology.* Philadelphia and London: Saunders.

11

PHYSIOLOGY

Physiology and Anatomy

Although the activities which distinguish living from nonliving things are more interesting than purely morphological studies, progress in physiology has always been slower than in anatomy. Whereas anatomical studies could be pursued with only the naked eye, a few simple tools, direct observation, and classification, physiology requires knowledge of chemistry and physics as well as imagination and the ability to reason. Progress in physiology waited on progress in other sciences and on the broader application of experimentation and analysis. Thus, although Aristotle, Galen, and Vesalius were interested in physiology—in elucidating the relationship of structure to function—they could not push their work very far beyond vague ideas of "cooking," "fermentation," "animal heat," and "spirits" or "faculties." It was not until the seventeenth century that new ways of dealing with the dynamic functions of the body were realized. Physiology was retarded not only by the lack of experimental techniques, but also by the lack of a guide in the form of an overall theoretical framework that would help rather than hinder the investigator.

By the end of the seventeenth century, a new concept of the cosmos was firmly established. Through the work of Galileo, Kepler, and Newton, the universe emerged as a great, law-bound machine, with the earth merely one of the planets circling the sun. Gravity, not crystalline spheres and angels, kept the heavenly bodies moving in their orbits. Although the revolution of ideas was not as deep or extensive in biology, it was not without effect. Two different ways of exploring vital phenomena emerged with new clarity in this period. These might be called the "metric experimental," which is best exemplified in the work of William Harvey, and the "rational-philosophical," which was used by the philosopher and mathematician René Descartes.

288

Harvey represents the greatest triumph in physiology of his time, and perhaps up to the nineteenth century. His method was the culmination of the work of Aristotle, Galen, and Vesalius. Harvey said that through vivisections he was led to truth—structures and activities of the living body gave him clues to explain vital phenomena in a new way. As Foster said, Harvey did not "appeal to any knowledge or to any conceptions outside the facts of anatomy and the results of experiments."[1] Indeed, there was little of fundamental value in chemistry and physics to guide the biologist when Harvey began his research. Harvey had essentially no tools or instruments that had not been available to his famous predecessors, but he was infused with the scientific philosophy that animated Galileo. Yet Harvey stands alone, towering over all physiologists of his time and well beyond.

The career of Santorio Santorio, who sat so patiently in his balance and measured what could be measured, proves that measurement and patient experimentation are not enough where the fundamental question is improperly phrased and the possible conclusions merely trivial. Quantitation alone gives no insight into vital phenomena without a proper theoretical basis for investigation. Physiology in the seventeenth century, except for exposition of the circulation of the blood, was essentially stymied for lack of development in the basic sciences. Successful physiologists in the seventeenth century dealt with problems of a strictly mechanical or anatomical nature. Digestion and metabolism, muscle and nerve action are problems requiring an understanding of chemical and electrical phenomena. Recognizing the power of mechanical explanation in Harvey's work, other physiologists tried to force vital phenomena to fit grotesque or trivial mechanical analogies. The jaws were explained as pincers, the stomach as a retort, the veins and arteries as hydraulic tubes, the heart as a spring, the muscles and bones as a system of cords and pulleys, the lungs as bellows, the viscera and kidneys as simple sieves and filters.

Mechanical analogies, though obviously strained or patently false, were preferred to chemical explanations because mechanics was making remarkable strides at the time, while chemistry was still mired in an embarrassment of ancient myths, obscure alchemical techniques, and outright quackery. Yet during the second half of the seventeenth century, important progress was to occur in chemistry. Indeed, shortly before his death Harvey handed on the torch of physiological research to Robert Boyle, first of the modern chemists.

Physiologists divided themselves into warring camps, choosing to call themselves either "iatromechanists" or "iatrochemists." The iatromechanist believed that all functions of the living body could be explained on physical and mathematical principles. The iatrochemists, in contrast, believed all these problems could be explained as chemical events. Unfortunately for progress in physiology, these views were held to be irreconcilable with each other; partisans maintained that life must be *either* a chemical or a physical phenomenon.

Mechanism and Vitalism

Physiological studies from the most ancient times to the present have been
guided implicitly or explicitly by a philosophical framework that has been
either "mechanistic" or "vitalistic." The mechanistic philosophy asserts that
all life phenomena can be completely explained in terms of the physical-chemical
laws that govern the inanimate world. Vitalist philosophy claims that the real
entity of life is the "soul" or "vital force" and that the body exists for and
through the soul, which is incomprehensible in strictly scientific terms.

Controversies between mechanists and vitalists have often been very violent,
but seemed to be fading away by the end of the nineteenth century as new
studies and discoveries blurred the distinctions between them. Realizing that
these were but vague generalizations at best, scientists began to believe that
physiology could abandon them altogether. It is interesting that today when
many life phenomena are being analyzed at the molecular level, new challenges
to the mechanist or analytical approach have been raised, forcing some scientists to hold forth on their philosophical assumptions as well as their experimental results.[2]

Sanctorius (Santorio Santorio) (1561–1636)

Although he did not make any original contributions to the new chemistry
and physics of his time, Sanctorius did set an example of patient and quantitative medical research. Little is known of his life except that he studied
medicine at Padua, graduated in 1582, travelled a great deal, and then settled
into a medical practice in Venice. He was appointed to the Chair of Theoretical Medicine at Padua in 1611 and was highly regarded as a teacher. In
1629 he resigned his position to return to private practice and research in
Venice.

Sanctorius is best remembered for his researches on variations of body
weight measured after eating, drinking, during sleep, at rest, during exercise,
and in illness. In 1614 he published his results as a series of aphorisms in a
small book titled *De medicina statica aphorismi*. Obviously quite popular, the
book was translated into several languages, and went through more than thirty
editions in Latin. The problem which obsessed Sanctorius was akin to the
modern concept of metabolism. He called the problem "insensible perspiration"—a process in which volatile substances supposedly left the body. Each
aphorism in the book was presented as a deduction from his careful measurements. Only a bare outline of his actual experimental methods was given, but
it is apparent that he carefully weighed himself, all the food and drink he ingested, and all his excreta. For some 30 years he spent as much time as possible weighing himself by sitting in a chair suspended from a steelyard.

In his preface he wrote: "It is a new and unheard of thing in Medicine that anyone should be able to arrive at an exact measurement of insensible perspiration. Nor has anyone, either Philosopher or Physician, dared to attack this part of medical inquiry. I am indeed the first to make the trial, and unless I am mistaken I have by reasoning and by the experience of thirty years brought this branch of science to perfection." Thinking others might share his interests and patience, he suggested they might use a similar balance to live a life "according to rule," by taking meals sitting on a balance that would be adjusted to warn when the proper amount of food had been consumed.[3]

While it is hard to imagine anyone taking the suggestion of living on a balance seriously, some of his inventions had clinical significance. In 1612 he described a rather cumbersome thermometer which could measure body temperature. In 1625 he elaborated on the use of the clinical thermometer in studying diseases. He also invented a hygrometer to measure humidity, a method for bathing a patient in bed, and a means of comparing pulses. Without accurate means of measuring small intervals of time, physicians could only speak of the strength or rhythm of the pulse. By using pendulums, where the length of the string was adjusted to match the beat of the pulse, Sanctorius was able to compare pulse rates.

The spirit of invention and investigation was highly developed in Sanctorius, even if his deductions were not commensurate with the pains he took in his work. No doubt he would have been surprised to know how his work was either laughed at or forgotten while the work of a man who cared little for mere measurement proved a great inspiration to biologists. This was the mathematician and philosopher, René Descartes.

René Descartes (1596–1650)

Although Paracelsus and his followers clung to chemical theories of life, the course of the history of physiology was diverted to the mechanistic philosophy mainly through the writings of René Descartes. While Descartes himself never made an original physiological discovery, and dissected only to support his preconceived ideas, he provided his followers with a complete mechanistic system. His writings on physiology were not meant as particular contributions to that science, but were part of a general system of philosophy.

Descartes was born into a wealthy family, although his early life was marred by illness. His mother died when he was born, and his own "pulmonary weakness," seemed to doom him to an early grave. His early education was at a Jesuit school, where most of the teachings impressed him as mere quibbling. Mathematics was the only reality he found in all his studies. Although convinced that he could erect a philosophical framework which would unite all the physical and natural sciences, when Descartes heard of the persecution of

Galileo (1633), he suppressed his own *Treatise on the World,* which owed too much to the censored system of Galileo to be safe. For a time, Descartes published works on safer subjects. Some of his most important and original works were not published until after his death.

Even a superficial introduction to philosophy will include Descartes' system and his famous phrase, "I think, therefore, I am." This was the one idea that Descartes was sure of as he practiced his method of philosophy by systematically doubting everything. After his death, parts of his great treatise, *On the World,* were published. For physiology the most important works were *On Man* and *On the Formation of the Fetus.*

In his scientific work Descartes was generally correct in his broad premises, sometimes brilliantly anticipating later advances, but in the particulars his work abounds with errors. His method involved determining whether ideas passed his criteria of truth—clearness and distinctness. Such ideas were "self evident" and became the premises by which other truths were established. Although Descartes acknowledged the importance of observation, he remained a rationalist who subordinated the crude facts of observation and experiment to the test of reason and his *a priori* principles. Descartes favored Plato's abstract world of mathematics, not the quantitative approach of Harvey and Sanctorius. Descartes preferred to begin with the abstract and proceed in a rational way to truth, unimpeded by the complexity, even perversity, of nature. The first reality was his own existence as a thinking entity. His universe was a great mechanical system in which God was the "first cause" of all motion. Matter had the essential qualities of extension, divisibility, and motion. "Give me motion and extension," Descartes said, "and I will construct the world." He rejected the atomic theory of Democritus and Lucretius and allowed for no empty space in his universe.

Animals were treated as automata with all physiological functions explained in purely mechanical terms as the motions of material corpuscles and the heat generated by the heart. The Galenic doctrine of the three spirits was replaced by one where only the animal spirits remained, transformed into a "fluid" utilized in the brain and nerves. Even man emerged from Descartes' exposition as another "earthly machine"—one which differed from animal automata by virtue of the rational soul which inhabited and governed man. Serving as the agent of thought, governing volition, conscious perception, memory, imagination, and reason, the rational soul was the only entity exempted from a purely mechanical explanation.

The purpose of Descartes' physiological treatise on man was to extend his concept of the universe as a machine to the explanation of man as also a machine working in accordance with physical laws.[4] To do this, he picked through the anatomical knowledge of his time, dissected on his own, added certain unproven "self evident" ideas, and concocted a theory of the action of the body as a machine. Descartes described the ingestion and digestion of food in physical terms—with the expected analogies to grinding, pulverizing, and sieves. He was enthusiastic about Harvey's recent work on the circulation—the perfect example of a mechanical system—but misunderstood or misrepresented much of it. Descartes saw the heart as a "heat machine" rather than a pump. Thus, in direct opposition to William Harvey, Descartes continued the ancient idea that the action stroke of the heart was not the contraction phase, but the expansion step.

According to Descartes, the heart contains "one of those fires without light . . . which renders it so fiery and hot" that as soon as the blood enters either of the two chambers or cavities of the heart, it immediately expands and dilates. "And the fire in the heart of this machine that I am describing," wrote Descartes, "serves no other purpose than to dilate, warm and subtilize the blood that falls continually drop by drop through a passage from the vena cava, into the right cavity whence it is exhaled into the lung . . . ."[5] The obvious test of the heart as heat machine was certainly a technically simple one even in the seventeenth century, since the thermometer was available. But it

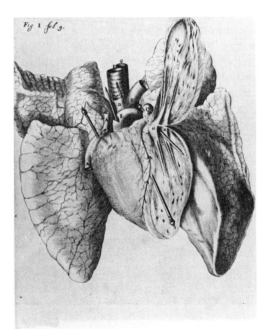

Descartes' figure of the heart. [From J. F. Fulton and L. G. Wilson (1966) *Selected Readings in the History of Physiology*. Courtesy of Charles C Thomas, Publisher, Springfield, Illinois.]

was not Descartes' way to subject such self evident ideas to crude experimental verification.

In contrast to the fiery heart, the lung tissue was described as delicate, soft, and freshened by the air. In the cool tissue of the lung the blood vapors condensed and then fell "drop by drop into the left cavity of the heart." By thus using the work of Vesalius and Harvey, it was only too simple for Descartes to restructure their findings to fit his purposes in discussing blood circulation, digestion, respiration, and certain other physiological phenomena. The nervous system presented more of a challenge, but Descartes valiantly attacked this problem. According to Descartes, all other systems were subservient to the nervous system and labored to supply the material basis of nervous energy. The nervous system in turn served to carry out the commands of the rational soul. Therefore, the strongest and subtlest parts of the blood are carried to the cavities of the brain by arteries taking the most direct route from the heart. Those special parts not only nourish and sustain the substance of the brain, but also "produce there a certain very subtle wind, or rather a very lively and very pure flame, which is called the 'animal spirits.'" Through conduits in the brain the animal spirits are able to enter the nerves. By their tendency to enter some nerves rather than others, the animal spirits "are able to change the shapes of the muscles into which these nerves are inserted and in this way to move all the members."[6]

Although Descartes continued to speak of "animal spirits," he transformed them into a special subtle fluid subject to the general physical laws applicable to fluids so that it flowed through the nerves in a purely mechanical way. According to Descartes, the nerves were tubes in which valves governed the direction of flow of the nervous fluid, just as the venous valves governed the flow of the blood. This is a very clear example of how Descartes was willing to describe structures that suited his theory although there was not a shred of anatomical evidence for their existence.

Although Descartes wrote some hundred years after Vesalius, and could have benefited by the progress made in that interval, in some ways his views are retrogressive, particularly concerning the nervous system. Vesalius had performed many vivisections to understand the relationship of nerve and muscle and had compared the brain of man to many other species. Vesalius knew that functional differences in brain capacity could not be revealed through dissection alone. Vesalius trusted the book of nature and ridiculed "the figments of men who have never studied the handiwork of God." Descartes had studied the "handiwork of God," but had embellished on His Creations in his exposition of the nervous system. He made daily trips to the butcher shop to see animal anatomy and did dissections "especially of heads of different animals to explain what imagination, memory, etc., consist of." To provide a clear account of the functioning of the brain and nervous system,

Descartes included various parts and activities in the system which did not exist, although his discussion seems to refer to items of general knowledge.

A major problem in his system was explaining how the rational soul could interact with the earthly machine. According to Descartes, the pineal gland was the seat of the soul and the only site of direct interaction between immaterial soul and the corporeal machine. The pineal gland was chosen because it is located at the base of the brain, is a single rather than a paired organ, and because it was supposed (erroneously) to be an organ present in man but absent in lower animals. Descartes believed it was very soft and mobile, and a reservoir for animal spirits.

The animal spirits flowed along the nerves and thence to the muscles which made the limbs move. Descartes compared the nerves, muscles, and tendons of the body with the engines and springs which moved the devices in the grottoes and fountains of the French royal gardens. The gardens were rather like an early version of Disneyland where mechanical monsters, gods, and goddesses moved about when hidden plates were depressed by the visitor. In Descartes' anatomical system, nerves not only possessed valves, but also delicate threads along the length of their cavities which went from brain to sense organ. The tiniest motion along the thread thus exerted a pull upon the site in the brain where the thread originated, opening up the orifices of certain pores on the internal surface of the brain, allowing animal spirits to flow into the muscles, causing the machinery to move. Except for ideas, all physiological functions of the body were as mechanical as the workings of a clock or mill, or any automaton. These ideas are reminiscent of Plato's description of men as the "puppets of the Gods."

Descartes even explained thoughts and emotions in mechanical terms. Ideas were defined as "impressions received by animal spirits as they leave the pineal gland," but he was not always consistent as to whether ideas were differentiations impressed upon matter or the soul alone. Memories seem to have been physically recorded—rather like permanent press—into the fabric of the nerves. Since man is a combination of soul and body, emotion is experienced both in the soul and in the machinery of the body. For example, the joy experienced in the soul has its counterpart in the body where the blood becomes finer and is more easily expanded in the heart, more easily exciting the nerves.

Certainly if judged from the standpoint of exact anatomical and physiological knowledge Descartes must plead guilty to inventing a scheme based as much on imagination as anatomical knowledge. Yet he was widely read, imitated, and honored. He challenged scientists to treat man's physical and mental aspects in the same manner as all other scientific problems. Nicolaus Steno saw in Descartes the first man "who dared to explain all the functions of man, and especially of the brain, in a mechanical manner."[7]

At first Descartes' philosophy was criticized by the Church, but eventually, despite the basically materialistic nature of his system, a reconciliation was effected. Like many other systems of thought, Descartes' began as heresy and ended as dogma. The mechanistic approach guided many scientists more experimentally inclined than Descartes. Muscle action was a favorite problem of iatromechanists like Giovanni Borelli.

Giovanni Alfonso Borelli (1608–1697)

Like Descartes, Borelli favored the mechanical mode of explanation for physiological processes, but Borelli differed from Descartes in several important ways. If Descartes may be regarded as the founder of iatromechanism as a philosophy, Borelli may be regarded as the founder of iatrophysics as an experimental science.

Little is known about Borelli's early life. He was born in Naples in 1608 and his father was probably an officer who had served with Philip III of Spain. At a very early age Borelli's genius for mathematics was recognized. By 1640 he was appointed to the Chair of Mathematics at Messina. Inspired by the work of Galileo, he obtained the consent and financial support of the university to take a prolonged leave to study with Galileo and Evangelista Torricelli in Florence. Unfortunately, Galileo died that same year. Borelli returned to Messina and in 1649 published his first work, a description of the epidemic raging in Sicily. Obviously he was already interested in medical problems, but it is doubtful he ever studied medicine formally. In 1656, Borelli was invited to the Chair of Mathematics at the University of Pisa by Ferdinand, Duke of Tuscany. Here Borelli carried out the best part of his researches. Although his introductory lecture is said to have been interrupted by loud booing and laughter from his student audience, Borelli continued to teach mathematics. With Marcello Malpighi, he began to study anatomy.

Borelli helped to make the University of Pisa famous for both mathematical and medical sciences and as a center for propounding the new experimental approach to natural science. Borelli and Francesco Redi founded the Academia del Cimento. Although Borelli was one of the leaders of the Academy, some of the other members found his personality almost unbearable. He was described as morose and argumentative, stubborn and jealous of others, one who became enraged when contradicted or criticized. After 12 years at Pisa, and many quarrels with his colleagues, Borelli asked to leave the university because of the bad climate and his need for more leisure and quiet. In 1668 he returned to the University of Messina where he indulged himself in literary and antiquarian studies, investigated an eruption of Mount Etna, and continued to work on the problem of animal motion. In 1674 he was accused of involvement in a political conspiracy to free Sicily from Spain. Forced into exile, he fled to Rome where he became part of ex-Queen Christina's unofficial court of intellectuals. Borelli served as her physician and lectured at Christina's academy. In 1676 his great work, *De Motu Animalium,* was almost finished, but further misfortunes attacked the aged Borelli. Cheated out of all his money and property, he died before his book was published. An introduction was added by an "ecclesiastic dignitary" who commended Borelli for upholding the authority of the Church in his lectures on astronomy. In physiological matters, too, Borelli remained a faithful son of the Church by acknowledging that all the mechanical phenomena described in his book were ultimately governed by the Soul.

Borelli's great physiological treatise was an attempt to thoroughly apply mathematical and mechanical principles to the study of muscular work. He regarded physiology as a "part of physics" which he expected to "ornament and enrich by mathematical demonstrations." The book, he wrote, "contains an analysis of the mechanics of contraction: how muscles act at joints, what tension they must develop in order to overcome a given mechanical disadvantage, etc. It also contains interesting observations upon reciprocal innervation and tonus." The first part of this work dealt with the movements of individual muscles and groups of muscles treated geometrically in terms of mechanical principles.[8] Articulations were similarly treated. Borelli dealt with posture in man and animals and different kinds of movement such as walking, running, swimming, and flying. The second part was a study of the cause of contraction in muscles. Animal movements were divided into external movements, such as those carried out by the skeletal muscles, and internal movements, those of the heart and viscera, and the fluid parts of the body. To understand the movements of animals, Borelli's protocol involved an examination of the structure, parts, and visible action of muscles. The method of study involved a progression from the simplest element of the motor system, the independent muscles, up to the more complicated organs and organ systems, and finally the power of movement of the organism considered as a whole.

Great faith in his methods enabled Borelli to attack the study of muscular mechanics as a problem which "could be solved by the almost direct application of known mechanical methods, and which called for little special research beyond the mere determination of the necessary data." Studies of muscle were being carried out by microscopic observation of their fine structure. Borelli's friend Malpighi was a leader in this field, but Nicolaus Steno had also formulated a newer conception of muscle structure through microscopic studies. Borelli and Steno agreed that the important fundamental fact of muscle contraction was the part played by the fleshy muscular fiber and that the fibers of the tendon are merely passive agents which do not take part in contraction.[9]

Heart action particularly intrigued Borelli. More experimentally inclined than Descartes, he recognized that the heart was a muscular pump rather than a heat engine. A simple experiment confirmed this. Measuring the temperature of the heart and other internal organs in a vivisection experiment on a deer, he proved that the temperature of the heart was not significantly different from other parts of the body. Borelli likened the ventricles to a wine press or piston and noted that during systole (contraction) the walls of the ventricles obliterate the cavities from which the blood has been driven. He concluded that in contracting the heart muscles increase in bulk and that this was a general phenomenon in muscle action. The hardening and tension apparent during muscular contraction must then be due to inflation of the muscular substance by something flowing into the muscle, or due to a sudden fermentation in the muscle triggered by animal spirits travelling from the brain through the nerve and to the muscle. Trying to formulate a strictly mechanical explanation, Borelli suggested that muscular fibers are chains of rhombs and that contraction was due to inflation caused by the sudden insertion of a number of wedges.

In trying to unravel the interaction of nerves and muscles, Borelli first dismissed theories involving an incorporeal influence, spirits, or air. If air rushed into the cavities of muscles like the wind experiments should show this. A simple experiment in which the muscles of a living animal were divided lengthwise while the animal was held under water refuted this hypothesis. If some "spiritous gas" had entered the muscles it should "burst forth from the wound, and ascend through the water." Inflation of the muscles could not be caused by the influence transmitted by the nerves alone, but must result from something in the muscles themselves. When the nerves transmit some influence from the brain to the muscle, within the muscle, wrote Borelli,

> there takes place something like a fermentation or ebullition, by which the sudden inflation of the muscle is brought about. . . . That such an action is possible is rendered clear by innumerable experiments which are continually being made in chemical elaborations as when spirits of vitriol are poured on oil of tartar; indeed all acid spirits when mixed with fixed salts at once boil up with a sudden fermentation. In like manner, therefore, we may suppose that there takes place in a muscle a somewhat similar mixing from which a sudden fermentation and ebullition results, with the mass of which the porosities of the muscle are filled up and enlarged, thus bringing about the turgescence and inflation.[10]

In conducting experiments to determine the weight that different muscles can support, Borelli found that the muscles on the two sides of the jaw could support a weight of more than 300 pounds when acting together. He assumed that the force of contraction of all healthy muscular tissue would be the same for a given unit of bulk. Thus, he extrapolated from his measurements of some exterior muscle to the forces that presumably could be generated by the heart and other internal muscles. While his basic ideas were good, his data were too crude to yield correct values. Indeed, accurate mathematical analyses of the work of muscles were not conducted until the late nineteenth century.

Even digestion was treated as a primarily mechanical process. Borelli was very impressed by the strong muscles of the stomach and their ability to crush foods. This was especially striking in certain birds where "the crushing, erosion, and trituration of food is effected by the muscular stomach itself. . . ." Borelli conducted certain experiments to prove this. He introduced "glass globules, or empty vesicles, and leaden cubes, similarly hollowed out, pyramids of wood and many other things" into the stomachs of turkeys. The next day he found "the leaden masses crushed and eroded, the glass pulverized and the remaining ingesta in the same condition." Trying to quantitate the force of the stomach muscles, he compared the force needed to break the shell of certain nuts by the molar teeth of man and glass vesicles of comparable strength that were crushed into powder in the stomach of the turkey. Arguing that the action of the teeth and the fleshy stomach are similar because both "act by pressure like a winepress, and overcome the same resistance," he concluded that the motive powers of the two were equal. Since he had already shown "that the absolute force of the muscles which close the human jaw represents a power greater than that of a weight of 1350 lbs.," he calculated the force of the turkey's stomach as "not less than the power of 1350 lbs."[11]

Borelli admitted that certain chemical reactions play a role in physiological processes. He realized that birds of prey digest their food "by means of a certain very potent ferment much in the same way as corrosive liquids corrode and dissolve metals." Some of his more extreme followers refused to admit that such a digestive juice was present. Extreme iatrophysicists argued that the only purpose of the stomach was to triturate food into chyle.

Even as Borelli was writing his great study of muscles, other scientists through simple experiments were proving that muscles do not increase in volume when contracting. Both Jonathan Goddard and Francis Glisson demonstrated that when a muscle is contracted under water the water level does not rise.

Francis Glisson (1597-1677) and Irritability

Glisson was interested in various medical and physiological problems, such as rickets, the fine structure of the liver, and the physiological property of "irritability." It was probably Harvey's work that inspired Glisson to study medicine. In 1634 Glisson became Doctor of Medicine and in 1636 was appointed Regius Professor of Physics. Glisson was a member of the interesting

group of intellectuals who met at Gresham College and later founded the Royal
Society. When the civil war broke out, Glisson had to leave Cambridge which
came under Royalist influence. Glisson was a staunch Presbyterian. Although
Glisson held the Regius Professorship until his death, he spent little time there.
Although he apparently gave no course of lectures, he did visit Cambridge on
occasion to examine candidates for the degree of Doctor of Medicine. Despite
his negligent attitude towards his professorship, he petitioned the University
for five years' salary in arrears for the time he spent entirely at Colchester.

During his studies of the liver, Glisson tried to explain why the bile is not
discharged into the intestines continuously, but only when needed. He found
experimentally that the gall bladder and biliary duct discharge more bile when
they are irritated. This occurs, he argued, because they have the capacity to
be irritated. This doctrine of "irritability" became very influential in physiology, largely through the work of Albrecht von Haller who revived Glisson's
concept. While Glisson used the term "irritability" to describe a broad range
of phenomena, Haller tried to confine the concept to the property that caused
muscles to contract in response to an external stimulus. Other workers extended the concept to mean any kind of change in a living organism whether
movement, conformation, growth, etc.

Nicolaus Steno (1638-1686)

The life and work of Nicolaus Steno are of interest not only for his many
contributions to anatomy and geology, but because they bear witness to the
tensions built up in a man of considerable genius by the conflicting calls of
his science and his religion. Steno, son of a prosperous goldsmith, was born
and educated in Copenhagen. Apparently the victim of a restless nature,
Steno traveled widely and studied in many places. During a period in Amsterdam he discovered the parotid salivary duct, named the ductus Stenorzi-
anus in his honor. While in Paris, he made important studies on the anatomy
of the brain. In Florence he was supported by the Grand Duke of Tuscany
who encouraged his research and appointed him to a hospital post which took
up little of his time. Steno was elected to the Accademia del Cimento founded
by the Duke's brother Leopold de'Medici. It was during this period that
Steno carried out his famous examination of the shark which led him to re-
consider the old question of "stone tongues," (fossilized sharks' teeth) and
develop novel and prescient ideas about geology and the meaning and formation of fossils.

Although Steno's father had been a very zealous Lutheran, Steno converted
to Catholicism after a time of spiritual crisis. Previous efforts by Bosseut in
Paris to convert Steno had failed because he had been too busy with science
to be interested in theology, but after returning to Florence at the age of 36,

he underwent a sudden change. While at the pharmacy attached to the Santa
Maria Novello, a holy brother selling drugs talked to Steno about remedies for
the soul, causing him to meditate on science and religion. Within a year Steno
gave up science and took holy orders. This seems to have upset his intellectual
balance, although it led to many worldly advantages and promotions. Steno
was sent as Vicar Apostolic to various countries to win converts. Zealous in
carrying out his religious goals, Steno even tried to use his reputation as a
scientist for this purpose. With Church approval, he taught anatomy in Copenhagen, although it was a "heretic" university. But religion, not science, was
his passion now and he soon resigned to devote himself to the duties of the
priesthood. Becoming increasingly ascetic, he ruined his health, and died at
age 48.

Before his religious conversion, Steno's ambition had been to apply the
methods and philosophy of Galileo to the biological problem of muscular contraction. While his approach was primarily geometrical, his analysis of the
nature of contraction confirmed the fact that the apparent swelling of a working muscle was due to shortening of the fibers and did not involve an increase
in volume. Some of Steno's ideas were probably influenced by another eccentric figure, Jan Swammerdam. Having met at Leyden, the two travelled
together to Paris in 1665. Swammerdam had conducted experiments to show
that muscles do not swell during contraction and had demonstrated that a
frog can swim for some time after its heart has been cut out. Following up
this kind of evidence, they conducted experiments that showed movement of
the feet and tail of a tortoise continued for 24 hours after the head was cut
off. Furthermore, an isolated heart could continue to beat for some time.
Steno noted in experiments on dogs that muscles convulsed when he dissected
nerves or pressed on them strongly. Some things about muscles, he argued,
should be quite clear from experimental evidence—such as the fact that the
heart is a muscle. He was particularly impressed with the way the heart continued to beat *in vitro* although no new blood flowed in and no new "spirits"
could enter the nerves coming from the brain. He realized that the voluminous writings on the "spirits" or "fluids" that were presumably involved in
muscle action merely were an attempt to disguise a profound ignorance.
"Many people talk of the animal spirits, the more subtle part of the blood,
the juice of the nerves," he wrote, "but these are mere words signifying
nothing." [12]

Use of the microscope seems to have influenced Steno's work as much as
the new experimental spirit. Having studied the microstructure of tissues, he
saw the muscle as a collection of motor fibers. Each motor fiber was in turn
a complex of very minute fibrils arranged lengthways—with a middle part
which differed from the ends in its consistency, thickness, and color. The
only part involved in contraction was the fleshy part of the motor fibers,

which became shorter, firmer, and more corrugated on the surface. The tendinous parts did not change during muscle contraction. Thus, ancient misconceptions about muscle and tendon were finally corrected in the seventeenth century. The Hippocratic writers, who confused tendons and nerves, had assumed that tendons caused motion and that muscles were passive, fleshy material.

It is strange that the man who could enunciate such a clear defense of the value of science would abandon the pursuit of scientific truth. Steno wrote:

> To those who decry the value of science I would give as an answer this demand that they should ask their own consciences and see what solid basis there is for all those dogmas which they pronounce with such bold ease when they explain the symptoms of apoplexy, paralysis, convulsions, prostration of strength, syncope, and other diseases affecting animal movements, on what foundation they rest when they apply remedies for removing these evils, with the result that they do away not with the paralysis, not with the convulsions, but with the paralytic or the convulsed man.[13]

Physiology in the Eighteenth Century

Although physiology was still a mixture of speculation and experimentation in the eighteenth century, the theoretical and experimental questions being probed were reaching a new level of sophistication. Reflecting a partial assimilation of the chemical revolution, physiologists made significant progress in the study of digestion and respiration. Much of this work was initiated and promoted by teachers at the leading medical schools—such as Hermann Boerhaave at the University of Leyden, who taught chemistry, physics, botany, ophthalmology, and clinical medicine.

Hermann Boerhaave (1668-1738)

During his lifetime, Boerhaave was honored as an intellect profound as Newton. It is difficult to understand quite why, since he is all but forgotten today. Yet according to one story illustrating Boerhaave's international reputation, a letter from Asia addressed "To the Greatest Physician in the World" was delivered directly to Boerhaave.[14] Boerhaave's influence derived more from his teachings and writings than his own research. Indeed, he made no major original discovery and in some of his disputes he was quite wrong. For example, he denied that the gastric juice was acid. Boerhaave was an eclectic concerning physiological theories—seeing that anatomy, physics, and chemistry

were each useful in their place. His *Elements of Chemistry* was still lauded
some 80 years after publication as "the most learned and most luminous
treatise on chemistry that the world has yet seen."

Albrecht von Haller was inspired by the teachings of Boerhaave, but surpassed his master in contributions to experimental physiology.

Albrecht von Haller (1707–1772)

Haller was a man of many talents, or at least interests, for in addition to his
scientific work he was a poet, biographer, and statesman. Physiology owes a
great debt to Albrecht von Haller—who carried out an enormous number of
experiments despite an almost paralyzing melancholy, bad health, and a profound disgust for the pain caused by his vivisections.

Obviously a prodigy, Haller had mastered Greek and Hebrew, literature,
and science by the time he was 10 years old. By 15 he had written an epic
poem and several tragedies. Travels in Belgium and England expanded his
education, as did his studies with the mathematician Bernouilli in Basel. At
19, he obtained his medical degree and became the favorite pupil of the great
physician and teacher, Hermann Boerhaave. In 1736, at the instigation of
George II of England, Elector of Hanover, Haller was appointed professor at
the new University of Göttingen, where he served as Professor of Anatomy,
Botany, and Medicine. He established a botanical garden and anatomical
theatre, founded a scientific society, spread the teachings of Boerhaave, and
published many scientific works. Despite success in his research and teachings,
he became increasingly depressed and homesick. Always the victim of poor
health, he used opium almost constantly. After seventeen years at Göttingen,
he refused all academic appointments and returned to Bern (1753). The last
24 years of his life were devoted to writing, bibliographical research, and
service to the state. He became a member of the Swiss council, was active in
public health matters, and founded a state orphan asylum. Despite his involvement in scientific and charitable works, he became increasingly melancholy and sick. He died with his finger on his pulse telling his friends: "The
artery no longer beats."[15]

Among his many contributions to physiology was the encyclopedic *Elements of Physiology*. His research and literature review were so extensive that
Magendie later complained that whenever he thought he had performed a new
experiment, he always found it had already been attempted or described by
Haller. *Elements of Physiology* conveniently summarized the state of physiology at the time in a concise and accessible form, although it ran to eight
volumes.

Haller and his students extended anatomical knowledge and linked it to
physiology by experiment, and by applying dynamic principles to physiological

problems. His studies concerned the form and function of various organs and organ systems—such as muscles, nerves, blood vessels, the circulatory and respiratory system, the formation of bones, and the development of the embryo. He was also interested in the actions of drugs and often used himself as a guinea pig.

Haller is most remembered for his fundamental studies of the "irritability" of muscle and the "sensibility" of nerves. Reviving Francis Glisson's concept of irritability, Haller narrowed the definition of the term. He showed that the organs of the body are partly irritable and partly non-irritable. The task of the physiologist was to determine which parts belonged in each category through vivisection experiments. The irritable parts were defined as those which contract when touched, while the sensible parts were those which conveyed a message to the mind when they were acted on. The tendons, bones, cerebral membrane, liver, spleen, and kidney were found to be insensitive. The property of irritability in muscles was shown to be due to the nerves. For example, Haller showed that the diaphragm could be made to contract by irritating several nerves.

In *Elements of Physiology* Haller discussed the nature of the contractile force which he believed was present in the plant as well as the animal kingdom. Through the contractile force "the elements of fibres are brought nearer to each other." According to Haller, this was not only the cause of cohesion in general, "but is rendered manifest by the fact that a fibre drawn out lengthways when let go very soon returns to its previous length, and never lays aside the effort to become shorter until it has so returned to its previous length." [16] After discussing various general phenomena, Haller focused on the special contractile force which is specific to muscles. In a living animal, or in one that had just died he often found spontaneous contractile movements in muscle tissue. This contractile property can also be induced by applying some stimulus such as pinching, pricking, or certain chemical agents.

Haller argued for a fundamental difference between the "living contractile force" and spontaneous contractions since "the two agree neither in the laws which govern them, nor in their duration, nor in their seat." He called the special force in living beings the "vis insita" (the inherent force). The tissues in which this force resided were "irritable." He argued that the property of irritability is different from that of feeling or sensing. "There are many parts which feel but which are not irritable," he wrote, "and in particular a nerve, which is above everything sensitive, and yet possesses no contractile force except that common one found as stated above even in dead things." Irritability, he concluded, is the definitive characteristic of muscle fibers. While others had called this special force the "vital force" Haller rejected this because he found the special contractile force of muscles to survive for some time after death. He preferred to call it the "force inherent in or proper to muscle." He found

that another force existed in muscle fiber which he called the "vis nervosa."
This was carried to the muscles from the brain by the nerves and had the
power to initiate muscular contractions. It too could not be called the vital
force because it was also found for a time in dead animals—either cold or
warm-blooded.

Although Haller realized just how difficult it would be to understand the
brain and nervous system, he dealt with them in a remarkably restrained and
modern spirit. Always he tried to make experiment his guide: "As the nature
of the brain and of the nerves is one and the same, so are these alike in function. In treating of them we will so far as possible make use of experiments,
nor will we at first at least go beyond the testimony of our senses." Experiments indicated that the nerves alone serve as the instruments of sensation so
that only those parts of the body served by nerves experience sensations. The
nerves also serve to elicit the power of contraction from the muscles which
are the only instruments of movement. Irritating different parts of the brain
in living animals determined the sensitivity by the movements which resulted.
Haller was interested in the question of whether specific parts of the brain had
particular properties and functions. While trying to answer this question by
evidence from various pathological processes and experiments on animals, he
realized the question was too complex to be answered satisfactorily at the
time. He was willing to enter into "conjectures" based on probable arguments
and extrapolations from the rather limited data available.

Examining the views of the ancients and his contemporaries, Haller rejected
most of their speculations on the nature of the nervous system. He refuted
the idea that the nerves act as solid bodies like elastic strings which convey
vibrations. Instead, Haller argued that the nerves work by means of a special
subtle fluid which is not like an albuminous solution, or spirituous—like alcohol, or acid, or combustible, or aerial—like the "ether" of Newton, or
electrical material. Having satisfied himself as to what the nervous material
was not, he concluded that it was

> an element of its own kind unlike everything else. An element, too
> subtle to be grasped by any of the senses, but more gross than fire, or
> ether or electric or magnetic matter, since it can be contained in chan-
> nels and restrained by bonds and moreover is clearly produced out of
> and nourished by food. What forbids, since light is something different
> from fire, and the material of the magnet differs from both, and air and
> ether are unlike all the rest, what forbids that there should be this
> element of its own kind known to us only by its effects?[17]

Therefore, the nerves must be hollow and by anology the fibers of the brain
must also be hollow. Only one kind of nervous fluid existed and it was

responsible for sensation, movement, and the preservation of life. Haller rejected the idea that the soul was diffused over the whole body. He concluded that "the soul has nothing in common with the body other than sensation and movement." Because both sensation and movement seemed to have their source in the medulla, that must be the seat of the soul.

Judgments of Haller vary all the way from praise for the greatest physiologist of his time to criticism of his inaccuracies and his basically reactionary view of life. Doubtless his major contribution was to the spirit and method of physiological research as a new field of "animata anatome"—vitalized anatomy.

Julien Offroy de La Mettrie (1709-1751)

The life and work of Julien de La Mettrie offer a sharp contrast to the pious Haller. La Mettrie was quite willing to let his materialistic scientific theories conflict with or contradict Christian dogma, despite the fact that he was himself a priest. La Mettrie's father was a rich merchant who demanded that his son study for the priesthood despite the latter's desire to be a poet. A friend advised La Mettrie that "a mediocre physician would be better paid than a good priest." Thus, La Mettrie became interested in medicine and like Haller studied with Boerhaave at Leyden. Later serving as a surgeon in the French army, he saw "one percent glory and 99 percent diarrhea." Demonstrating his talent for arousing conflict, La Mettrie antagonized the ultraconservative medical faculty in Paris by translating Boerhaave's work into French. The Paris faculty had opposed Vesalius and Harvey and were maintaining their reactionary stance by opposing Boerhaave. La Mettrie increased their antipathy by publishing a series of pamphlets satirizing his learned opponents. While suffering from a violent fever, La Mettrie realized that the clearness of his thinking varied with the severity of the illness. This proved to him that thought is a function of the brain, dependent on physical conditions. He expressed these ideas in his controversial work, *A Natural History of the Soul.*

According to La Mettrie, we cannot know what the soul truly is, but neither do we know what matter is. Since we never find a soul without a body, to study the properties of the soul, we must study the body, and to do this, we must investigate the laws of matter. The natural history of the soul traced the evolution of an "active principle" in matter through plants and animals to man. These materialistic views were clearly in disagreement with Christian dogma. The theologians were ready to do battle with the heretical La Mettrie and the chaplain of his regiment had him dismissed. La Mettrie had also made enemies of the physicians with his book on *The Politics of Physicians,* which satirized their greed, competition, and scheming. With his practice and reputation in ruins, La Mettrie sought refuge in Holland, where he amused himself by writing another attack on the physicians. Feeling relatively safe in Leyden, La Mettrie published another unorthodox book, *Man, the Machine* in 1748. Without Descartes' veneer of piety, he discussed man as a machine whose actions were entirely due to physical-chemical agents. He based his work on experiments with animals where he had observed that peristalsis continued after death and isolated muscles could be stimulated to contract. He reasoned that if this were true for animals, it must also be true for man—because both were essentially the same in terms of bodily composition. Attempting to get away from the mind-body dualism of Descartes, La Mettrie argued that even the mind must depend directly on physiochemical processes. Substances such as opium, coffee, alcohol, and other drugs affect both body and mind—thought, mood, imagination and will. Diseases attack mind as well as body. Likewise, he argued, diet can affect character. For example, eating red and bloody meat causes a savage character.

Although his heretical book had been published anonymously, it was quite obvious who the author was. Fortunately, La Mettrie was able to find refuge at the court of Frederick II of Prussia where he carried on a medical practice, wrote, and lectured at the royal court. This peaceful period lasted only a short time. La Mettrie had called death the conclusion of a farce, and proved this with his own peculiar exit from life. A feast was given in his honor by a grateful patient whom La Mettrie had cured of a serious illness. Bragging about his capacity for enjoying all of life's pleasures, La Mettrie ate an enormous quantity of a truffle pastry. He fell ill immediately and soon died in great pain. Did he die of simple overeating, indigestion, or some septic poison in the truffles? This we do not know, but Voltaire said it was a great occasion, since, for once, the patient had killed the doctor. This abrupt and unusual death increased La Mettrie's notoriety. The Church maintained that no one could die in peace without its blessings, which La Mettrie certainly had not had. Thus, the theologians saw in his death a fitting reward for the author of such heretical and materialistic views of the nature of man.

In his opposition to Cartesian dualism, La Mettrie paved the way for more modern biological theories. He denied Descartes' claim that man was essentially different from animals. For La Mettrie, man himself was essentially a variety of monkey, superior mainly by virtue of the power of language. Animals were not merely machines, although their ability to reason was less developed than the human level. La Mettrie thus narrowed the gap between the physiology and mental powers of man and the animals.

Possessed of a rather sardonic humor, La Mettrie dedicated one of his books to the pious Albrecht von Haller as a practical joke. As another jest, he added to the attacks on his book by publishing a pretended and mocking critique of his own called *L'Homme, plus qu'une machine* (*Man, More than a Machine*). For the most part, his books were more polemical than scientific. In place of Church dogma, he wanted to substitute a "natural" system of ethics based on principles obtained by direct observation of life.[18]

Other eighteenth century scientists were more interested in elucidating particular physiological problems than in elaborating great philosophical systems. Some of these were quite ingenious in terms of the experimental approach they took. Among the problems attacked with ingenuity and simple equipment was the regulation of body temperature.

Regulation of Body Temperature

When thermometers were developed it became possible to experiment on another physiological problem—the regulation of body temperature. The ancients had qualitatively noticed that body temperature was rather constant in healthy people, but varied in pathological states. It was also apparent that no matter what the ambient temperature, certain species are warm to the touch when alive, but become cold when dead. Other species were relatively cold, or varied in temperature and activity depending on ambient temperature. Sanctorius had made his own thermometer and found that, except for fevers, man's body temperature remains very constant. Where to measure body temperature was a problem which Dr. George Fordyce solved by placing the thermometer under the tongue.

Charles Blagden (1748-1820) and John Hunter (1728-1793) undertook an interesting series of experiments which proved that body temperature was constant at a broad range of ambient temperatures. Small rooms were maintained at temperatures of 110-120°F, 180-190°F, even 260°F, while outdoors the temperature was about 32°F. The scientists worked in these hot rooms, noted their own reactions, and measured body temperature. As controls, nonliving materials and foods were tested—e.g., egg, meats, wines, water. From these experiments Blagden concluded that temperature regulation was a fundamental characteristic of life.[19]

The Chemical Approach to Life

Jean Baptista van Helmont (1577-1644)

Van Helmont was an intellectual disciple of that picturesque exponent of alchemical theory, Theophrastus Bombastus von Hohenheim, better known as Paracelsus (1493-1541). Like Paracelsus, van Helmont believed that life was essentially a chemical phenomenon. Unlike Paracelsus, van Helmont has an unarguable place in the history of chemistry for his pioneering studies on gases and fermentation. It was van Helmont who introduced the term "gas" to replace the Paracelcian word "chaos" (which indicated what typically happened in the alchemist's laboratory when substances being heated released great quantities of gas). For Paracelsus the workings of nature could be explained in terms of "visible matter and invisible forces." A spiritual force called the "archeus" governed matter and its transformations. In living systems, all physiological processes were ruled by the internal archeus—whose operations determined health and disease, life and death.

Many scientists regarded Paracelsus as little better than a quack and his ideas as remnants of ancient mysticism, but van Helmont revived and reformed these chemical explanations of life. As intellectual heir to Paracelsus, van Helmont also inherited the smoldering resentment of established medical tradition for the new science of chemistry. Finding his studies of philosophy wholly empty and unsatisfying, van Helmont refused to take the degree of Master of Arts which he saw as a symbol of scholasticism rather than a sign of learning. Rejecting philosophy, he turned to systematic botany, then law, and finally medicine. At the age of 22 he took the degree of Doctor of Medicine and then spent the next 10 years travelling through Europe—seeing Switzerland, Italy, and England. After marrying a wealthy woman, he settled near Brussels and devoted the bulk of his time to chemical experiments, practicing medicine as a work of charity. Although van Helmont was a very devout Catholic, he came into conflict with the authorities for a work published in 1621, *De Magnetica Vulnerum Curatione*. The Church saw his naturalistic exposition of magnetic cures as a challenge to orthodox interpretations of cures as miracles. Van Helmont published little else during his lifetime, but was allowed to have his own home serve as his prison. Just before his death, van Helmont instructed his son Franciscus Mercurius (1614-1699) to publish his manuscripts. *Ortus Medicinae* (dedicated to Jehovah) was first published in 1648, but was little noticed until translated into English, French, and Dutch at about the same time as Descartes' work *On Man* first appeared.

Van Helmont tried to reconcile the chemical view of life with a vitalist philosophy. However, his chemical researches were of an exact and quantitative nature in which the idea of the indestructibility of matter was implicit. For example, he proved that a metal could be dissolved in acid and then recovered without loss of weight. Intrigued by the transformation of liquid water into an "air," or vapor, under the influence of heat and the way this vapor could again be transformed into liquid water, van Helmont planned and executed an experiment to prove that water is the source of all things. In a large earthenware vessel he placed two hundred pounds of dried earth and a willow tree that weighed five pounds. After watering the tree with pure rain water for five years, he found that the tree weighed 169 pounds and 3 ounces. The weight of the soil was virtually unchanged. Van Helmont was satisfied that his experiment proved water was the primary element. The experiment certainly proves something. It proves that merely being quantitative will not illuminate the secrets of nature where theory throws a false light.

To confirm his belief that earth could be formed from burning vegetation, which in turn had come from water, van Helmont carried out an experiment in which he burned 62 pounds of charcoal. Only one pound of ash remained and he assumed that the rest of the material had been driven off into the air in the form of a spirit which he named *gas sylvestre.* This same gas was released by burning organic matter, during fermentation of beer or wine, from certain mineral waters, and from shells and limestone treated with acid. While he seems to have recognized that different gases existed, he lacked the special apparatus required to collect them for further study and characterization. His use of the term gas is nearly modern. Moreover, these studies of gas led him to reject Paracelsus' concept of the three elements—sulfur, mercury, and salt. Instead he argued there are only two elements—air (the natural atmosphere) and water. Water in his system meant everything that is not air.

Like Paracelsus, van Helmont believed that all physiological phenomena could be explained as chemical actions. The chemical processes in the body were governed by a hierarchical series of *archaei* in various organs. Subordinate to the *archaei* was an entity which he named *"blas."* There was a special *"blas humanum"* for all specifically human functions and other kinds of *blas* for common physiological processes.

Van Helmont based his physiological system on ferments and their action. This modernized the ancient suspicion that events in the fermentation of wine were somehow related to vital phenomena in the animal body. He regarded all changes in the body—including digestion and nutrition, movement, gestation—as due to the action of ferments. The *blas* or *archaeus* acted on matter through the ferments rather than directly. Digestion was explained by a scheme with six conversions—six digestions or concoctions—which transformed food into living flesh. Ignorant of the role of saliva, he postulated that the first stage of digestion took place in the stomach by means of a ferment which comes from the spleen when needed. The ferment in the stomach was acid but its action was greater than that of common acids such as vinegar. The acid chyle prepared in the stomach then passed into the duodenum and acquired a salty nature. Very complicated ferments were at work here in this second digestion. He believed that the bile contained another ferment. The work of the acid ferment of the stomach ceased upon reaching the duodenum. The third digestion was caused by the liver. This was the prelude to conversion of chyle into crude blood. The refuse of the food after the nutritious chyle was absorbed, passed to the caecum where another ferment converted it into feces.

In the heart and arteries the fourth digestion took place. The darker and thicker blood of the vena cava became lighter in color and more volatile. This actually corresponds to the change from venous to arterial blood. Van Helmont did not clearly distinguish this change from his fifth digestion in the arteries which changes the blood into vital spirit. A peculiarity of his view of the heart was that it had minute pores in the septum through which *"spiritus vitalis*" could pass from the left side to the right side, but through which blood could not pass. He still believed that blood passed through pores from the right side to the left side of the heart. The last digestion he wrote "takes place in the kitchens of the several members, for there are as many stomachs as there are nutritive members. In this sixth digestion . . . a ferment innate in each place cooks its food for its life."[20] That is, all tissues take what they need from the blood and transform nutrients into new tissue components.

Although his theory was obscure and permeated by out-moded mysticism, the idea that physiological problems could be solved by chemical experiments was valuable. Furthermore, the analogy between fermentation and physiological processes was a fruitful one which later workers with more exact chemical knowledge could successfully exploit.

Franciscus Sylvius (1614–1672)

An outstanding iatrochemist and teacher, Sylvius was the intellectual heir of van Helmont. Although he taught and practiced medicine, much of his work was purely chemical rather than medical. For this research he established his Laboratorium at the University of Leyden which seems to have been the first of its kind. Here he established himself as the founder of a vigorous program of chemical and experimental approaches to physiology. He helped to establish Harvey's new concepts in Holland, as well as other advances in anatomy and mechanical principles applied to physiology. He always encouraged his students by reminding them of how much remained to be discovered and by providing them with problems for their inquiry. Steno recalled being inspired by the teachings of Sylvius.

The properties of acids, bases, and salts fascinated Sylvius. He recognized that many salts are formed from the union of acids and bases and that volatile alkalis can be found in plants. Studies of the chemistry of acids and alkalis were transformed into a theory which explained disease as the result of an excess of "acridity"—acid or alkali. Therapy was designed accordingly. While Sylvius was very intrigued with van Helmont's concept of fermentations in the body, he rejected the mystical *archaei*. Digestion, he taught, was a fermentation initiated by the saliva, then proceeding under the influence of the bile and the pancratic juice (discovered by Regnier de Graff in 1664 while working with Sylvius).

A very important and bold insight expounded in Sylvius' work is the idea that the chemistry of living things is the same as the chemistry of nonliving things. Theoretically then it should be possible to reproduce in the laboratory chemical events supposedly peculiar to the living body. Sylvius tried to rid iatrochemistry of the spiritual and mystical tendencies of Paracelsus and van Helmont. This was to make the chemical approach to physiology a more

respectable one. Like Borelli, Sylvius was the founder of a distinct and disputatious school of thought. While Borelli tried to explain physiological
phenomena in mathematical and mechanical terms. Sylvius rested his case on
the new science of chemistry, a science much advanced since the time of Paracelsus, but still lacking the power and exactness of the physical sciences. Both
had great confidence in their methods and theories, and both thought they
had the exclusive key to the explanation of physiological phenomena.

Digestion

Although iatromechanists tried to explain digestion in purely mechanical terms,
this was one process where chemical explanations had an obvious advantage.
Several seventeenth century scientists had studied the physiology of digestion,
including Regnier de Graaf (1641-1673), who investigated the pancreatic juices,
and Réné Antoine Ferchault Réaumur (1683-1757), who obtained samples of
the gastric juices of his pet hawk. In the eighteenth century, Lazzaro Spallanzani (1729-1799), whom we have already discussed for his experiments in
generation, extended these very preliminary attacks on the mechanism of
digestion.

   Spallanzani began his studies of digestion by testing the action of saliva on
foods so that he could explain the mechanism to his students at Pavia. He
went on to verify Réaumur's work on the digestive powers of gastric juice and
also proved that it prevented putrifaction. Since there had been few experiments conducted on man, and Spallanzani viewed arguments based on data
from other species as probable but not conclusive, he valiantly swallowed
tubes and bags containing different foods despite discomfort, vomiting, and
the apprehension that such materials could cause a fatal obstruction in the
alimentary canal.

   Quite independent of Spallanzani, Edward Stevens published a thesis in
1777 which contained one of the first descriptions of the isolation of human
gastric juice and the *in vitro* study of its properties. Little is known about
Edward Stevens other than the date when he defended and published his
thesis. An English translation of a part of Stevens' thesis was appended to the
English translation of Spallanzani's *Dissertations relative to the natural history
of animals.* Stevens was fortunate in finding a human subject to participate
in his studies. He conducted his experiments at Edinburgh "upon an Hussar,
a man of weak understanding, who gained a miserable livelihood, by swallowing stones for the amusement of the common people."[21] After 20 years of
swallowing stones, this "human stone-swallowing regurgitator" might have enjoyed swallowing the hollow silver spheres which Stevens prepared. These
delicacies were divided into two cavities and perforated on the surface with
holes through which needles could be admitted. Various foods were placed in

the spheres and the action of the digestive juices on them was compared. Knowledge of human digestion was not substantially improved until 1833 when William Beaumont (1785-1853) published the results of studies conducted through the gastric fistula of Alexis St. Martin.

The Phlogiston Theory

About the same time that Robert Boyle (1627-1691), the "skeptical chemist," was undertaking the housecleaning that chemistry so desperately needed, another scientist in Germany was formulating a strange theory which would affect the course of chemistry until the end of the eighteenth century. This was the doctrine of phlogiston, derived from alchemical theory by Johann Joachim Becher.[22]

Johann Joachim Becher (1635-1681)

Becher's father, a Protestant minister, had been ruined by the Thirty Years War. Despite his lack of a formal education, Becher forged for himself a career that brought him academic and financial success, alternating with periods of exile and infamy. He seems to have won his medical doctorate by marrying a privy counsellor's daughter. Although his chemical knowledge was self taught, he was appointed Professor of Medicine at the University of Mainz in 1666. Later he became physician to the Elector of Bavaria. Becher was as interested in economics as chemistry and took a position at the Commercial College in Vienna. In 1678 various disputes in Vienna forced him to take refuge in Holland, but in 1680 he again had to change residence, this time to England. While in Holland, Becher had persuaded the city of Haarlem to buy his secret process for turning silver into gold. He promised that with hundreds of pounds of silver he could transform tons of sand into mountains of gold. The failure of his demonstrations forced him to seek refuge in England.

According to Becher, bodies consist of air, water, and three kinds of earth: "terra pinguis" (fatty), "terra mercurialis" (mercurial), and "terra lapidia" (stony). The earths correspond to the three principles of Paracelsus: sulfur, mercury, and salt. During combustion the terra pinguis was released as fire. Becher's crude, pseudoscientific theory of combustion had the virtue of dissociating the chemical study of actual substances from the alchemical notions of "essenses" and "potential qualities." Although Paracelsus had not precisely meant his principle of combustability—"sulfur"—to be equated with actual sulfur, confusion was inevitable. Becher argued that many substances without sulfur could burn and thus combustion did not involve sulfur at all. Combustion was defined as the breakup of a body with the loss of its more volatile components.

It was Georg Ernst Stahl who transformed terra pinguis into phlogiston.

Georg Ernst Stahl (1660-1734)

In Stahl's hands the doctrine of phlogiston became a framework that held together many bewildering chemical phenomena and guided chemical research for a century. Unfortunately, the popularity of Stahl's theories obscured some of the really valuable insights Robert Boyle and other members of the Royal Society had gained in their pioneering studies of combustion and respiration.

Like Becher, Stahl was the son of a Protestant minister, but his education was in a more formal and conventional mold. Stahl studied medicine at Jena and taught there from 1683-1693. In 1693 the new university of Halle was founded. Friedrich Hoffmann (1660-1742), a distinguished physician and founder of one of the famous medical "systems" of the period, requested that Stahl be appointed to a professorship at the new university. Hoffmann taught chemistry, physics, anatomy, surgery, and medical practice. Stahl's responsibilities included botany, physiology, dietetics, pathology, and materia medica, but he also included chemistry in his courses. Soon, he and Hoffmann became bitter rivals, vilifying each other in the epic battles of vitalist and mechanist. (Hoffmann saw the body as a hydraulic machine driven by a hypothetical fluid called the "nervous ether" which circulated in the nervous system.)

In 1703 Stahl published an edition of Becher's *Physica Subterranea* and appended his own exposition of the mechanism of combustion. He spoke of a potent agent—"fire, flaming, fervid, hot"—which intervenes in combustion. "Phlogiston" was defined as "a material principle and as a constituent part of the whole compound, the material and principle of fire, not fire itself."[23] While it is easy to ridicule the phlogiston doctrine by reducing it to the trivial core that things burn because they are burnable, some judgments of Stahl are more kindly, calling him "one of the first to formulate a rational system of chemistry on a structure of experimental observations."[24]

The definition of chemistry which Stahl formulated in his *Fundamenta chymiae* (1723) sounds quite modern and is a definite break with the objectives of alchemy:

Universal chemistry is the Art of resolving *mixt, compound,* or *aggregate* Bodies into their *Principles*; and of composing such Bodies from those *Principles.*

It has for its *Subject* all the *mix'd, compound,* and *aggregate Bodies* that are *resolvable* and *combinable*; and *Resolution* and *Combination,* or *Destruction* and *Generation,* for its *Object.*[25]

The principle of fire was found in the vegetable, animal, and mineral kingdom, but was more abundant in the first two—which left little residue when they burned. According to phlogiston theory, when things burn they lose phlogiston. Studies of combustion, oxidation, respiration, and photosynthesis were stimulated by phlogiston theory. Although flaws and false predictions were soon embarrassingly evident, the theory was not discarded for one hundred years.

One glaring discrepancy was the gain in weight that occurred when metals were burned. Many ingenious rationalizations were put forth to save the theory. Since phlogiston could not be trapped and analyzed there was no compelling reason to assume it was an ordinary substance with weight. Like heat, light, magnetism, electricity, or ether, it might be a "subtle fluid." Some chemists argued that instead of being attracted to the earth like other elements, phlogiston tended to rise. This caused an increase in the weight of metal oxides and a weight loss when the oxides were reduced.

Stahl's physiological theories were in many ways a return to the mysticism of Paracelsus and a retreat from the iatrochemistry of Sylvius, who tried to prove that events occurring in the living body were really ordinary chemical events. For Stahl, all chemical changes in the living body, no matter how they seemed to resemble ordinary chemical events, were of a fundamentally different nature. In the living body all chemical changes were controlled by the sensitive soul, the *"anima sensitiva."* This sensitive soul pervaded all parts of the body and governed their function. Rejecting Descartes' dualism and description of the body as a machine, Stahl contended that physiological phenomena in the living body obeyed laws entirely different from those governing the inanimate world. Thus, all living things, no matter how simple, obeyed a different set of laws from all nonliving things, no matter how complex. In denying the Cartesian duality of mind and body, Stahl substituted the duality of living and nonliving worlds.

Taken as a whole Stahl's system was regressive and put physiology even further behind the physical sciences of his time. Phlogiston chemists implicitly accepted the alchemical notion that substances are composed of matter plus some intangible principles, essences, or spirits which could be separated from the material entity by appropriate techniques.

Pneumatic Chemistry

Phlogiston theory retained the loyalty of some of the most eminent eighteenth century chemists—Joseph Black, Henry Cavendish, and Joseph Priestley. Intelligent men, skillful in experimentation, and innovative in many ways, they were aware of difficulties with the theory, but did not feel compelled to reject it.

While the connection between the processes of respiration and combustion had been suspected long ago, it was not until chemistry joined the modern

sciences at the end of the eighteenth century that the problem could be solved
in terms of chemical reactions. Solving the physiology of respiration required
the confluence of knowledge of the circulation of the blood, the movements
of respiration, the microanatomy of the lungs, and the chemistry of the gases.
By the end of the eighteenth century, the isolation and characterization of
gases led to a revolution in chemistry. The pneumatic trough, invented by
Stephen Hales, made this unprecedented burst of chemical discovery possible.

Stephen Hales (1677–1761)

Neither a physician nor professional scientist, the inventor of the pneumatic
trough was a parish priest, Stephen Hales, the Perpetual Curate of Teddington.
Between sermons he kept himself busy with a series of experiments on the
hydraulics of the vascular system, the chemistry of gases, the respiration of
plants and animals, and the measurement of blood pressure. He was also interested in sanitation and the value of ventilation. In addition to many papers
read to the Royal Society, he published *Vegetable Staticks* (1727) and
Statistical Essays, containing Haemastaticks (1733).

By separating the reaction vessel from the collection vessel and collecting
gaseous products over water, Hales made it possible to separate and store gases.
While he was able to isolate several distinct species of gas, including carbon
dioxide (from mineral waters) and oxygen (from nitre), he did not fully
realize the qualitative differences among gases. Hales assumed that, whatever
the source, he was merely dealing with ordinary air, somewhat modified by his
experiments. He called all gases "air," but saw air as a "volatile Proteus among
the chymical principles."[26]

Applying more rigorous qualitative methods to the study of plants and animals, Hales returned to van Helmont's plant growth experiment. Not content
to assume that water alone entered into plant material, but actually measuring
the amount of water taken up by roots and given off by leaves, Hales concluded that, in addition to water, something material is supplied by the air to
plant growth. Air, he reasoned, must contain some "secret food of life."

Joseph Black (1728–1799)

After studying languages and natural philosophy, Black became interested in
science and medicine. In 1754 he received the doctorate of medicine for a
dissertation entitled, "On the acid humour arising from food and on magnesia
carbonate alba." Stones formed in the human bladder and gravel passed in
the urine were of interest to medical researchers at the time. A heated
controversy was raging over the use of caustic agents as solvents for "the
stone." Prime Minister Walpole and his brother had been treated by the

secret remedy of Mrs. Joanna Stephens. Satisfied with the results, the Walpoles paid £5000 for her recipe which was published in the *London Gazette,* June 19, 1739. When he discovered "fixed air," Black was testing the major component of Mrs. Stephens' powders and pills—calcined snails.

According to phlogiston theory, limestone or chalk become caustic quick lime by taking up phlogiston. When quick lime was slaked it gave off phlogiston. Black discovered that when chalk was calcined or burned it gave off a gas which Black called "fixed air." Driving "fixed air" through a clear solution of lime water, caused the formation of "mild lime" which precipitated out of solution, providing a simple test for the presence of "fixed air." Black proved that this same "fixed air" was released during fermentation, was present in expired air, and was produced by burning charcoal. Fixed air (carbon dioxide) was deadly to animals, and would extinguish a flame. At first Black thought that the part of the atmospheric air that was unsuited for respiration must be his "fixed air." But the experiments of Dr. Daniel Rutherford showed that there was another noxious gas in the atmosphere—it was later identified as nitrogen.

Van Helmont had discovered "fixed air" before and called it "gas sylvestre." But Black could certainly consider his work far superior to van Helmont's since it was quantitative and rigorously experimental rather than speculative. His work stimulated other British chemists to study the chemical nature of gases.

Henry Cavendish (1731–1810)

A wealthy and eccentric private scholar, Henry Cavendish made important studies of the composition of gases, electricity, and geophysics. Although he studied at Cambridge, he took no degree because he objected to the strict religious tests applied to candidates for degrees at the University. After studying physics and mathematics in Paris, he settled in London. A large inheritance allowed him to devote himself to scientific experiments, using the finest instruments and reagents available. Otherwise he had no interest in money. Described as the "richest of the learned, and the most learned of the rich," he left behind a fortune of about one million pounds and many remarkable unpublished works.

In 1766 Cavendish published in the *Philosophical Transactions* of the Royal Society "Three Papers, containing Experiments on Factitious Air." Factitious air was defined as "any kind of air which is contained in other bodies in an unelastic state, and is produced from thence by art." By dissolving zinc, iron, or tin in vitriolic acid, Cavendish produced a new "factitious air" which he called "inflammable air." For a time, he thought this gas—now called hydrogen—must be phlogiston itself. Careful specific-gravity measurements clearly differentiated between "inflammable air" and other gases. Exploding a mixture of hydrogen and oxygen, Cavendish proved that water was not an element, but a compound of hydrogen and oxygen. Studies of atmospheric gases in 1873 proved that the composition of the atmosphere is constant at different times and places.[27]

Carl Wilhelm Scheele (1742-1786)

Oxygen was discovered almost simultaneously by Carl Scheele and Joseph Priestley, but both being staunch phlogistonists could not fully appreciate the nature of this new gas. Conducting a series of experiments to prove that air is made up of two components, "foul air" and "fire air," Scheele determined the ratio to be one part "fire air" to three parts "foul air" (volume/volume). Scheele believed that the function of "fire air" was to absorb phlogiston given off by burning substances. When air became saturated with phlogiston, it could no longer support combustion.

"Phlogiston," wrote Scheele, "is a true element and quite simple principle. Phlogiston cannot be obtained by itself since it does not separate from any substances unless there is another present with which it is in immediate contact." Scheele and Priestley knew that "fire air" was used up in combustion and that the remaining air had less weight and volume than the original mixture. This was ingeniously explained by postulating that the combination of air and phlogiston produced a compound so subtle it passed through the pores of the glass and dispersed into the atmosphere.[28]

Joseph Priestley (1733-1804)

Although mainly remembered as an experimental chemist, Priestley was also a minister, theologian, author, and educator.[29] During his lifetime, he was more famous as a radical theologian and political thinker than as a scientist. He wrote about 150 books, most of them religious or educational. He began his experiments on gases in the public brewery that adjoined his home in Leeds. The gas which bubbled out of the beer making vats was, of course, Joseph Black's "fixed air." Priestley discovered that although the gas was insoluble in water, it produced a very pleasant effervescent beverage which he likened to the best mineral waters. Priestley had invented soda water, which the College of Physicians once recommended to the Lord of the Admiralty as a possible cure for the sailors' disease—scurvy.

In July 1791 a mob burnt down Priestley's house during a series of attacks on dissenters and radicals. Priestley barely escaped the riots with his life, but his home, library, laboratory, apparatus, and many manuscripts were destroyed. He reluctantly decided he must emigrate to America and finally settled in Northumberland, Pennsylvania where his house still stands. (His laboratory has been partially restored as a bicentennial project of the American Chemical Society.)

An enthusiastic and independent experimenter, Priestley was curious about everything, but unsystematic in his researches. He often said that if he had known any chemistry he would never have made any discoveries. The role of chance loomed very high in his estimation, for he said "more is owing to what we call chance . . . than to any proper design, or pre-conceived theory in this business."[30] Despite the apparent lack of plan or system, Priestley was a careful observer. Among the gases he discovered were ammonia, sulfur dioxide, carbon monoxide, hydrogen chloride, nitric oxide, and hydrogen sulfide. Humphry Davis said of Priestley, "No single person ever discovered so many new and curious substances."

It did not seem possible to prepare "air" that was purer than the best common air, but this was what happened in 1774 when Priestley extracted a new "air" from mercuric oxide, using a burning lens with a diameter of twelve inches and a twenty inch focal distance. The new air allowed a candle to burn with a very vigorous and enlarged flame. Priestley tested the effects of his new air on mice, plants, and himself. Mice survived in this new air longer than in a similar limited quantity of ordinary air. On breathing this air himself he felt his breath "peculiarly light and easy." Predicting that it would have usefulness in medicine, he prophesized that it might be dangerous too: "As a candle burns out much faster in this air than in common air, so we might live out too fast . . . the air which nature has provided for us is as good as we deserve."

Studies of gases led Priestley to investigate the problem of respiration and the relationship between plants and animals. He became interested in ways to restore air damaged by breathing to a useful state. After many failures, he found that air that had been "injured" by animal respiration or by the burning of candles could be restored by plants. This surprised him greatly at first. Since air was vital to life, he expected both plants and animals would affect air in the same manner.

Unfortunately, Priestley saw all of his discoveries in terms of phlogiston theory. He never accepted Lavoisier's oxidation theory, but clung to his original name for his new gas—"dephlogisticated air." Priestley believed that air became saturated with phlogiston from combustion, respiration, and other chemical processes which render it "unfit for inflammation, respiration, and other purposes to which it is subservient." Plants, in contrast, take up phlogiston and purify the air. Changes in the color of the blood were also explained in terms of phlogiston. Dark venous blood, laden with phlogiston, released phlogiston in the expired air. Bright red dephlogisticated arterial blood was, therefore, capable of absorbing phlogiston from body tissues. These changes could be reproduced in vitro. Blood exposed to dephlogisticated air became as bright as arterial blood. Since Priestley visualized respiration only in terms of the turnover of phlogiston, he failed to appreciate Black's observation that fixed air was exhaled.

Priestley acknowledged that "the phlogiston theory is not without difficulties. The chief of them is that we are not able to ascertain the weight of phlogiston." Yet he did not see this as a reason for rejecting the theory because no one could "pretend to have weighed light or the element of heat, though we do not doubt that they are properly substances capable by their addition or abstraction, of making great changes in properties of bodies, and of being transmitted from one substance to another."

Antoine-Laurent Lavoisier (1743–1794)

In life style and approach to science, there could be no greater contrast than that between Joseph Priestley and Antoine Lavoisier. While Priestley worked unsystematically following his insatiable curiosity rather than a plan, Lavoisier approached chemistry as a physical scientist who had mapped out a protocol for the complete reformation of chemistry at the outset of his career. Politically, Priestley was a leading radical, forced to seek refuge in the new world. Lavoisier, although a reformer and humanitarian, lost his life because of his ties to the ancient regime of France.

Lavoisier was the son of a wealthy lawyer. While preparing himself to follow his father's profession, he became interested in the natural sciences. At the age of 21 he submitted the first of many memoirs to the Royal Academy of Sciences. Four years later he was elected a member of the Academy and decided to devote himself to scientific researches. To support himself in a manner that would not interfere with his experiments, he became an assistant-farmer in the private tax-farm used to collect government taxes. He completed his preparations for his chosen life-style by his marriage to the 14 year old Marie Anne Pierrette Paulze in 1771. Marie was the perfect wife; she illustrated his scientific works, assisted him in the laboratory, kept his notes, translated the works of English chemists into French for him, and entertained his famous visitors.

In 1789 Lavoisier was quite satisfied that he had succeeded in creating a revolution in chemistry. His *Traité elementaire de chimie,* the first really modern textbook of chemistry, was published in that year. Translated into English, German, Dutch, Italian, and Spanish, it became a best seller and major influence on the direction of chemistry as theory and technique.

Astute as he was in science, Lavoisier did not seem to recognize the danger of the social and political revolution erupting all around him. As a tax collector and opponent of Jean Paul Marat, Lavoisier was in grave danger. The hated tax-farmers were arrested, tried, and guillotined. The mathematician

Antoine Laurent Lavoisier with Mme. Lavoisier and other associates studying human respiration. (© 1959, Parke, Davis and Company.)

Joseph Louis Lagrange said the day after Lavoisier was executed: "It took but a moment to cut off that head; perhaps a hundred years will be required to produce another like it." But Coffinhal, President of the Revolutionary Tribunal said: "The Republic has no use for savants." In one of those ironies of history, the day that Lavoisier was executed for his role in a reactionary system, Priestley was on a ship bound for America, marked as a radical. Madame Lavoisier later married another chemist, Benjamin Thompson (1753–1814), an American Tory who was awarded the title Count Rumford. Their marriage was stormy and short-lived.

Lavoisier's chemical work was not especially remarkable for originality, but was special in terms of its precision, planning, and accuracy. In praising Lavoisier, Liebig said: "He discovered no new body, no new property, no natural phenomenon previously unknown. His immortal glory consists in this—he infused into the body of science a new spirit." On November 1, 1772, Lavoisier deposited a sealed letter with the Secretary of the Academy of Science. This letter, which discussed some preliminary experiments on combustion and a general summary of his tentative conclusions, was to serve

him as a claim to priority against future publications—provided all went as he
expected. In 1772 he wrote in his notebook that the work he planned to
carry out "seemed destined to bring about a revolution in physics and chemistry." The revolution would be accomplished by repeating previous work
"with new safeguards in order to link our knowledge of the air that goes into
combination or that is liberated from substances, with other acquired knowledge and so to form a theory." It was impossible to carry out this plan until
1772, when Lavoisier met Priestley and learned about his "dephlogisticated
air." Lavoisier saw the implications of this new gas, repeated Priestley's experiments and prepared the gas he named "oxygen." It is possible to regard
Lavoisier as the *real* discoverer of oxygen.[31]

By 1777 Lavoisier had rigorous proof that "eminently respirable" air is
converted into "fixed air" by both combustion and respiration and was ready
to exorcise the spirit of phlogiston from the corpus of chemical theory. In
1783 the Lavoisiers conducted a ceremony reminiscent of the way Paracelsus
had treated the books of the ancient authorities. Mme. Lavoisier, robed as a
priestess, celebrated the hundredth anniversary of the death of Becher by
burning the books of Stahl and the phlogiston theorists and chanting a
requiem for phlogiston. "It is time to bring chemistry to a more rigorous
way of reasoning," Lavoisier wrote, and "strip off the merely hypothetical."
Pointing out the path chemistry must follow in the future, he did not expect
older scientists to accept his views for their minds had been "creased into a
way of seeing things" so that they could "rise only with difficulty to new
ideas." But he took comfort in his belief that all young chemists read and
followed his ideas: "All young people adopt the new doctrine, and from this
I conclude that the revolution in chemistry is accomplished."

As early as 1777, Lavoisier had published a paper on respiration called
"Experiments on the Respiration of Animals and on the Changes which the
Air undergoes in passing through the lungs." With a sure understanding of the
nature of the gases involved, Lavoisier lucidly explained respiration as a slow
combustion or oxidation. Unlike Priestley, Lavoisier saw clearly that respiration used oxygen and released carbon dioxide. To investigate the physiochemical basis of physiological phenomena, Lavoisier collaborated with Pierre Simon
Laplace (1749-1827) in the design of an experimental system which could
quantitatively measure the production of animal heat. With the ice-calori-
meter, respiration and combustion could actually be compared in quantitative
terms—rather than as metaphor. In his "Memoir on Heat" Lavoisier concluded
that one may consider "the heat released in the conversion of pure air to fixed
air by respiration as the principal cause of the maintenance of animal heat."[32]

Unfortunately, the French Revolution truncated Lavoisier's career just as
the point where he was learning to apply the new chemistry to physiological
phenomena.

Physiology in the Nineteenth Century

The first half of the nineteenth century was a period rich in innovative researches and discoveries. It is a period often compared with the Renaissance in terms of scientific creativity. With a newly matured chemistry and a firmly grounded physics, physiology too could develop into a true science. Casting off vague "virtues," "faculties," "spirits," and "vital forces," scientists searched out physicochemical explanations of life phenomena.

Jöns Jacob Berzelius (1779-1848)

After completing his medical studies with a thesis on the therapeutic effects of galvanism (negligible, according to Berzelius), he found that he preferred the study of chemistry to the practice of medicine. At the beginning of the century, John Dalton (1766-1844) had proposed a new atomic theory based on weights and measurements rather than ancient philosophical convention. Recognizing the value of Lavoisier's and Dalton's work, Berzelius carried out the tedious analytical work that transformed Dalton's theory and clumsy symbols into a workable system. (After an argument about the use of Berzelius's "horrifying symbols" Dalton suffered the first of a series of strokes.) Berzelius prepared a table of the 50 elements then known, determined the atomic weights of almost all of them, discovered many new elements, began a chemical journal, developed new analytical methods, corresponded with all the important chemists of his time, and trained so many chemists that almost all the important nineteenth century chemists were either his students or students of his students.

Since Berzelius began his studies of natural science with medical training, it is not surprising that his interests included organic substances such as bile, blood, and feces. In 1812 he began to analyze these complex materials and found that even here the laws of chemical combinations applied. Among his accomplishments were the discovery of pyruvic acid and the development of the concepts of isomerism and catalysis. A prolific and fluent writer, Berzelius unleashed a torrent of articles and books which helped unify and direct chemical studies for almost a century.

In some respects Berzelius retained a vitalist view about the origin of natural products, but he recognized that the use of the term "vital force" was merely a cover for ignorance. "We know," he wrote, "after this explanation, as little as we knew before." Above all, Berzelius brought the chemicals of life within the purview of the atomic theory. Inevitably, the first man to synthesize an organic chemical in the laboratory was one of the Berzelius' students—Friedrich Wöhler.

Friedrich Wöhler (1800–1882)

The son of a veterinary surgeon, Wöhler described himself during his student days as distinguished neither by special zeal nor broad learning. Since a very early age, his chemical experiments and mineral collections had distracted him from his assigned schoolwork. At Heidelberg University the chemist Leopold Gmelin was so impressed with Wöhler that he persuaded him to do research and forego attending chemistry courses as a waste of time. After Gmelin's colleague, Fritz Tiedemann, suggested the possibility of a chemical approach to physiology, Wöhler began an investigation of the transformation of substances in urine, often using himself as guinea pig.

Work on cyanic acid began while he was a medical student and may be regarded as preliminary to his later famous studies of urea. After graduation, Wöhler followed Gmelin's example and went to Stockholm to study with Berzelius. By evaporating a solution of ammonium cyanate, Wöhler produced the isomeric compound—urea. This was the first time that a chemical produced by living beings had been synthesized from materials which were available, at least in principle, from nonliving matter. The synthesis of urea *in vitro* is often cited as the event which abolished the distinction between organic and inorganic compounds. Rather cautiously, Wöhler drew attention to this aspect of his discovery as "an example of the artificial production of an organic, so-called animal, substance from inorganic substances." In a letter to Liebig he was less restrained in language. "I can no longer contain my chemical water," he wrote. "I can make urea without kidney of man or dog."[33]

Although the preparation of urea is often seen as proof of the theoretical equivalence of inorganic and organic chemicals, this event was actually a more significant landmark in isomer studies. Not until the synthesis of acetic acid by Hermann Kolbe (1818–1884) in 1845 was there a clear case for the rejection of vital action as prerequisite to the production of organic chemicals. It took many other demonstrations to really overthrow vitalism in chemistry. Marcelin Berthelot (1827–1907) is generally credited with giving the old theory its deathblow, but certainly Wöhler was the first to seriously injure the concept.

Animal Chemistry

Prior to 1750, nutrients were regarded as necessary for the animal as lubricants for the muscles and joints and as replacements for parts worn out by the wear and tear of daily life. Chemists assumed that these components were present in foodstuffs and were directly assimilated. By 1800 some rather complex organic compounds had been separated and described—including lactose, starch, chlorophyll, glycerol, and tarataric, lactic, uric, prussic, oxalic, citric, and malic acids. Some chemists, such as Lavoisier and Michel-Eugene Chevreul

(1786-1889), had been impressed by the naturalists use of the species concept and had attempted to apply it to chemical species.

Although organic chemistry seemed to be overwhelmingly complicated, chemists expected that knowledge of nutrients and the composition of organisms would provide a scientific basis for agriculture and nutrition. Believing that empirical formulae adequately characterized organic chemicals, chemists still found themselves unable to deal with all the chemical species found in foodstuffs. They turned instead to an analysis of bulk foods, body fluids, solids, and excrements. A simple means of dealing with such complexity was to regard foods as a mixture of saccharine, oleaginous, and albuminous materials (i.e., carbohydrates, fats, and proteins).

Chemists and physiologists differed as to the proper approach to elucidating the vital phenomena of the body. Despite chemistry's new pride in its sophistication, chemical explanations of life had a difficult time in working their way into nineteenth century physiology and medicine—even as they had in the time of Paracelsus. Many physiologists saw "animated anatomy"—vivisection—as the only way to unravel the complex phenomena of life. Test tube chemistry, in their view, could contribute nothing to understanding life. "If life be chemical," wrote Charles Caldwell, Professor of Medicine at Transylvania University, "man is degraded to the level of a laboratory."[34]

The successes and failures of the chemist, Justus von Liebig, and the physiologist, Claude Bernard, illustrate the consequences of this dispute.

Justus von Liebig (1803-1873)

Struggling to create a community of chemists in Germany, Liebig conceived it his mission to prove that chemists could deal with the world of living beings as well as the inanimate world and that chemical science would revolutionize nutrition, agriculture, and industry.

Quite early in his life, Liebig revealed his passion for chemical experiments. His father was a pharmacist and dealer in chemicals. As a youth, Liebig carried out syntheses and experiments in the little laboratory in his father's shop. Liebig's career and work fell into two phases. He was first satisfied with classical organic chemistry, developing and improving methods of analysis and identification, and his teaching. In the fourth decade of his life he turned to more complex problems of agricultural chemistry and the chemistry of living things. This new interest brought him his greatest fame and triumphs, but also revealed his weaknesses as a thinker and his growing tendency to formulate grandiose theories from a shaky and meager base of data.

Rejecting all evidence that yeast is a living organism, Liebig was contemptuous of the fermentation studies of Schwann and Pasteur. Although Pasteur was not entirely correct, Liebig's theory that yeast was the product rather than cause of fermentation was wrong and retrogressive. According to Liebig, "putrefying" yeast communicated some kind of molecular vibration to the fermenting solution. Liebig believed that only green plants could build up complicated organic substances from the simple inorganic elements they took from the air and the soil. While plants were synthetic chemical factories, animals were degradative chemical processors which took the components they needed for their tissues from the plants and used the rest as fuel. While Liebig assumed that chemical transformation *in vivo* was essentially the same as that *in vitro,* he could not free himself totally from the old concept of the "vital force." Assuming that the conservation of matter was a law applicable to living things, he attempted to determine what chemical events occurred in the living body by measuring and balancing all the components that were ingested and excreted by the body.

In his approach to metabolic phenomena, Liebig is reminiscent of Santorio Santorio, and indeed he was also engaged in a futile effort. Claude Bernard later said that this attempt to deduce the invisible metabolic phenomena from such analyses of input and output was like trying to deduce what happened inside a house by measuring what goes in the door and out the chimney.

Liebig made other errors, such as his assumption that protein degradation alone supported muscular activity and that urea output could be used as an index of animal activity. His ideas about nutrition, especially about the role of proteins, stimulated food faddism such as that of Sylvester Graham—immortalized in the common name of the cracker.

Francois Magendie (1783-1855)

Experimental physiology owed much to the indefatigable, irascible, and ruthless Francois Magendie. Although his work did not culminate in any broad or all inclusive generalizations, he contributed many individual facts, a special spirit of inquiry, and virtually founded the field of experimental pharmacology with work on poisons and emetics. Magendie served as Professor of Medicine at the College de France, President of the French Academy of Sciences, and President of the Advisory Committee on Public Hygiene. In these offices and through his researches and private lectures, Magendie left his imprint on every branch of physiology. His *Formulaire* (1821) introduced the medical world to strychnine, morphine, iodides, and bromides.

The foundation of Magendie's work was vivisection; reports of his ruthless experimentation stimulated much antivivisectionist activity.[35] His goal was to establish physicochemical explanations of vital phenomena and he approached this through what was termed an "orgy of experimentation." Although he attempted a purely mechanical explanation of most vital phenomena, he seems to have allowed for a "vital force," at least in the nervous system (whose workings he thought were qualitatively different from other life processes). Studies of the nervous system, resulting in an acrimonious priority battle with Sir Charles Bell, included his most famous experiments. Magendie seems to be entitled to full credit for the discovery of the separate motor and sensory functions of spinal nerves. Nevertheless, his most important contribution to science was rescuing his student, Claude Bernard, from his turpor and mediocrity and inspiring him with a love for physiological science.

Claude Bernard (1813–1878)

Bernard surpassed his mentor in many ways. Even the egotistic Magendie had to acknowledge that Bernard's skill in vivisection was superior to his own and time has testified to his ability to formulate significant generalizations. The combination of these talents made Bernard the true founder of modern experimental physiology.

Born into a family of poor peasants, Bernard was fortunate in receiving some training from the parish priest in classical subjects. After more advanced studies, he taught language and mathematics at a Jesuit school while taking private pupils to earn more money. Financial difficulties forced him to become assistant to an apothecary at Lyons. Here he carried out menial tasks, such as sweeping the floors, washing the glassware, and delivering prescriptions. The first preparation that his employer allowed him to do independently was of shoe polish. Delivering prescriptions was the only enjoyable part of his work, for it allowed him to stop and watch the operations carried out at the nearby veterinary school.

Finding many aspects of medicine quite ridiculous, Bernard wrote a short play revolving around theriac. This was a popular medicine which contained at least 60 ingredients—so haphazardly compounded that no two batches were ever the same. The success of this play made him aspire to greater things and he began to work on an ambitious five-act historical drama. This so interfered with his involvement in the fine art of apothecary that his master terminated his contract. Arriving in Paris full of hope, Bernard was quite deflated when Saint-Marc Giradin, a literary critic at the Sorbonne, advised him to learn another profession if he wanted to eat. In 1834 Bernard entered the School of Medicine at Paris. Although he received an internship in 1839, he was hardly regarded as a brilliant student. Indeed, he ranked twenty-sixth out of 29 taking the examinations.

Bernard served as Magendie's assistant for many years, since he had difficulty getting a professorship himself. Trying to rescue himself from financial privations, he contracted an arranged marriage for the sake of a good dowry. His wife turned out to be a very pious lady who actively opposed vivisection.

Their quarrels exacerbated Bernard's bouts of ill health and attacks of gastri-
tis. From 1847, Bernard and Magendie shared the lectures at the College de
France. That same year, Bernard won the prize of the Academy of Sciences
for experimental physiology. In 1854 he became a member of the Academy.
Finally, in 1855 he succeeded Magendie in the professorship at the college.
Worn down by years of worry, quarrels with his wife, and the years spent in
the cold, damp cellar that was his laboratory, Bernard was forced to return to
his birthplace to try to regain his health at 47 years of age. During this period of rest he had time to reflect on the broader implications of his work.
The resultant book, *An Introduction to the Study of Experimental Medicine,*
became a landmark in the history of physiology and elevated Bernard from
scientist to sage.

Claude Bernard. [From Marcel Florkin, *A History of Biochemistry,* **Parts I and
II, Elsevier, Amsterdam, 1972 (vol. 30 of** *Comprehensive Biochemistry,* **edited
by M. Florkin and E. Stotz).**]

Shortly before his death, Bernard began a study he believed would throw new light on the problem of fermentation. Incomplete notes and speculations discovered after his death drove Pasteur into another frenzy of experimentation to refute these speculations. During his terminal illness, Bernard told his friends: "What a pity; it would have been good to finish it." France gave Bernard the funeral of a hero. The Chamber of Deputies voted him the first state funeral in honor of a scientist. The novelist Gustave Flaubert (1821–1880) described the event as more impressive than the ceremonies at the death of the Pope. Bernard's funeral, he said, "was religious and very beautiful."[36]

In the course of his researches, Bernard made more discoveries and contributed more insights into physiological phenomena than all previous physiologists. Among his discoveries were: the glycogenic function of the liver, the vasomotor nerves, the function of the pancreatic juices in digestion, and the nature of the action of curare, carbon monoxide, and other poisons. He wrote many papers and a 14-volume study of physiology, *Lessons in Experimental Physiology applied to Medicine*. Just as important were his conceptual contributions—notably the concept of the "internal secretion," the constancy of the internal milieu of complex creatures, and his scientific philosophy. For Bernard, vitalism and mechanism were essentially worthless philosophies—often obstructing scientific progress. Instead he promoted "determinism"—faith in the experimental method and its applicability to living organisms.

As early as 1854 Bernard had included a discussion of the principle of the constancy of the internal milieu in his lectures. The concept appears in his *Introduction to the Study of Experimental Medicine* and was further developed in his later writings. His doctrine of determinism, implicit in his earliest work, allowed him to develop this fundamental physiological principle. Bernard believed that there were always real physicochemical bases for all vital phenomena, no matter how diverse and mystifying they might appear. The guiding principle of research into vital phenomena must be to continue until a suitable physicochemical explanation has been reached. Bernard's goal was to completely banish that capricious agent, the "vital force," which resisted the laws applying to the inanimate world and made all acts performed by living organisms somehow beyond science.

In his *Introduction to the Study of Experimental Medicine* Bernard warned that experimenters must "doubt, avoid fixed ideas, and always keep their freedom of mind." But they must avoid the excesses of skepticism, for "we must believe in science, i.e., in determinism; we must believe in a complete and necessary relation between things, among the phenomena proper to living beings as well as in all others."[37]

Bernard was as concerned with establishing the superiority of the physiological approach to Liebig's chemical approach as he was with stamping out vestiges of vitalism. He realized that this was a difficult path and perhaps brooded that his work was overshadowed by more dramatic or glamorous scientists like Pasteur and Darwin. His image of the life of science expresses the difficulties quite well. "If a comparison were required to express my idea of the science of life," he wrote, "I should say that it is a superb and dazzlingly lighted hall which may be reached only by passing through a long and ghastly kitchen."[38]

While it was impossible to define precisely what "life" is, the scientist must not allow speculations to interfere with true scientific work—which is to analyze and compare the manifestations of life. If forced to define life in a single phrase, Bernard would say, "life is creation." He expanded on this phrase by defining a created organism as "a machine which necessarily works by virtue of the physico-chemical properties of its constituent elements." Although the term "vital properties" was often used, actually it was a phrase used for properties "which we have not yet been able to reduce to physico-chemical terms; but in that we shall doubtless succeed some day."[39] The way to understand life phenomena for Bernard was through vivisections and experiments. He argued that "the science of life can be established only through experiment, and we can save living beings from death only after sacrificing others."[40]

The kind of dedication demanded of the physiologist was a total one: "A physiologist is not a man of fashion, he is a man of science, absorbed by the scientific idea which he pursues: he no longer hears the cry of animals, he no longer sees the blood that flows, he sees only his idea and perceives only organisms concealing problems which he intends to solve."[41]

Glycogenic function of liver and the synthetic capabilities of animals

Bernard believed that his demonstration of the glycogenic function of the liver was his most important piece of work. To do this, he had to divorce himself from the prevailing theories of plant and animal metabolism and strike out in an independent direction. At the time, scientists believed that sugar in the blood of carnivores must have been supplied entirely from foods ingested. Animals, it was assumed, only use sugar which is synthesized ultimately by plants. Animals used foodstuffs synthesized by plants to support a combustion which supposedly took place either in the blood or the lungs. It was Claude Bernard who finally proved that animal blood contains sugar even when it is not present in the diet.

In tests of the theory that sugar absorbed from the food was destroyed in passing through the liver, or lungs, or some other tissue, Bernard put dogs on a carbohydrate diet for several days and then killed the animals during digestion. He found large amounts of sugar in the hepatic veins. For his control experiment, he fed dogs only on meat and was surprised to find much sugar in the hepatic veins, although not in the intestines. Before concluding that current theories were wrong, he did many additional experiments to ascertain the location of the tissue that supposedly served as the site of the destruction of carbohydrates. Through such experiments, Bernard discovered gluconeogenesis—the conversion of other substances into glucose in the liver. This was a major break with the idea that the liver could only absorb glucose from the blood—glucose which had to come from the diet. In addition, Bernard discovered: glycogen, the carbohydrate storage polymer of animals; glycogenesis, the synthesis of glycogen; and glycogenolysis, the breakdown of glycogen to glucose. Other experiments initiated to determine whether sugar is destroyed in passing through the lungs (to create animal heat) revealed that all tissues had ferments to use sugar which is not totally or exclusively destroyed in the lungs or any particular tissue.

These studies led to the concept of the "internal secretions" which were products transmitted directly into the blood instead of being poured out to the exterior of the gland or organ secreting them.

Internal and external environment

Although most French scientists had ignored Schwann's ideas of cell physiology and metabolism, with its emphasis on the importance of the nutritive medium bathing the cells, Bernard was perceptive enough to see its applicability to the fundamental problem of physiology—the relationship between the cells and their immediate surrounding environment. Bernard believed he was the first scientist to insist that complex animals have two environments. The animal experienced "an *external environment* in which the organism is placed, and an *internal environment* in which the elements of the tissues live. Life does not run its course within the external environment . . . but within the *fluid internal environment* formed by circulating organic liquid that surrounds and bathes all the anatomical elements of the tissues."

The great generalization drawn from this work was his dictum: "The constancy of the internal environment is the condition for free and independent life." Because the animal encloses itself in a kind of "hot-house," changes in external conditions do not perturb it. This freedom is the "exclusive possession of organisms which have attained the highest state of complexity or organic differentiation." Higher animals are not indifferent to their environment, Bernard explained, but are rather in close and intimate relation to it, so that their equilibrium is the result of compensation established as continually and as exactly as if by a very sensitive balance.[42]

Introduction to the Study of Experimental Medicine portrays Claude Bernard as the ideal scientist and sage, always lucid and rational, never at a loss for a hypothesis. A close examination of his research notebooks reveals

that the path to discovery was much more arduous and tortuous than his
polished publications admit.[43]   Although his skill in experimental surgery was
superlative, deficiencies in his mastery of chemical techniques seem to have
hindered him throughout his career. As Holmes shows, Bernard truly believed
in himself as a reformer of physiology and as a highly individualistic researcher
despite long years of struggle and obscurity during which immersion in his
research brought few successes, rewards, or recognition. Bernard's concept of
the constancy of the internal environment was not immediately appreciated,
particularly by his conservative French colleagues.

Elucidating the complex pathways of metabolism and the enzymes, hormones, and neural agents which regulate them still constitutes an enormous
and incomplete task. Bernard's work marks an epoch which divides modern
physiology—as biochemistry—from all that went before. After Bernard left
the field it seems a pause was necessary before his very shrewd and prescient
views could be fully exploited. As Bernard recognized late in life, one could
"do no better than to follow the indications of natural phenomena, using
theories as torches intended to illuminate the path and needing to be replaced
as they were consumed."[44]   In the twentieth century many researchers in the
United States, Britain, Germany, and France returned to the concept of the
constancy of the internal environment and allowed it to guide their work.
Lawrence J. Henderson (1878-1942) publicized Bernard's work in the United
States, encouraged the translation of *Introduction to Experimental Medicine*
into English (1927), and in his book *The Fitness of the Environment* explored
"The Characteristics of Life" in terms of the relationship between the internal
and external environment.

Walter Bradford Cannon (1871-1945) was especially important in promoting
understanding of physiological regulatory mechanisms—even beyond the small
academic world—through his voluminous, lucid writings. In 1926 Cannon
coined the word "homeostasis" to describe the conditions that maintained the
constancy of the interior environment. According to Cannon (1932), the
term homeostasis "does not imply something set and immobile, a stagnation.
It means a condition—a condition which may vary, but which is relatively
constant." The coordinated physiological processes which maintain the steady
state of the living organism, Cannon described as "so complex and so peculiar to living beings—involving, as they may, the brain and nerves, the heart,
lungs, kidneys and spleen, all working cooperatively." Some physiologists
shared Cannon's hope that the study of regulatory mechanisms in complex
living beings might lead to improved understanding of other kinds of organization "even social and industrial—which suffer from distressing perturbations."[45]

Since Bernard and Cannon made homeostasis the guiding principle of
physiological research, the concept has broadened considerably. New terms
such as feedback, loops, servomechanisms, transfer functions, and cybernetics

Walter Bradford Cannon. [From J. F. Fulton and L. G. Wilson (1966). *Selected Readings in the History of Physiology.* Courtesy of Charles C Thomas, Publisher, Springfield, Illinois.]

indicate that the concept is used not only by biologists, but also is applicable to problems studied by engineers, sociologists, economists, ecologists, and mathematicians. The concept is so fundamental that L. L. Langley believes: "Any paper published today, at least in physiology, which is worth the paper it is printed on, should further clarify a homeostasic mechanism."[46]

Notes

1. M. Foster (1970). *Lectures on the History of Physiology.* New York: Dover, p. 55.

2. For example,
   F. H. C. Crick (1967). *Of Molecules and Men.* Seattle: University of Washington Press;
   A. Koestler and J. R. Smythies (1969). *Beyond Reductionism,* Boston: Beacon;
   J. Monod (1971). *Chance and Necessity.* New York: Knopf.

3. Foster, p. 146.

4. R. Descartes. *Treatise of Man.* Translated by T. S. Hall (1972). Cambridge, Mass.: Harvard University Press.

5. Descartes, pp. 9–10.

6. Descartes, pp. 17–21.

7. Foster, p. 62.

8. Foster, pp. 67–68.

9. Foster, pp. 71–73.

10. Foster, pp. 74–75.

11. Foster, pp. 165–166.

12. J. F. Fulton and L. G. Wilson, eds. (1966). *Selected Readings in the History of Physiology.* Chicago: Thomas, p. 214.

13. Foster, p. 285.

14. H. E. Sigerist (1933). *The Great Doctors, A Biographical History of Medicine.* New York: Norton, p. 185.

15. Sigerist, p. 191.

16. Foster, p. 291.

17. Foster, pp. 296–297.

18. E. Nordenskiold (1936). *The History of Biology.* New York: Tudor, pp. 238–243;
T. S. Hall (1969). *History of General Physiology.* Chicago: University of Chicago Press, pp. 46–56;
A. Vartanian (1960). *La Mettrie's l'homme machine.* Princeton, N. J.: Princeton University Press.

19. L. L. Langley, ed. (1973). *Homeostasis: Origins of the Concept.* Philadelphia: Dowden, Hutchinson and Ross.

20. Foster, p. 141.

21. Fulton and Wilson, pp. 173–174.

22. J. R. Partington (1961). *A History of Chemistry.* New York: St. Martin.

23. Foster, p. 168.

24. H. M. Leicester and H. S. Klickstein, eds. (1968). *A Source Book in Chemistry 1400-1900.* Cambridge, Mass.: Harvard University Press, p. 58.

25. Leicester and Klickstein, pp. 61–62.

26. Foster, p. 232.

27. Leicester and Klickstein, pp. 134–153.

28. Leicester and Klickstein, pp. 101–112.

29. A. Holt (1931). *A Life of Joseph Priestley.* London: Oxford University Press.

30. Leicester and Klickstein, pp. 112–125; Foster, pp. 237–243;
I. V. Brown, ed. (1962). *Joseph Priestley: Selections from His Writings.* University Park: Pennsylvania State University Press;
J. Priestley (1803). *The Doctrine of Phlogiston established and that of the composition of Water refuted.* 2nd Ed. Northumberland, Pa.: Byrne.

31. D. McKie (1952). *Antoine Lavoisier.* New York: Schuman.
32. Leicester and Klickstein, pp. 154–180;
    E. Mendelsohn (1964). Cambridge, Mass.: Harvard University Press.
33. M. Florkin (1972). *Comprehensive Biochemistry.* Vol. 30, *A History of Biochemistry.* New York: Elsevier, pp. 251–253;
    Leicester and Klickstein, pp. 309–316.
34. Florkin, p. 175.
35. R. D. French (1975). *Antivivisection and Medical Science in Victorian Society.* Princeton: Princeton University Press.
36. J. M. D. Olmsted (1938). *Claude Bernard Physiologist.* New York: Harper, p. 245.
37. C. Bernard (1957). *An Introduction to the Study of Experimental Medicine.* New York: Dover, p. 35.
38. Bernard, p. 15.
39. Bernard, p. 93.
40. Bernard, p. 102.
41. Bernard, p. 103.
42. Bernard, Foreword by I. B. Cohen.
43. F. L. Holmes (1974). *Claude Bernard and Animal Chemistry.* Cambridge, Mass.: Harvard University Press.
44. Holmes, p. XV.
45. W. B. Cannon (1967). *The Wisdom of the Body.* 2nd Ed. New York: Norton, pp. 24–25.
46. Langley, p. 293.

References

Allen, G. (1975). *Life Science in the Twentieth Century.* New York: Wiley.
Beck, L. J. (1952). *The Method of Descartes; a Study of the Regulae.* Oxford: Clarendon.
Bernard, C. (1957). *An Introduction to the Study of Experimental Medicine.* New York: Dover.
Berry, A. L. (1960). *Henry Cavendish: His Life and Scientific Work.* London: Hutchinson.
Brooks, C. McC. (1962). *Humors, Hormones, and Neurosecretions; the origins and development of man's present knowledge of the humoral control of body functions.* Albany: State University of New York.
Brooks, C. McC., Kaizumi, K., and Pinkston, J. O., eds. (1975). *The Life and Contributions of Walter Bradford Cannon, 1871–1945.* New York: State University of New York Downstate Medical Center.

Brooks, C. McC., and Cranefield, P. F., eds. (1959). *The Historical Development of Physiological Thought*. New York: Hafner.

Brown, I. U., ed. (1962). *Joseph Priestley: Selections from his Writings*. University Park: Pennsylvania State University Press.

Cannon, W. B. (1932). *The Wisdom of the Body*. New York: Norton.

Cannon, W. B. (1915). *Bodily Changes in Pain, Hunger, Fear and Rage; an account of Recent Researches into the Function of Emotional Excitement*. New York: Appleton.

Cannon, W. B. (1923). *Traumatic Shock*. New York: Appleton.

Cannon, W. B. (1936). *Digestion and Health*. New York: Norton.

Castiglioni, A. (1947). *History of Medicine*. New York: Knopf.

Clark-Kennedy, A. E. (1929). *Stephen Hales, DD, FRS, and Eighteenth Century Biography*. Cambridge: Cambridge University Press.

Cohen, C. (1975). The Protein Switch of Muscle Contraction. *Sci. Amer. 233:* 36–45.

Coleman, W. (1971). *Biology in the Nineteenth Century*. New York: Wiley.

Crick, F. H. C. (1967). *Of Molecules and Men*. Seattle: University of Washington Press.

Crombie, A. C. (1959). Descartes. *Sci. Amer. 201:* 160–173.

Crowther, J. G. (1962). *Scientists of the Industrial Revolution*. Philadelphia: Dufour.

Descartes, R. (1662). *Treatise of Man*. Translated by T. S. Hall (1972). Cambridge: Harvard University Press.

D'Irsay, S. (1930). *Albrecht von Haller*. Leipzig: Thiene.

Eyles, U. A. (1954). Nicolaus Steno. *Nature 174:* 8–10.

Farber, E., ed. (1961). *Great Chemists*. New York: Interscience.

Foster, M. (1899). *Claude Bernard*. New York: Longmans, Green.

Foster, M. (1901). *Lectures on the History of Physiology during the Sixteenth, Seventeenth, and Eighteenth Centuries*. Reprint of 1901 edition. New York: 1970 Dover.

French, R. D. (1975). *Antivivisection and Medical Science in Victorian Society*. Princeton: Princeton University Press.

Fulton, J. F. (1949). *Physiology of the Nervous System*. New York: Oxford University Press.

Fulton, J. F., and Wilson, L. G., eds. (1966). *Selected Readings in the History of Physiology*. Chicago: Thomas.

Garboe, A., ed. (1959). *The Earliest Geological Treatise*. London: MacMillan.

Gibb, F. W. (1965). *Joseph Priestley: Adventurer in Science and Champion of Truth*. London: Nelson.

Goodfield, G. J. (1969). *The Growth of Scientific Physiology*. London: Hutchinson.

Gordon-Taylor, G., and Walls, E. W. (1958). *Sir Charles Bell, his Life and Times*. Edinburgh: Livingstone.

Gotch, F. (1908). *Two Oxford Physiologists*. Oxford: Clarendon.

Graubard, M. (1953). *Astrology and Alchemy; two fossil sciences*. New York: Philosophical Library.

Grmek, M. D. (1972). "A Survey of the Mechanical Interpretations of Life from the Greek Atomists to the Followers of Descartes," in *Biology, History, and Natural Philosophy*. Edited by A. D. Breck and W. Yourgrau. New York: Plenum.

Guerlac, H. (1953). John Mayow and the Aerial Nitre. *International Congress of the History of Science*, Actes VII. Paris: pp. 332–349.

Guerlac, H. (1954). The Poet's Nitre. *Isis 45:* 243–255.

Guerlac, H. (1957). Joseph Black and Fixed Air: A Bicentenary Prospective, with Some New or Little Known Material. *Isis 48:* 124–151, 433–456.

Gunther, R. T., and Franklin, K. J. (1932). *Early Science in Oxford*. Vol. 9. Oxford: Clarendon.

Hall, T. S. (1975). *History of General Physiology 600 B.C. to A.D. 1900*. 2 vols. Chicago: University of Chicago Press.

Haller, A. von. (1786). *First Lines of Physiology* (1786). Translated by W. Cullen (1966). New York: Johnson Reprint.

Hartog, P. (1941). The Newer Views of Priestley and Lavoisier. *Annals of Science 5:* 1–56.

Hawkes, E. (1928). *The Pioneers of Plant Science*. London: Sheldon.

Haymaker, W., ed. (1953). *Founders of Neurology*. Springfield, Ill.: Thomas.

Hemmeter, J. D. (1908). Albrecht von Haller. Scientific, Literary and Poetical Activity. *Johns Hopkins Hospital Bulletin 19:* 65.

Holmes, F. L. (1963). The Milieu Interieur and the Cell Theory. *Bulletin of the Institute of the History of Medicine 37:* 315–335.

Holmes, F. L. (1974). *Claude Bernard and Animal Chemistry. The Emergence of a Scientist*. Cambridge, Mass.: Harvard University Press.

Holt, A. (1931). *A Life of Joseph Priestley*. London: Oxford University.

John, H. J. (1959). *Jan Evangelista Purkyne, Czech Scientist and Patriot*. Philadelphia: American Philosophical Society.

Jorpes, J. E. (1966). *Jac Berzelius, His Life and Work*. Translated by B. Steele. Stockholm: Almquist and Wiksell.

Keynes, G., ed. (1949). *Blood Transfusions*. Bristol: Wright.

Kloaster, H. S. van. (1947). Jan Baptist van Helmont. *Journal of Chemical Education 24:* 319.

Koestler, A., and Smythies, J. R., eds. (1969). *Beyond Reductionism.* Boston: Beacon.

Langdon-Brown, W. (1946). *Some Chapters in Cambridge Medical History.* Cambridge: Cambridge University Press.

Langley, L. L., ed. (1973). *Homeostasis: Origins of the Concept.* Philadelphia: Dowden, Hutchinson and Ross.

Leake, C. D., ed. (1956). *Some Founders of Physiology: Contributions to the Growth of Functional Biology.* Washington, D.C.: American Physiological Society.

Leicester, H. M., and Klickstein, H. S., eds. (1968). *A Source Book in Chemistry 1400-1900.* Cambridge, Mass.: Harvard University Press.

Long, E. R. (1965). *History of Pathology.* New York: Dover.

Lowndes, R. (1878). *René Descartes: His Life and Meditations.* London: Norgate.

Lusk, G. (1922). A History of Metabolism. *Endocrinology and Metabolism* (edited by L. F. Barker) *3:* 1–78.

Maar, V. (1902). *Nicolaus Steno: Opera Philosophica.* Copenhagen: Glydendal.

Mahaffy, J. P. (1881). *Descartes.* Philadelphia: Lippincott.

McKie, D. (1952). *Antoine Lavoisier: Scientist, Economist, Social Reformer.* New York: Schuman.

McKie, D. (1942). The Birth and Descent of John Mayow: a Tercentenary Note. *Phil. Mag. J. Sci. 33:* 51–60.

McKie, D. and Heathcote, N. U. (1935). *The Discovery of Specific and Latent Heats.* London: Arnold.

Mendelsohn, E. (1964). *Heat and Life. The Development of the Theory of Animal Heat.* Cambridge, Mass.: Harvard University Press.

Monod, J. (1971). *Chance and Necessity.* New York: Knopf.

Moon, R. O. (1931). President's Address: Van Helmont, Chemist, Physician, Philosopher, and Mystic. *Proceedings of the Royal Society of Medicine 25:* 1–6.

Moulton, F. R., ed. (1942). *Liebig and After Liebig.* Washington, D.C.: American Association for the Advancement of Science.

Needham, D. (1971). *Machina Carnis. The Biochemistry of Muscular Contraction in its Historical Development.* Cambridge: Cambridge University Press.

Olmsted, J. M. D. (1938). *Claude Bernard, Physiologist.* New York: Harper.

Olmsted, J. M. D. (1944). *Francois Magendie, Pioneer in Experimental Physiology.* New York: Schuman.

Partington, J. R. (1961). *A History of Chemistry.* 2 vols. New York: St. Martin.

Partington, J. R. (1959). Some Early Appraisals of the Work of John Mayow. *Isis 50:* 211–226.

Partington, J. R. (1956). The Life and Work of John Mayow (1641–1679). *Isis 47:* 217–230.

Partington, J. R. (1936). Jean Baptista Van Helmont. *Annals of Science 1:* 359–384.

Patterson, T. S. (1931). John Mayow—in Contemporary Setting. *Isis 15:* 47.

Priestley, J. (1966). *A Scientific Autobiography of Joseph Priestley (1733–1804): Selected Scientific Correspondence.* Edited by R. E. Schofield. Cambridge, Mass.: MIT Press.

Ramsay, W. (1918). *Life and Letters of Joseph Black.* London: Constable.

Rolleston, H. D. (1932). *The Cambridge Medical School.* Cambridge: Cambridge University Press.

Rossiter, M. W. (1975). *The Emergence of Agricultural Science: Justus Liebig and the Americans, 1840–1880.* New Haven: Yale University Press.

Roth, L. (1937). *Descartes' Discourse on Method.* Oxford: Clarendon.

Sachs, J. von. (1909). *A History of Botany.* Oxford: Clarendon.

Scheele, C. W. (1952). The Discovery of Oxygen. *Alembic Club Reprints 8.*

Schiller, J. (1967). *Claude Bernard et les Problems scientifiques de son temps.* Paris: Editions du cedre.

Scott, J. F. (1952). *The Scientific Work of René Descartes.* London: Taylor and Francis.

Shenstone, W. A. (1901). *Justus von Liebig, his Life and Work.* New York: Cassell.

Sigerist, H. E. (1933). *The Great Doctors: a Biographical History of Medicine.* Translated by E. and C. Paul (1958). Garden City: Doubleday.

Smith, E. F. (1920). *Priestley in America, 1794–1804.* Philadelphia: Blakiston's.

Smith, H. M. (1949). *Torchbearers of Chemistry.* New York: Academic.

Soderbaum, H. G. (1934). *Jons Jacob Berzelius, Autobiographical Notes.* Translated by O. Larsell. Baltimore: Williams and Wilkins.

Stevenson, L. G. (1947). *Sir Frederick Banting.* Toronto: Ryerson.

Stillman, J. M. (1960). *The Story of Alchemy and Early Chemistry.* New York: Dover.

Strunz, F. (1907). *Johann Baptist van Helmont (1577–1644).* Leipzig: Deuticke.

Suner, A. P. (1955). *Classics of Biology.* New York: Philosophical Library.

Thorpe, E., ed. (1920). *Chemical and Dynamical Researches.* Monographs in Industrial Chemistry. Clayton (U.K.): William Margarine.

Thorpe, E., ed. (1894). *Essays in Historical Chemistry.* London: MacMillan.

Thudichum, Ludwig (1875–76). The Discoveries and Philosophy of Liebig, with Special Reference to Their Influence upon the Advancement of the Arts, Manufactures and Commerce. *Journal of the Society of Arts.*

Underwood, E. A. (1944). Lavoisier and the History of Respiration. *Proceedings of the Royal Society of Medicine 37:* 247–262.

Urdang, G. (1948). Berzelius and Pharmacy. *Journal of the American Pharmacological Society, Scientific edition 37:* 481–85.

Valery, P. (1947). *The Living Thoughts of Descartes.* Philadelphia: McKay.

Volhard, J. (1909). *Justus von Liebig.* 2 vols. Leipzig: Barth.

Walden, P. (1944). *Drei Jahr Tausende Chemie.* Berlin: Limpest.

White, J. H. (1932). *The History of Phlogiston Theory.* London: Arnold.

Wilson, G. (1851). *The Life of the Honourable Henry Cavendish.* London: Cavendish Society.

12

EVOLUTION

From Natural History to Biology

Because it is better suited to popularization and controversy among non-scientists, evolution overshadowed the more exact and rigorous nineteenth century biological sciences like physiology, cell theory, embryology, animal chemistry, and even microbiology. Without doubt the great synthesis of the old natural history and the new science of biology is epitomized by Charles Darwin's *Origin of Species.* The publication of this work in 1859 and the controversy that it ignited have long obscured much that was important in nineteenth century biology—both new concepts and a new level in experimental research. During the nineteenth century the "scientist" began to replace the "natural philosopher." More important, the professional scientist began to displace the amateurs, physicians, and clergymen who had previously found natural philosophy a hospitable vocation. Darwin has been described as the last and greatest of the natural philosophers.

The term "biology" was first introduced at the very beginning of the nineteenth century and was publicized by the writings of the French zoologist, Jean Baptiste Lamarck (1744-1829), and a German naturalist, Gottfried Treviranus (1776-1837). According to Lamarck's definition, biology is

> one of the three divisions of terrestrial physics; it includes all which pertains to living bodies and particularly to their organization, their developmental processes, the structural complexity resulting from prolonged action of vital movements, the tendency to create special organs and to isolate them by focusing activity on a center and so on.[1]

In the 1830s Auguste Comte (1798-1857), the French social philosopher, called biology one of the principle sciences of Positive Philosophy and specified

biologists as the only persons "competent to apply physical theories successfully to the rational solution of physiological problems."[2] Biologists were
ready to turn away from their traditional practices of purely descriptive
biology and fact-gathering, to studies of the active state of plants and animals
by means of observation and experiment. To understand the new desire of
biologists to observe and experiment we must review briefly the eighteenth
century passion for the accumulation and classification of masses of details
about plants and animals from all over the world. This is most clearly shown
in the life and work of Carl Linnaeus (1707-1778), the principal founder of
scientific taxonomy. Ironically, evolutionary theory appears as the fruition of
the work of taxonomists and natural philosophers who saw only stability and
design in nature. Believing that species were fixed at the creation, they believed their studies could only further enhance the glory of the Creator.

Some degree of transmutation had been acceptable to the church until
about 1600. Few species were known and the concept of species was poorly
defined. After all, if Adam was the father of all the human races, and all
races were of one species, some changes must have occurred since the creation.
Consideration of the size of the ark seemed to prove that Noah had assembled
far fewer pairs of animals than there were species on the modern earth. But
as Linnaeus and other scientists cataloged and scientifically defined species,
the church became more strict in its antievolutionary stance.

Since much of the scientific debate concerning evolution used the dialect
of systematic classification, a brief survey of that field is needed.

Catalogs and Classifications

As our knowledge of the world expands, we need to group individual objects
into general categories which can be used to communicate with others. Even
primitive people develop schemes for ordering objects in their environment into useful groups. These are usually based on easily understood, striking, and
important properties. Edible and inedible are obviously important categories,
as it is important to eat lettuce and not poison ivy. But some of these classification schemes are really quite sophisticated, even employing a form of
chemotaxonomy.

Beginnings of Classification

Plato and Aristotle were the first to attempt to deal with classification on
scientific lines. Plato's concept of the "forms" representing the "idea" of
each species was very influential, but Aristotle was more realistic, in that he
placed the forms inside nature. As discussed previously, he never developed a

The Lamia. [From Edward Topsell's *The History of Four-Footed Beasts and Serpents and Insects* (with an introduction by Willy Ley). (1967). Courtesy of Da Capo Press, New York.]

complete system of classification, but suggested ways of doing so. The only systematic terms in his work are "genus" and "species," but his concept of the "scale of nature" or "great chain of being" provided an intellectual framework for later naturalists.[3] Seeing himself as a pioneer gathering facts for future use, Aristotle argued that one must use many characteristics to discover natural groupings, rather than simple bifurcations as Plato did. If Aristotle's principles had been faithfully followed, taxonomy would have had a fairly good head start. Unfortunately, descriptions of animals were to become more and more popular, but less and less accurate. Natural histories were filled with strange and wonderful creatures, whose behaviors were even more bizarre than their appearance. Zoology degenerated to the point where separation of fact from fiction became a herculean task.

The search for systematic schemes of classification dominated the life sciences during the seventeenth and eighteenth centuries. Several factors contributed to this new preoccupation with classification, but the sheer bulk of new material was crucial. Current estimates put the number of species of living beings on earth at about 2,000,000. Aristotle described about 500 species of animals. Up until 1600 only about 6000 species of plants were known. By 1700, botanists had 12,000 new ones to sort through. Similar problems of "information retrieval" were facing the zoologist. Second, naturalists realized that their "science" lagged disgracefully far behind the physical sciences. Physicists had created an orderly universe, bound by natural

The Unicorn. [From Edward Topsell's *The History of Four-Footed Beasts and Serpents and Insects* (with a new introduction by Willy Ley). (1967). Courtesy of Da Capo Press, New York.]

laws. Certainly there must be laws that would create order from the chaotic mass of plant and animal species.

Another factor in the zeal for classification was an outgrowth of a long overdue "housecleaning" of natural histories. It was time for zoologists to institute a purge of the bizarre, exotic, and nonexistent species that cluttered

The Dragon. [From Edward Topsell's *The History of Four-Footed Beasts and Serpents and Insects* (with a new introduction by Willy Ley). (1967). Courtesy of Da Capo Press, New York.]

The Hydra. [From Edward Topsell's *The History of Four-Footed Beasts and Serpents and Insects* (with a new introduction by Willy Ley). (1967). Courtesy of Da Capo Press, New York.]

the pages of books purporting to be works of science. Many strange creatures had been cataloged in Pliny's ever-popular *Natural History*. Later writers continued to include these mythological beasts in revised natural histories. Travellers might hope to catch a glimpse of "wonder people" like the Monoculi, who have one eye in the middle of their forehead; or the Androgeni, who contain both sexes in one body; or the Ostomi, who have no mouths but subsist on odors. Among Pliny's unusual beasts was the fire-breathing chimera with the head of a lion, goat's body, and serpent's tail.

During the Middle Ages further corruptions of the *Natural History* were popular since animal stories were used to teach morals. St. Hildegard the Nun (1098-1179) used Genesis as a guide to organizing the plants and animals described in *Causes and Cures,* one of her medical works. Her work included a common error, a regression from Aristotle's classification, for she called the whale the first of the fishes. Even the greatest of the sixteenth century "encyclopedic naturalists," Conrad Gesner (1516-1565), included many dubious stories in his five-volume *Historia Animalium.*[4] Generally the animals were arranged alphabetically, although he sometimes grouped closely related animals. Species were intermingled indiscriminately wherever the alphabet placed them. According to Gesner, the bird of paradise flies so high that it has no need of feet, and the female has to lay her eggs in a hollow in the back of the male because no nesting sites were available to her. The tree goose was another peculiar bird which grew like fruit on a tree. Sometimes Gesner was skeptical about the animals included in his work, but apparently he could not resist including the basilisk—a lizard-like monster hatched by a serpent from the egg of a rooster. For each animal, Gesner included a description of its habits and behavior, means of capture, and uses as food or medicines. Many interesting illustrations accompanied the text. Gesner also planned a monumental *Historia plantarum,* but unfortunately it was never written. In the epidemic of plague that struck Zurich in 1564, Gesner, the Chief Town Physician, fell ill. He died the next year when the plague returned. Many other naturalists took up the task of describing and arranging the plants. Most herbals merely listed the plants in alphabetical order or by relative importance to man, but some botanists attempted to organize the plants according to overall affinities or by the properties of particular organs. However, the first naturalist to develop a taxonomic system applicable to plants and animals using the species as his unit, was John Ray. In the foundation of modern taxonomy he ranks second only to Linnaeus.

John Ray (1627–1705)

Sickly as a youth, Ray often compounded his own herbal remedies. His mother probably stimulated his interest in the medical uses of plants inasmuch as she was the local herbalist, or healer. Ray was ordained and held many college offices, but lost his positions during the reign of Charles II for refusing to sign the Act of Uniformity. Constantly fighting poverty, Ray spent the rest of his life studying natural history in Scotland, Holland, Germany, Italy, and France. Ray's writings are extensive and very diverse, including sermons, religious essays, handbooks on the classics, and treatises on folklore and natural science.

In his *General History of Plants,* Ray summarized the entire body of botanical knowledge then available. Beginning his work with the pious intention of completing the task of Adam—naming the plants and animals—Ray went on to the more complex labor of describing, classifying, and interpreting the meaning of form and function in living beings. Morphology, behavior, instinct, interactions with the environment had to be reconciled with a simple anthropomorphic, teleological viewpoint. In nature, Ray found order and purpose in accord with God's own design. These ideas were developed in Ray's most influential work, *The Wisdom of God manifested in the Works of the Creation* (1691). As indicated by the popularity of Ray's work, science and theology have not always been at war. The period from the publication of Ray's *Wisdom of God* until 1859 when Darwin published *Origin of Species* saw the flowering of "natural theology." Scientists and clergymen shared the comfortable conviction that studies of nature, God's "universal and public manuscript" would always be in harmony with God's written word—the Bible.

In Nature, Ray found a stable and perfect world created by God, but in his descriptive work, religious convictions rarely interfered with objectivity. Since he treated the species as the fundamental unit of classification, Ray analyzed the concept of species with the greatest thoroughness and insight up to that time. Plants were regarded as the same species if they produced plants similar to themselves through their seeds. That is, a breeding test should serve as the basis of the definition of species. He wrote in 1686:

> After a long and considerable investigation, no surer criterion for deter-
> mining species has occurred to me than the distinguishing features that
> perpetuate themselves in propagation from seed. Thus, no matter what
> variations occur in the individual or the species, if they spring from the
> seed of the one and the same plant, they are accidental variations and
> not such as to distinguish a species.[5]

Trivial differences would not be the basis of species. For example, different colored flowers were not different species any more than different colored calves. Color variations in plants could be preserved by asexual methods, cuttings, but not through the seeds.

Because God had filled the entire Scale of Nature, there were no great gaps between living forms. Careful studies would reveal the intermediate types and fill in the Scale of Nature as God intended. Ray stated that the number of species does not change because God rested on the seventh day from all his work—which was the creation of new species. However, Ray did not regard the stability of species as absolute; he had observed that plant species, at least, can be varied through "degeneration" of the seeds. Ray included exam-ples of this phenomenon, such as the production of leaf-cabbage from the seed of cauliflower, and certain items of folklore such as the degeneration of cultivated grains into common weeds such as wheat to *Lolium* and maize to other kinds of weed. Thus, although all species were descendants of pairs originally created by God, there was some flexibility of type. Species were not absolutely fixed in their properties.

Natural Theology promised that God always preserved the species. Al-though individuals were allowed to perish, extinction was impossible. There-fore, Ray was troubled by the possibilities suggested by fossils. He comforted himself with the belief that man was God's most recent creation even if fossil plants and animals suggested a greater antiquity of the earth.

Studies of the natural history of the earth and its living creatures grew in popularity—with the blessings of the theologians. Natural Theology promised a new way to understand God. A transition from "bible studies" to "field and stream" appealed to many people. There were pleasant, outdoor working conditions, no mind-bending training nor studies of ancient languages, no ex-pensive tools, and the puzzle posed by nature was an exhilarating challenge.

Nature studies could reveal God's plan—obviously carefully contrived and skill-fully executed, considering the short time available for its unfolding.

The relationship between the work of Ray and Linnaeus is not easily un-derstood. While some say that Ray's work "falls far short of Linnaeus,"[6] Ray's biographer, Charles Raven, feels that Ray had a truer appreciation of the real task of the scientist than Linnaeus, who defined the fundamentals of botany as classification and nomenclature, thereby limiting science to systematology.[7]

Because of Ray's religious and political problems, poverty, caution, and his reactionary retention of Latin instead of the vernacular, he was destined to be overshadowed by Linnaeus in the eyes of history. It is possible that Ray's ideas, if further developed, could have led to natural systems of classification that would have yielded a developmental systematics more compatible with evolutionary ideas than the Linnaean system.

Carl Linnaeus (1707-1778)

Botanist, entomologist, zoologist, physician, and mineralogist, Linnaeus was above all the founder of modern taxonomy. The task of biology for Linnaeus was to complete the assignment of Adam: to name the plants and animals, to contemplate the handiwork of God, and to marvel at His creations. His energy, ability and temperament were those exactly suited to the intellectual priorities of his time. Samples, specimens, and seeds were sent from all over the world to Linnaeus, even through naval blockades. The greatest joy for many amateur scientists was to have Linnaeus name a new species they had uncovered with a Latinized version of their own name.

In all, Linnaeus dictated four autobiographies. Each is a different version of his life, but he emerges quite clearly as a very temperamental individual. One fact is not in question—Carl was literally brought up in a garden. His family tree can be traced directly to a tree. His father was a peasant who took the name Linnaeus from a linden tree near his home. Because of its age and size the tree was regarded as almost sacred by the country people. First child in a large family, Carl shared his father's interest in botany and natural science, but Carl hated school. He thought classical studies were worthless and found his schoolmasters crude and cruel. However, his physics teacher encouraged him with books by naturalists and physicians and urged him to study medicine rather than for the priesthood. His other teachers advised his father not to waste money on Carl's education as he would never amount to anything; the boy should be apprenticed to a tailor or carpenter. Instead Carl went on to the University at Upssala where his poverty led to bouts of catarrh, scurvy, toothache, headache, rheumatism, gout, stones, and hunger.

In 1732, Linnaeus had the opportunity to make an extensive field trip to Lapland to collect natural science data from some 4600 miles of wild and unexplored regions in Lapland. Carefully noting his expenses, he purchased a measuring stick, telescope, magnifying glass, knife, paper and plant drying equipment. In all, the expedition cost about $100. During the trip he collected not only articles of natural science, but also anthropological data and met his future wife, the patient Sara-Lisa Moraeus, to whom he was engaged for eight years. His future father-in-law would not allow the marriage to take place until Linnaeus became a physician. Linnaeus went to the University of Harderwijk in Holland because there was no medical program in Sweden. Apparently medical school was not too difficult, because in a matter of months he obtained his doctor's degree.

Linnaeus in Lapp costume. [From Heinz Goerke (1973). *Linnaeus* (translated from the German by D. Lindley). New York: Charles Scribner's Sons. Courtesy of Wissenschaftliche Verlagsgesellschaft, Stuttgart, Germany.]

More important than his physician's license was the publication of the first edition of *Systema Naturae* in 1735. The book brought him instant fame and paved the way for his successful career as arbiter of species. Returning to Sweden, Linnaeus was rewarded with a university appointment and many honors. Yet he complained that all the recognition bestowed on him prevented him from undertaking new expeditions. Practically a national monument, he was too valuable to be risked. Many of his young disciples travelled for him and some died on such missions. "The death of many whom I have induced to travel has turned my hair grey and what have I gained?" he asked. "A few dried plants, with great anxiety, unrest and care." The effects of poverty and overwork undermined his health. After 1770 he suffered a series of strokes which led to paralysis and the loss of mental powers. When Linnaeus died in 1778, he was buried with all the honors proper to royalty. Quarrels over possession of Linnaeus' collection followed his death. Unfortunately for Sweden, his collection was sold to an Englishman. Legend has it that Sweden sent a warship out to recapture it, but the British ship was too fast.[8]

Linnaeus in 1775. [From Heinz Goerke (1973). *Linnaeus* (translated from the German by D. Lindley). New York: Charles Scribner's Sons. Courtesy of Wissenschaftliche Verlagsgesellschaft, Stuttgart, Germany.]

Linnaeus had formulated the basic principles of his life's work by the age of 25. In the following years his work was often revised and his ideas developed and changed. Different editions of his works contain different statements about species. However, many critics of Linnaeus quote those passages which support their own ideas, as if the man never changed his mind. Well aware that his system was not "natural," Linnaeus emphasized the virtues it undeniably had: It was practical and easy to use; plants and animals were grouped into species, genus, order, class, and kingdom. The binary nomenclature standardized by Linnaeus became the universally accepted method of naming plants and animals.

For Linnaeus, naming things was the very foundation of science. The complete title of the famous *Systema Naturae* continues . . . *or The Three Kingdoms of Nature systematically proposed in classes, orders, genera, and species.* As well as taking up the task of Adam, naming things, Linnaeus also assumed the burden Aristotle took up in *Historia Animalium,* that of knowing living things as they exist. "The first step in wisdom," he wrote, is to know the things themselves; this notion consists in having a sure idea of the objects; objects are distinguished and known by classifying them methodically and giving them appropriate names . . . classification and name-giving will be the foundation of our science." His best work was, doubtless, to consist of botanical studies. Here, again, he saw the fundamentals of science as "the division of plants and systematic name-giving, generic and specific."[9]

Linnaeus felt that his most important work concerned procreation in plants, a study which revealed "the very footprints of the Creator." Although many today find the subject somewhat less than inflammatory, Linnaeus reveled in poetic descriptions of the "celebration of love and nuptials in plants, in the bridal bed provided by the petals."[10] The descriptions of plant sexuality utilized the language of human concerns, such as "pure marriage" and "adultery." Plants were divided up into 24 classes according to the number and character of the stamens. These classes were divided into orders by the number and character of the pistils. Therefore classification required careful attention to the reproductive organs and sex life of the plant.

Some judgments of Linnaeus' taxonomy were so harsh, it appeared the young scientist would indeed be "philosophically and botanically annihilated."[11] For a time, Linnaeus almost gave up botany for medical practice. His use of plant sexuality was also viewed as a shocking and inappropriate method of classification to teach to tender young minds. His anthropomorphic approach to plant classification made the study of innocent flowers seem quite risqué. (In "The Loves of the Plants," Erasmus Darwin made Linnaeus' allusions to promiscuity, marriage, and adultery more explicitly pornographic.)

Although Linnaeus was best known for his botanical work, he did make valuable contributions to zoology, including the use of the binomial system for nomenclature, and his great personal influence on zoological science through friends, associates, and pupils. In the first edition of the *Systema Naturae* (1735), Linnaeus deplored the underdeveloped status of zoology: "Zoology, the noblest part of Natural History, is much less worked up than the other two parts." This was a poor state of affairs since "animals are the highest and most perfect works of the Creator."[12] Linnaeus was weaker as a zoologist than in botany. Unfortunately, he lost his colleague and zoologist collaborator, Artedi, when that young man fell into a canal and drowned one night at the age of 30.

Linnaeus has been called an arrogant man who wrote as if personally present at the Creation.[13] But while taxonomy came to stand for fixity and a static view of nature, Linnaeus had discreet doubts about this during his career and had even written to a friend about the problem of the conservative religious authorities who could take away one's job for expressing wrong opinions. The canonical biological knowledge of the time held that evolution of life forms could not have occurred since all species were fixed entities created by God. In his *Classes Planterum* (1738), Linnaeus stated that "there are as many species as there were originally created diverse forms." Thus, during the early years of his long career, Linnaeus apparently believed that species were fixed entities and, therefore, he could not have had any sympathy for evolutionary speculation. In his observation of the Three Kingdoms, in the 1735 *Systema Naturae,* his fourth proposition is that all species were created by God, and that there are no new species.

As there are no new species (1): as like always gives birth to like (2): as one in each species was at the beginning of the progeny (3): it is necessary to attribute this progenitorial unity to some omnipotent and omniscient Being, namely *God,* whose work is called *Creation.* This is confirmed by the mechanism, the laws, principles, constitutions and sensations in every living individual.

The first proposition, an examination of the process of generation, also led to the conclusion that species were created and fixed as such.

1. If we observe God's works, it becomes more sufficiently evident to everybody, that each living being is propagated from an egg and that every egg produces an offspring closely resembling the parent. Hence no new species are produced nowadays.

Generation has increased the number of individuals, but not the number of species.

2. Individuals multiply by generation. Hence at present the number of individuals in each species is greater than it was at first.[14]

It seems rather odd that Linnaeus did not attach much importance to varieties of organisms, since he worked primarily with plants, where varieties of any given plant species are very easy to find and identify. Linnaeus considered that the "archetype" was the most important measure of a species, and that variations were only shadows of the archetype. These views seem to reflect the transmission and influence of the Greek ideas about species previously discussed, now mixed with Christian natural theology.

Linnaeus classified plants on a strictly morphological system, refining the technique that is now known as the typological species concept, but problems soon arose with Linnaeus' species concept. Joseph Kolreuter (1733-1806), produced evidence that plants could be hybridized. A change in Linnaeus' thinking is found in his *Systema Vegetabilium* (1774) in which he wrote:

> Let us suppose that the Divine Being in the Beginning progressed from the simpler to the complex; from few to many; similarly that He in the beginning of the plant kingdom created as many plants as there were natural orders. These plant orders He himself, therefrom producing, mixed among themselves until from them originated those plants which today exist as genera.
>
> Nature then mixed up these plant Genera among themselves through generation of double origin and multiplied them into existing species, as many as possible (whereby the flower structures were not changed), excluding from the number of species the almost sterile hybrids, which are produced by the same mode or origin.[15]

Using the old device of casting unorthodox ideas as a thought-game, Linnaeus admitted that species were not static.

Close inspection has lead various authors to try to rehabilitate Linnaeus from his reputation as a strictly antievolutionist, tedious classifier. Evidence found in his writings suggests a streak of heterodox thinking in Linnaeus which was suppressed because of unwillingness to offend the powerful religious interests of his age. When Linnaeus said, "every genus is natural and was in the beginning of things created such," it seems logical that species would have experienced a similar origin. However, later on Linnaeus discussed certain cases of species which seem to be the daughters of others. Linnaeus noticed a sportiveness in his own garden, where varieties kept popping up spontaneously and "abnormal" plants were derived from common ones. In his own mind he puzzled over the distinction between true species as made directly by the

Creator and the varieties which all gardeners and practitioners of animal husbandry knew. He even removed from later editions of the *Systema Naturae*
the statement that no new species can arise. However, varieties were regarded
as trivial, the products of changes in climate and environmental conditions.

That Linnaeus did not wish to encounter ecclesiastic disapproval, is evident
in a letter he wrote to a trusted friend, J. C. Gmelin, in 1747. He felt restrained in his published writings by the need to avoid offending the Lutheran
and orthodox ecclesiastics who ruled the "destinies of all seats of learning."
Linnaeus was probably not as rigid, unimaginative, and sure of himself as his
less creative followers became.[16]

Evolutionary Speculation

Because historians of science have collected precursors of Darwin with all the
zeal of somewhat unbalanced butterfly collectors after a rare variety, it is
difficult to say where a discussion of evolutionary theory should start. As we
have seen, before Aristotle Greek natural philosophers had postulated bold
theories of evolution in which man was treated as just another part of natural
development. A partial list of evolutionists would include at least Anaximander and Empedocles, among the Greeks; Benoit de Maillet, Buffon, Lamarck, Etienne Geoffroy St-Hilaire, and Maupertuis, among French Evolutionists; Erasmus Darwin, W. C. Well, J. C. Prichard, W. Lawrence, Patrick
Matthew, Robert Chambers, and A. R. Wallace, among the English. Some of
these are well known and others languish in well-deserved obscurity.

To call attention to some of the difficulties as well as the brilliant flashes
of insight that marked the development of evolutionary science we will survey
the work of some of these forerunners of Darwin, beginning with the French
author, Benoit de Maillet.

Benoit de Maillet (1656-1738)

A controversial book entitled *Telliamed: Or Discources Between an Indian
Philosopher and a French Missionary on the Diminution of the Sea, the
Formation of the Earth, the Origin of Men and Animals, etc.,* first appeared
in 1748. The author, Benoit de Maillet, had died 10 years before his book
was given to the public. De Maillet had been a diplomat and traveler with an
interest in paleontology and theology, but he was not really a scientist. Even
though he had suppressed the book during his lifetime, de Maillet disguised his
heretical ideas by using the device of a mythical foreign sage, named Telliamed.
This heathen (the author's name spelled backwards) could safely express
heretical or unorthodox ideas, while his interrogator, the respectable, but

curious French missionary, knew his religion would not allow him to hold such ideas. The book became popular in France and England, bringing theories of cosmic and organic evolution to many who would not be reached by purely scientific works.

De Maillet attempted to link cosmic and biological evolution in an all embracing cosmology reminiscent of the pre-Socratic Greeks. As in the ancient cosmologies, the world was eternally experiencing cycles of creation and destruction, with changes occurring through the effect of chance. The seas were all-important to this scheme and were the agents responsible for changes in the earth's crust as well as the site of origin of life forms. As in Anaximander's theory, all land animals were derived from creatures that had lived in the sea. Living beings developed only in response to favorable environmental conditions and adapted as the environment around them changed. As the seas began to recede, living beings adapted to the changing conditions until they were finally able to go up onto the land. From this point, all the familiar land organisms developed. De Maillet believed that "seeds of life" served as the points of origin for new species and that

the number of species of trees, plants, and flowers has increased with the extension of emerged lands, and that none existed before the appearance above water of the summits of the first mountains, it is equally certain that their number will increase from one century to the next as long as the diminution of the sea gives us new lands.[17]

Although de Maillet believed that species were always changing, he was faced with the problem of the observed fixity of species as opposed to the rapid metamorphoses of animal life. The explanation he adopted was the one used by later, better known students of evolution; that is, changes in species are never directly observed because they take place over such long periods of time. Incorporated into his theory was the transmission by heredity of newly acquired characteristics. To his great credit, de Maillet recognized the true nature of fossils, the successive deposition of geological strata, and anticipated calculations of the great age of the earth.

Since the time of Galileo, cosmic changes had been amply documented. The once immutable heavens and incorruptible celestial bodies were subject to change just as much as the earth. A bold intellect like de Maillet extended the findings of changes in the heavens to the idea of changes in the earth and its organic beings. Like Buffon, de Maillet was fascinated by the idea of "organic atoms" such as those seen under the microscope. Such minute living entities could be the primeval filaments or "seeds of life" which evolved into ever more complex creatures. De Maillet's theory is not one of sustained progress and hope of improvement. He believed that the process of evaporation from the seas was still occurring at a rate of about three or four inches per 100 years. When all the waters evaporated, there would be a period of universal volcanic action and the earth would burn up and become a sun.

This imaginative and unorthodox book exerted a strong influence on others, notably Cuvier and Buffon. While Cuvier vigorously rejected such ideas, Buffon seems to have followed many of de Maillet's arguments in his own works.

Pierre Louis Moreau de Maupertuis (1698–1759)

Maupertuis has been described as one of those scientists "who are more capable of formulating brilliant ideas than of testing them in a laboratory."[18] Although he was often on the verge of an important discovery, he never quite formalized the idea, tested it, or put it into a form that convinced his contemporaries. Yet a modern reader will be struck by the clarity and directness of his arguments in the light of present knowledge. While mostly remembered as a mathematician, especially for his studies of the properties of curves, Maupertuis was also interested in the problem of heredity and the origins of species. His speculations about evolution and inheritance were not generally appreciated, but his transformist theory is today regarded as an anticipation of the modern concept of mutation. Maupertuis demonstrated the fallacy of one of the commonly accepted ideas of his time—preformationism—by the process of simple reasoning from readily available observations.

In a remarkable little book, *The Earthly Venus,* Maupertuis promised to approach the great problem of life "only as an Anatomist" and "not as a Metaphysician." "Do not be angry," he warned the reader, "if I say you were a worm, or an egg, or even a kind of mud."[19] Although not much of an experimentalist himself, he advocated animal experiments to study the question of generation. A review of the ovist and spermist theories of generation quickly revealed the inadequacies of both camps. A consideration of the history of the subject brought up the question of how such erroneous theories could flourish after the great William Harvey had promoted epigenetic theory. Maupertuis concluded that Harvey had only "beclouded the issue." Since the time of Harvey, scientists ought to have learned more because they had acquired a powerful new tool—the microscope. But the new tool had led scientists even further from a true theory of generation. According to preformationist theories only one parent could contribute heritable traits to the offspring, but Maupertuis pointed out many contradictions to this idea, including the general observation that children could resemble either or both of their parents. A more striking case of mingling of traits involved the mulatto offspring of black and white parents. In the animal kingdom, the offspring of the donkey and the mare was well known as a composite of both parents. Here the parental types were so dissimilar that the reproductive organs of the mule were quite useless. How could such a composite arise if the fetus were preformed in the egg or sperm?

Although chemical theory was primitive at the time, Maupertuis suggested that chemical affinity was the principle that guided inheritance. He believed that the seminal fluid of each species has many particles suitable to form more animals of that species. These speculations led to the hypothesis that "each part furnishes its germ." This idea was later revived by Charles Darwin as his theory of Pangenesis. Maupertuis offered a possible test of his hypothesis. Why not try for many generations to mutilate some animal and see if the amputated parts eventually diminished or disappeared? Clearly, Maupertuis had anticipated the experiment associated with Weismann. The development of a new species from a variety might be enhanced by mating the same types for several generations. Like Darwin, Maupertuis believed that new varieties arise by chance, but he did not rigorously exclude the possible influence of climate and food. Thus, the heat of the Torrid Zone might somehow be more favorable to the "particles" that compose black skin than those for white skin. Given enough time selective matings might eventually produce new races of giants or dwarfs.

Maupertuis' ideas on species and evolution influenced a better known French scientist and author, Georges Buffon.

Georges Louis Leclerc, Comte de Buffon (1707–1788)

Born into a family with all the advantages of wealth and power, Buffon readily acquired a reputation as a charming and elegant man. A youthful love affair led to a duel which forced Buffon to leave France. Escaping from the rigidities of the Ancient Regime, Buffon enjoyed the excitement of the new experimental work of English scientists and the physics of Newton. In London Buffon studied mathematics, physics, and botany. When he finally returned to France, Buffon was said to have affected British mannerisms, even in his speech. By translating Newton's *Fluxions* and Hales' *Vegetable Staticks* into elegant, readable French, Buffon helped to bring British science to France. Although his first interest had been physics, Buffon decided to make biology his special field. A large inheritance freed him from the nuisance of having to earn a living and allowed him to devote his life to science. He began his work at 6 AM every day, but interrupted his studies twice a day to have his hair dressed and powdered.

In 1739 Buffon became an associate of the French Academy of Science and was appointed Keeper of the Jardin du Roi. Buffon made the "garden" into a valuable research center. His work resulted in the publication of a multivolume compendium of natural history which became a best seller in an increasingly literate culture. Even Louis XV, not known for intellectual curiosity in general, encouraged Buffon's work. Buffon's writings induced many young scientists to study biology rather than physics or mathematics. In the 44 volumes of the *Natural History,* Buffon presented a complete picture of the cosmos—from the stars and solar system to the earth and its organic and inorganic constituents.

In his extensive and elegant writings, Buffon managed to touch on many dangerous and forbidden topics so delicately that he nearly concealed his unorthodox views. Taken as a whole, his writings contain all the elements of the Darwinian synthesis, but Buffon assembled facts that support the theory of evolution while simultaneously denying that he was attempting to prove such a theory. Either Buffon did not realize the import of his ideas or he neatly camouflaged heretical tendencies by pointing out how the facts might lead to an evolutionary interpretation if one did not know what Genesis and the True Church taught. Charles Darwin[20] called Buffon "the first author who in modern times has treated [the origin of species] in a scientific spirit . . . ." Yet "his opinions fluctuated greatly at different periods," making Buffon a very uncertain ally in Charles Darwin's battle for scientific acceptance.

George Louis le Clerc, Comte de Buffon. [From A. Meyer (1939). *The Rise of Embryology.* **Stanford, California: Stanford University Press.]**

Despite his cryptic style, Buffon aroused the suspicion of orthodox scientists and theologians. He was called before the Syndic of the Sorbonne, the Faculty of Theology, on June 15, 1751. He was told that certain parts of the *Natural History* contradicted religious dogma and had to be withdrawn. The offending passages dealt with the age of the earth, the derivation of the planets from the sun, and the idea that truth can only be derived from science. Buffon acknowledged the infallible authority of the church and promised to subdue his heretical proclivities (at least in public). Despite the displeasure of the church and more orthodox thinkers, Buffon continued his writings—in a cautious but provocative style—and grew in popularity and influence.

Completely won over by Newtonian science, Buffon struggled to extend the concept of the rule of natural law to the inanimate and animate objects on the earth. Seeing before him the great complexity and diversity of Nature, he ridiculed the oversimplifications and arbitrary arrangement that flawed Linnaean taxonomy. At one stage of his thinking Buffon denied the reality of species and argued that genus and species are not real things but merely names for convenient mental divisions of a reality too complex for present science. Genus and species were no more "real" than the lines of longitude on maps, for in nature only individuals exist. In his *Histoire naturelle* (1753), Buffon wrote: "There only exist individuals and *suites* of individuals, that is to say, species." The idea of species was both artificial and mischievous since even for classification, one could only approach truth by increasing the number of divisions made "since in reality individuals alone exist in nature."[21] The source of Buffon's belief that increasing the number of divisions increases accuracy may have been Newton's calculus for after Buffon's pilgrimage to the land of Newton, he had returned to France a devout believer.

Later studies of breeding behavior forced Buffon to revise his ideas, for in the mule he saw an operational test of membership in a species. Species were groups of animals which could interbreed to produce fertile offspring and not sterile mules. Buffon's appreciation of the breeding test made his view of species very similar to the modern biological species concept. Through scientific experiments on hybrid fertility, Buffon believed species could be shown to be Nature's only "objective and fundamental realities." Species could then be defined as "a whole, independent of number, independent of time; a whole always living, always the same; a whole which was counted as one among the works of the creation, and therefore constitutes a single unit in the creation."[22]

Buffon, like Linnaeus, originally viewed species as fixed, but subsequent research undermined his faith in the stability of species. Delving into comparative anatomy, he found evidence of imperfect, useless, and rudimentary organs. Why should such structures exist if all species had been perfectly designed and were unchanged since the Creation? Although he presented evidence that species had changed, he also stated that species are separated by a chasm that cannot be bridged. In some passages, Buffon seems to have found a useful compromise: species were neither entirely mutable nor fixed, but could assume a number of forms. Yet Buffon also said that each species was an individual creation. The first individual of each species served as a model for its descendants. Some of the ambiguities and contradictions may reflect Buffon's more cautious nature taking over to protect him from charges of heresy.

Buffon was particularly interested in the development of the earth as a place suitable for living creatures. This theme was explored in two of his works, "Theory of the Earth" and "The Epochs of Nature." Buffon broke with the biblically based estimates of the age of the earth. Such calculations had fixed the Creation at 4004 B.C. Buffon's time scale was on the order of 100,000 years. Such a figure is certainly modest in comparison with current estimates, but for the eighteenth century it was a radical departure from prevailing beliefs. A theory that entailed any degree of geological and biological evolution was incompatible with a 6000 year time limit. Yet Buffon equivocated enough to maintain a superficial orthodoxy by the conventional devices of giving God the role of initiator of all change and presenting his theory as a hypothetical concept. Therefore the theory did not conflict with truth as taught by the church. To maintain a proper link with Scriptures, Buffon introduced the stratagem of treating the seven days of creation as seven epochs of indeterminate length. The origin of the earth marked the first epoch. Buffon thought the earth and other planets were formed from the sun after a collision with a comet. Rotating and cooling for some 3000 years, the incandescent gases were transformed into a spheroid. During the second epoch the earth hardened and congealed into a solid body—a process requiring about 30,000 years. During the third epoch the envolope of vapors around the earth became covered by a universal ocean. The tides were a major factor in the development of the earth during this 25,000-year period. Proof that the seas had once been so extensive could be found in the stratified nature of geological formations and the fossil marine shells found on mountains. Volcanic activity was the major influence during the 10,000 years of the fourth epoch as the waters began to subside. Once the cool land surface was exposed, vegetation began to grow. During the fifth epoch, which lasted some 5000 years, land animals began to appear. The entire land mass was originally connected in Buffon's theory. Elephants and other tropical animals wandered through the areas which are now northern and cold. During the sixth epoch the continents separated from each other and in some 5000 years assumed their present conformation and position on the globe. Finally, man appeared in the seventh epoch. Diversity among human races was the result of exposure to particular climates, soils, ideas, manners, and tastes. During recent times, new varieties of animals had been formed by man's own selective efforts. In nature, new varieties could develop through migration and segregation.

Although studies of plants and animals for classification suggested "degeneration" from a common ancestor, Buffon dutifully stated that "all species issued
fully formed from the hands of the Creator."

Although his theory of the earth was largely speculative, Buffon did try to
estimate the duration of his various epochs by experimentation. Spheres made
of identical size, but different materials (such as iron, copper, brass, granite,
limestone) were heated in a blacksmith's furnace and placed on wire holders.
With a watch, Buffon measured the time that elapsed until he could touch the
sphere without being burned. Calling upon his excellent mathematical abilities,
he then calculated for a given size and composition planet how many years
would have elapsed before the red-hot earth was cool enough for life to
appear. Certainly Buffon's calculations seem naive compared to modern
methods which suggest that the earth was formed more than four billion years
ago. "Epochs of Nature" must not be judged in terms of modern science, but compared with Archbishop Usher's calculation of the date 4004 B.C. for the creation—
a profundity enshrined in the official English Bible in the seventeenth century.

With his provocative speculations and elegant writing, Buffon helped to
spread new ideas and stimulate further scientific studies. Even in fields where
he had no expertise of his own he inspired others to follow. Although not an
anatomist, he recognized the advantages of the study of comparative anatomy
and relied on Louis Jean Marie Daubenton (1716–1799) to assist him in composing the comparative anatomy section of the *Natural History*. But it was
Buffon's spirit that inspired Cuvier. Bichat and Lamarck were also followers
of Buffon. Buffon tried to treat man as a part of nature, who could be
studied by the methods of natural science. He recognized many physical
characteristics shared by man and beast, but regarded man's intellect as incomparable to that of the other animals. Buffon's denigration of the native
animals and peoples of the New World even stimulated a rebuttal from
Thomas Jefferson. Jefferson prepared special comparative tables in his *Notes
from Virginia* to show American superiority in the size of indigenous mammals.
He noted that there were "18 quadrupeds peculiar to Europe," but America
had 74. Finally, "of 26 quadrupeds common to both centuries, 7 are said to
be larger in America, 7 of equal size, and 12 not sufficiently examined." To
prove the falsity of Buffon's pronouncement that mammals degenerated in
size in the New World, Jefferson went to great trouble and expense to have
the skeleton of a seven-foot moose and the bones of American caribou, elk,
and deer sent to Paris to prove the "immensity of many things in America."[23]

In 1788, just 10 years after the death of Linnaeus, Buffon died. His son's
life was terminated by the guillotine the next year during the Reign of Terror
and the grave and monument of the scientist were desecrated by the revolutionaries. Evolutionary philosophy was carried forward by a man who had
enjoyed Buffon's patronage—Jean Baptiste Lamarck.

Jean Baptiste Pierre Antoine de Monet,
Chavalier de Lamarck (1744–1829)

By the turn of the century scientists were ready to promote the idea of
species transformation more consistently and openly than Buffon. Lamarck is
the best remembered, and probably least read of these, as well as one of the
most unusual and most gratuitously maligned of the actors in the drama of
evolutionary thinking.

   Although he was the youngest of eleven children, deaths among his
brothers eventually conferred the family title on him. Unfortunately for
Lamarck there was no fortune to go along with it. As a young man with no
prospect of an inheritance, Lamarck was sent, against his wishes, to a Jesuit
school to become a priest. Lamarck had no choice in the matter while his
father lived, but at 16 his father's death brought him just enough money to
buy an old horse. Rejecting the career that had almost been forced upon him,
he rode away to join the army in Germany near the end of the Seven Years'
War. According to an account by his son (probably somewhat inflated), all
the officers of his company were killed in battle and Lamarck assumed the
command. His bravery was so outstanding that the Field-Marshal promoted
him immediately. When only 22, Lamarck was forced to leave the army because of chronic inflammation of the lymphatic glands of the neck (supposedly induced when another soldier lifted Lamarck by the head). Living in Paris,
Lamarck variously studied medicine, botany, and music, barely able to support
himself on his pension, writings, and work as a bank clerk. During this period
he met Rousseau, with whom he discussed natural history during their long
walks together. Some of his ideas about the inheritance of acquired characteristics may have been influenced by Rousseau.

   Lamarck's interest in natural science grew out of his studies of the flora
along the Mediterranean. In 1778 he published *Flore francaise,* a much admired key to the determination of species, which went through several editions. Impressed by this study, Buffon helped to pave the way for Lamarck's
scientific career. Lamarck became assistant in the botanical department of the
King's natural history museum and travelled widely through Europe, sometimes with Buffon's son, to study and collect botanical specimens. While the
French Revolution was destroying the family of his patron, Lamarck was
elevated to the Chair of Zoology by the National Convention. The Jardin du
Roi was renamed the Jardin des Plantes. In the process of reforming national
institutions, the National Convention created a number of professorships, including two in zoology. In the absence of more qualified candidates, Lamarck,
a botanist, and Geoffrey St-Hilaire (1772-1844), a minerologist, became instant, improvised zoologists. Dividing up the animal kingdom between themselves, Geoffrey lectured about the vertebrates and Lamarck turned his attention to the invertebrates. Already nearly 50 years of age, Lamarck transformed

himself into a zoologist and a pioneer in a new field. Despite progress in his
scientific career, Lamarck remained poor, modest, and retiring to the end.
Married and widowed four times, he saw most of his children die, and became
blind in his old age. He died in Paris in December 1829 and was buried in a
pauper's grave—his bones thrown in a trench to mingle with the skeletons of
other nameless unfortunates.[24]

In many ways Lamarck could serve as the prototype of the "crackpot"
scientist. Many of his interests were seen as eccentricity bordering on
madness, such as his hobby of weather forecasting. Although his contemporaries regarded him mainly as a systematist, he did not want to be remembered
for such dry and lifeless, although valuable, work. His views on transformation of species were generally ignored or ridiculed until after the publication
of Darwin's *Origin of Species.* Georges Cuvier, a younger but more successful
colleague, succeeded in burying Lamarck's work in ridicule with his vicious
"eulogy" at the death of his former patron. Charles Lyell introduced
Lamarck's ideas to the English speaking world in the 1830s, giving them a
fair treatment although he disagreed with Lamarck's conclusions. Few critics
took the trouble to consult Lamarck's original elucidation of the problem of
transmutation. Relying on second-hand misinformation, they created a figure
of ridicule.

Charles Darwin gave Lamarck credit for being "the first man whose conclusions on the subject [Origin of Species] excited much attention." Indeed,
Lamarck "did the eminent service of arousing attention to the probability of
all change in the organic, as well as in the inorganic world, being the result of
law, and not of miraculous interposition." Darwin admitted he had not
studied the original works, but got his information from the *Life* written by
Geoffrey Saint-Hilaire's son. Further denigrating Lamarck's importance,
Darwin continued: "It is curious how largely my grandfather, Dr. Erasmus
Darwin, anticipated the views and erroneous grounds of opinion of Lamarck
in his 'Zoonomia'. . ."[25] However, searching for precursors of Darwin, both
Haeckel and Samuel Butler referred to Lamarck as an important pioneer of
evolutionary thought.

Becoming a zoologist late in life, a self-taught scientist, without any systematic scientific training, Lamarck indulged interests so broad that any
particular area was necessarily lacking in depth. His most brilliant ideas were
needles in a haystack of vague and foolish fantasies. This is most apparent in
his *Memoirs on physics and natural history, based on reason independent of
any theory.* In attempting to formulate a general theory of existence based
on physics, chemistry, and physiology, Lamarck revealed the extent of his
ignorance. Opposed to the theories of Lavoisier, he clung to ideas not much
different from the chemical theories of the ancient Greeks. It is best to ignore
Lamarck's tresspasses into physics and chemistry, and let him speak only as a

biologist. Indeed, Lamarck coined the word "biology" and was quite successful in his work on invertebrates, a group of animals not well known when he
began his systematic studies.

The idea of the inheritance of acquired characteristics is so much associated
with Lamarck that many people have the mistaken impression that he invented
it. Actually, what Lamarck did was to apply a very old assumption about
heredity in an original way as the mechanism of evolution. He believed that
as the earth changed over long periods of time, geographic and climatic alterations produced new influences on plant and animal life. Given time without
limit, transmutation of species occurred in accordance with four "laws" of
evolution. Unfortunately, Lamarck dictated his final views on evolution in
1815 when he was old, ill, blind, and probably not very sharp of mind. His
earlier work, *Zoological Philosophy* (1809), explained evolution through the
cooperation of two factors.[26] The first was a fundamental innate tendency to
evolve towards increasing complexity of structure. Spontaneous generation
produced the lowliest forms of life—worms and infusorians. The tendency to
evolve into higher life forms was inherent in these creatures. The inheritance
of acquired characteristics was the second factor driving evolution from simple
to complex. If the environment was stable, evolution would be driven by the
first factor alone. When the environment changed, animals were forced to new
kinds of exertions which eventually produced modifications. Lamarck believed
that the increased or decreased usage of a part caused proportional changes in
strength and resulted in increased or decreased size. This change in size applied to muscles and perhaps to some organs. Heredity preserved characteristics acquired by the species through long usage.

In his later work, *Invertebrate Zoology,* Lamarck elaborated four "laws."
The most important and misunderstood was the second law which essentially
stated that "the production of a new organ in an animal body results from a
new *need* which continues to make itself felt." Under new conditions, animals
have a physiological need for new organs. The French word "besoin" was
often mistranslated into English as "want" or "desire" rather than "need" as
Lamarck had used it. Lamarck did not suggest that the animal caused the
changes through its conscious wishes or desires, but his work was misunderstood and misrepresented by critics and admirers alike who never took the
trouble to determine what Lamarck really meant.

Some Romantics and Transcendentalists found the idea that "desires"
determined evolution quite inspiring. As Emerson would have it:

And striving to be man, the worm
Mounts through all the spires of form.[27]

When all living animals were known, Lamarck believed they could be arranged in a linear series with some small collateral branches. Such an arrangement would reflect the "natural order" of animals in nature. Although finding this natural order was the task of the systematic zoologist, in practice zoologists and botanists were forced to use arbitrary taxonomic systems. Such systems were useful, even necessary, but Lamarck insisted that taxonomists recognize the arbitrary nature of their work. The true "natural order" would range from the lowest monad to the highest creature on the scale—man. If all species were included, the gradations between neighboring species would be nearly imperceptible. Apparent gaps in the scale represented undiscovered species. Lamarck's scale of nature has been compared to the periodic table of the elements where apparent gaps served as stimuli to chemists to find new elements.

Lamarck hinted at the relationship between man and ape, but his views as to the relationship between man and the scale of animals was generally obscure. Although in intellect man was clearly different from the "brutes," if only structure was considered man could be classified as one family of mammals containing six varieties—Caucasian, Hyperborean, Mongoloid, American, Malayan, and Negro. Although the great apes were not well known, Lamarck seems to have suspected a common origin for man and ape. He even proposed a scheme in which an ape might have learned to stand more erect in order to see further. The new posture could have caused heritable structural modifications. Other modifications and advances would accompany changes in needs and behavior. Lamarck began his "Observations with regard to Man" quite boldly with the assertion that if man were distinguished from the animals only by organization "it could easily be shown that his special characters are all due to long-standing changes in his activities and in the habits which he has adopted and which have become peculiar to the individuals of his species."[28]

Although the anatomical structures of man and the apes were similar, it was possible to explain the physical and mental superiority of man as the advantage passed along by heredity after countless years of increased use of these powers. It was even possible that the higher apes could be brought up to a higher level of intelligence through training and usage. After all, Lamarck pointed out, few men in all history achieve real intelligence and the rest remain in a state of bestial ignorance. Like Descartes, Buffon, and other unorthodox thinkers, Lamarck protected himself by casting such daring ideas in the form of a game or hypothesis merely for argument's sake rather than as a theory to be taken seriously. Although man could have evolved from a simian ancestor, Lamarck lamely concluded that these were merely "reflections which might be aroused, if man were distinguished from animals only by his organization, and if his origin were not different from theirs."[29]

Etienne Geoffroy Saint-Hilaire (1772–1844)

Less well known than Lamarck, Geoffroy St-Hilaire also turned the energies of
his second career towards highly speculative theories. Like Lamarck, Geoffroy
had been destined for the priesthood by his father. However, he was allowed
to indulge his interest in natural science, chemistry, and crystallography and
became a minerologist. In 1792, he was appointed to a Chair of zoology
along with Lamarck. Despite his lack of zoological training, he became quite
successful in his studies of the vertebrates. Geoffroy was able to assemble a
splendid collection by accompanying Napoleon Bonaparte on his expedition
to Egypt. Like Lamarck he was hurt professionally through rivalry with
Georges Cuvier. Ironically, it had been Geoffroy who had recommended his
nemesis to his first professorship.

Nature philosophy shaped much of Geoffroy's speculations. His philosophy
of biology was founded on the belief in a common fundamental type. "There
is," he declared, "philosophically speaking, only a single animal."[30] Although
he made wild and impossible comparisons among various types of animals, his
theory of the "archetype" served many comparative anatomists as a workable
alternative to either a strict creationist or evolutionary framework. His son,
Isidore, adopted and extended many of his ideas. Lamarck may have wanted
to turn the old Ladder of Creation into a moving staircase, but there were too
many gaps for other scientists to climb along with him. Despite his pioneering
studies of the invertebrates, he was unable to find a way to connect them
with the vertebrates. Geoffroy Saint-Hilaire attempted to bridge this chasm
by inventing a "bent vertebrate" from the Cephalopods.

According to his son, Geoffroy St-Hilaire began to suspect as early as 1795
that modern species were "degenerations" of a primeval organism. However,
not until 1828 did he publish his belief that species had not been fixed at
their point of origin. Changes in species were ascribed to their conditions of
life. In spite of this glimpse of evolutionary possibilities, Geoffroy did not
believe that species were presently undergoing modification.

Erasmus Darwin (1731–1802)

If only for being the grandfather of both Charles Darwin and Francis Galton,
Erasmus Darwin would rate a place in a history of biology, but he is quite
capable of standing on his own merits as an enthusiastic if amateur natural
philosopher. His circle of friends was as full of prominent names as his family.
Although Erasmus Darwin remained a country physician all of his life, he was
a member of the Lunar Society; associate of Josiah Wedgewood, the potter
(Charles Darwin's uncle and father-in-law); James Keir, the chemist; and James
Watt, the engineer, whom he introduced to another friend, Matthew Boulton,

the industrialist. Erasmus Darwin was an inventor as well as an author. He turned his keen observations of man and nature into ingenious, poetic, and sometimes bizarre speculations about the nature of life.

His personal life was as colorful as his writings.[31] Married twice, he produced five children in the first marriage, seven in the second, and two illegitimate daughters in between. The venerable patriarch and all his issue lived together in the appropriately named "Full House." The good doctor was tall, enormously fat, scarred from smallpox, and a great talker despite his stutter. As the result of a fall from one of his own inventions (a carriage), he walked with a limp.

More radical, if less systematic, a thinker than his grandson, Erasmus proposed a theory of the transmutation of species in which all organisms had descended from some primordial filament. His views were aired in *Zoonomia* (1794), *Phytologia* (1800), the *Botanic Garden* (1789) and the *Temple of Nature, or Origin of Society* (1803). Although his work might have been forgotten by history if not for the fame of his grandson, at the time of their publication these books were sufficiently exciting to earn a place on the *Index of Prohibited Books*.

Zoonomia was translated into several languages and was especially popular among the German natural philosophers. It was read twice by Charles Darwin and probably had more effect on his own thinking than he would admit. On first reading, Charles greatly admired the *Zoonomia,* but ten to 15 years later when he reread it he was "much disappointed; the proportion of speculation being so large to the facts given."[32] Charles denigrated his grandfather's work, but also came close to accusing Lamarck of stealing ideas from Erasmus. Despite Charles Darwin's insinuation that Lamarck used the ideas of Erasmus, it is more likely that both had been inspired by Buffon's idea of mutability and more mundane facts derived from plants and animal breeding. Both Erasmus Darwin and Lamarck believed that organisms could attain various modifications of structure through their own efforts and that these were passed on to their offspring.

The natural philosophy expounded in *Zoonomia* began with the view that "spirit" and "matter" are the foundations of nature. "Life" was defined as the result of a special force called "irritability" in accordance with the physiology of Albrecht von Haller. According to Erasmus Darwin the conditions of the parents' life affected the development of their progeny. New varieties which had been obtained among domesticated animals provided proof for this assertion.

All the elements that were to appear in *Origin of Species* in 1859 were present at least in germ in the works of Erasmus Darwin, along with suggestions for future research. Erasmus argued that time extended back through millions of ages. He discussed rudimentary structures and vestigal organs as the "wounds of evolution." Although he referred to the "struggle for existence" and sexual selection, he explained the mechanism of change in terms of "degeneracy" and the influence of the environment. Also noted were studies of homologous and analogous organs, heredity and disease, and the dangers of inbreeding. Unlike Charles, who only dealt with the transmutation of complex species created through some unknown mechanism, Erasmus boldly derived all living forms from some "primal filament." Through the study of "similarity in features of nature" we would realize that "the whole is one family of one parent." The driving force that created myriad forms was "the power of acquiring new parts, attended with new propensities, directed by irritations, sensations, volitions and association; and thus possessing the faculty of continuing to improve by its own inherent activity and of delivering down these improvements by generation to its posterity, world without end!" Although he recognized the struggle for existence in the phrase, "One great Slaughter-house the warring world," he also suggested that sympathy should exist among all life forms because of their relatedness through descent. Even "imperious man" who "styles himself the image of God . . . arose from rudiments of form and sense, an embryon point, or microscopic ens!"[33]

Lamarck and Erasmus Darwin were intrigued by the idea circulating among eighteenth century intellectuals that unlimited progress and organic change were possibilities. Fears generated by the excesses and terror of the French Revolution suppressed interest in such radical ideas. Stability in society and nature was perceived as more desirable than limitless, unpredictable change. Georges Cuvier, a scientist who won great acclaim for his authoritarian theories and strict organization of animate beings, rejected and ridiculed the evolutionary theories of his predecessors.

Georges Leopold Chrétien Frédéric Dagobert, Baron Cuvier (1769–1832)

Georges Cuvier, preeminent comparative anatomist and founder of palaeontology, was an implacable opponent of Lamarckian ideas in general and evolutionary theory in particular. Although he remained opposed to the possibility of the transmutation of species, Cuvier's great renovation of comparative anatomy was to become instrumental in assembling the data which eventually was turned to support the theory of evolution.

A child prodigy, able to read at the age of four, Cuvier entered the Academy of Stuttgart when 14 years old and soon won recognition for his prodigious memory and disciplined application to his studies. His biology professor, Karl Friedrich Kielmayer, believed the study of comparative anatomy would lead to the discovery of the archetype of all living creatures. Such knowledge would create order among the great jumble of species. When

Cuvier became a tutor in Normandy at the age of 18, he began to bring order
to the study of marine animals. Fortunately for Cuvier, his diagrams of these
creatures so impressed Geoffroy St-Hilaire that Cuvier was appointed assistant
to the Professor of Comparative Anatomy at the Museum of Natural History
in Paris in 1795. Like Lamarck and Geoffroy St-Hilaire, he began his new
post singularly unprepared. Although his duties included training medical
students, Cuvier shared with Galen the handicap of never having dissected a
human being. But, unlike his two senior colleagues, Cuvier soon became
famous, successful, rich, and a favorite of the Emperor Napoleon. In addition
to his research and teaching, he took on the reform of the French educational
system as "inspecteur general" in the department of education. Cuvier's interest in scientific organization and education found expression in his position
as Chancellor of the Imperial University and as Councillor of State during the
reign of Napoleon and Louis Phillippe. As a member of all the major scientific societies of Europe, he made his influence felt throughout the continent.
For his researches on the animal kingdom he was honored as a "second
Aristotle."[34]

A disciplined, energetic and ingenious researcher, Cuvier produced numerous books and scientific papers. While Cuvier reformed the taxonomic relationships of many different groups of animals, his most exciting studies were
of fossil mammals and reptiles. Particularly interesting fossils had been discovered in the tertiary formations near Paris. To understand the function and
form of fossil animals, Cuvier studied the osteology of living creatures for
comparative purposes. A true understanding of the meaning of the fossilized
remains of plants and animals was long in coming. Agricola (Georg Bauer
Agricola, 1494-1555) had used the term "fossil" in *De Re Metallica,* published in 1555, to refer to anything dug out of the earth. Those things which
resembled animal or plant forms were regarded as sports of nature, a kind of
art form that Nature practiced rather capriciously to amuse and mystify man.

Cuvier engaged in a thorough and systematic excavation for fossils in the
limestone deposits near Paris. Understanding fossil forms was quite difficult
at the time because they were rare, often incomplete and in poor condition.
Most collectors of fossils were too ignorant about the structure of living beings
to know with certainty that particular bones did not correspond to living
forms. Cuvier turned his knowledge of extant forms and his "correlation
theory," which encompassed a complete appreciation of the functional relationships between organs and their functions, to the analysis of fossils. It was
his expertise in reconstructing the forms and functions of animals from bits
and pieces that led to the creation of modern paleontology. Often, his ability
to reconstruct the whole animal from a fragment amazed his contemporaries
and dismissed the superstitious nonsense that had grown up around puzzling
specimens. For example, Cuvier proved that fossil bones identified as the

skeleton of a man who had died before the great flood of Noah were really the remains of a gigantic extinct salamander.

Although Cuvier's study of fossils led him to ponder the nature of changes in the earth and the differences between species past and present, he did not find the transmutation of species a necessary conclusion. Cuvier defined species as "certain forms, which, from the origin of things, have perpetuated themselves" within certain limits. Varieties existed as "accidental subdivisions of species." Although anatomy was his major work, he believed that the definition of species should go beyond morphological characteristics and include a genetic or historical component. "Generation being the only means of ascertaining the limits to which varieties may extend," he wrote, "species should be defined—*the re-union of individuals descended one from the other, or from common parents, or from such as resemble them as strongly as they resemble each other*."[35] He realized that the definition was strict, and that the experimental test implied—mating—could not be done in all cases.

Recognizing discontinuities between populations of extinct and living species, Cuvier explained changes in the animal world in terms of "catastrophies" or "revolutions" which had decimated whole populations of living things in prehistoric times. Both the physical evidence of the local geological strata and the recent political and social events in France may have influenced Cuvier's concept of sudden, violent transitions. The sharp lines of demarcation between various geological strata, the inversion and disorder of strata at some sites, and the great numbers of fossils in some strata seemed proof of sudden deaths of life forms and rapid changes in the surface of the earth. When mammoths were found frozen into the ice in Siberia, this was taken as evidence for sudden and violent changes.

Many cultures have legends about catastrophies, particularly floods, which took place in the dim past. By the end of the eighteenth century evidence had accumulated that the earth was quite old and that there must have been many periods of creation and destruction. In his preface to *Research on Fossil Bones* ("Discourse on Revolutions of the Surface of the Globe"), Cuvier furnished a dramatic account of the catastrophies that had established the geology of our own epoch, which dated back about 6000 years. Trying to avoid unscientific speculations, Cuvier focused his attention on the comparative anatomy of fossils and extant species, and the way in which species had been destroyed. Cuvier's time scale was quite limited in comparison to that of Lamarck and other evolutionists. He denied that any human remains were true fossils—predating the last revolution. Any suspicious human bones were explained away as the accidental admixture of modern human bones with true fossils. Despite Cuvier's staunch resistance to evolutionary theories, advances in geology and paleontology led others to question the fixity of species and even the antiquity of human origin.

Robert Chambers (1802-1871)

Like Benoit de Maillet, Robert Chambers is best classified as an amateur naturalist, an enthusiastic, self-taught lover of science who pulled together the strands of evolutionary thought and gave people a scientific-sounding alternative to the Special Creation theory. Chambers was the author of a much maligned book called *Vestiges of the Natural History of Creation*.[36] Not until 1884, well after his death, was his authorship revealed. By that time, it was possible for scientists to discuss evolution openly—indeed Darwin had published *Descent of Man* in the year of Chambers' death.

Robert and his brother William came to an interest in the problem of "variety in nature" very directly. Both brothers were complete hexadactyls, that is hands and feet had six digits. Both were operated on to correct this anomaly, but Robert's convalescence was long and painful.

Although *Vestiges* does not seem very dangerous or particularly impious today, it was quite controversial when first published in 1844. Chambers' contemporaries saw the book as a "work of black materialism that threatened to cut away the foundations of all morality and religion." The book was carefully read by Darwin, Wallace, and T. H. Huxley, as well as Disraeli, Lincoln, Florence Nightingale, and Frances Cobbe. Francis Darwin noted that his father had carefully read and marked passages in his copy of *Vestiges*. T. H. Huxley wrote that the "author knew no more science than could be picked up in *Chambers' Journals.*" Clergymen and naturalists denounced the book so loudly that they roused public interest in the question of evolution as never before. Adam Sedgwick said he did "loath and detest *Vestiges.*" Further, he declared that the book was so stupid, he knew the anonymous author must be a female. Others suspected that Prince Albert was "Mr. Vestiges." While scientists and clergymen competed in villifying the poor author of *Vestiges,* the public devoured the book. It was to go through 10 editions in 10 years and was translated into several languages. Chambers had taken extreme precautions to keep his authorship a secret, but he could not resist replying to some of his critics in 1845. He attempted to defend his ideas in *Explanations by the Author of Vestiges.*[37]

Vestiges was a layman's guide to scientific theories concerning all development—the universe, plants and animals, man and society. Although Chambers was certainly an amateur, his general knowledge of many subjects, such as geology and astronomy, was quite good. The distance between professional scientist and interested amateur at the time was not so great as it would become with increasing specialization. Some of Chambers' worst errors were in his discussions of biology where his background was especially weak. Chambers believed that the processes of "cosmogony" were not fixed at one moment in creation, but that the universe is "still and at present" developing.

The universe was run according to "Laws" for which physicists had established a sound mathematical basis. Seeing a unity in the development of the cosmos and of life, Chambers actually believed his arguments were quite pious and did more to glorify the Creator than did the idea of constant intervention in fashioning new species of all creatures from elephant to dung beetle. For Chambers, God was the "First Cause to which all others are secondary and ministrative, a primitive almighty will, of which these laws are merely the mandates . . ." Something of an expert on geology, Chambers dealt quite well with the composition and history of the earth. His discussion of the fossil record foreshadowed many of the problems Darwin would face. The incompleteness of the fossil record was a difficulty, but evolutionists could point out that more discoveries would surely be made in the future and that conditions in the past had not always favored the formation of fossils. Even with these provisos an unprejudiced observer had to admit that the fossil record clearly showed an increasing degree of complexity and organization.

"That God created animated beings," Chambers wrote, "is a fact so power-fully evidenced, and so universally received, that I at once take it for granted." What he would not grant was that God would constantly be involved in the creation of the myriad forms of life. He could do this more economically through the operation of natural laws which are the "expressions of His will." If not in complete harmony with the letter of the scriptures, Chambers argued his account of organic evolution was completely compatible with the spirit.[38] In many details, Chambers displayed excessive naiveté and a tendency to rely on bizarre, unproven anecdotes to support his theory. In the formation of crystals he saw the mechanism of the growth of plants. In the pattern of frost on the window he saw a meaningful analogy to vegetable forms. Even in the amalgam of silver and mercury he saw the shapes and organization of shrubs. He used very poor examples of "spontaneous generation" as support for his theory, including the *in situ* formation of intestinal worms, mushrooms produced in a mixture of cow and horse dung, and white clover from lime. One of his favorite examples was the "electrical generation" of insects from inorganic chemicals in the laboratory of Crosse and Keekes in 1842.

Realizing the importance of recent progress in chemistry, Chambers argued that since chemists had synthesized urea and alantoin, there was no proof for any special chemical peculiarity of organic compounds. His discussion of cell theory indicates that Chambers appreciated one of the most important new biological generalizations of his time. He saw the cell as the basis of life and utilized this insight as evidence for evolution. Chambers saw the cell as the "meeting point between inorganic and organic—the end of mineral and be-ginning of vegetable and animal." Thus, the problem of the creation of life was reduced to the question of the creation of the first primitive cell. Chambers suggested that some "chemico-electrical operation" had caused the formation of the first cell.[39] In contrast to Chambers, Charles Darwin was never too sure about the truth and usefulness of cell theory.

In refuting Special Creation, Chambers adroitly assembled studies of comparative anatomy, particularly evidence of rudimentary and vestigal organs. Such "abortive or rudimentary organs," he argued, appear as "blemishes or blunders" incompatible with the idea of a perfect and efficient Creator. While Chambers seemed quite confident that his evidence proved evolution had occurred, he was unable to provide a compelling argument as to the mechanism driving evolution although he was acutely aware of variations in nature. Chambers fell back on the old argument that environmental influences affect the improvement or degeneration of individuals and species. Citing the well-known fact that adult bees can change the embryonic development of their kind to produce a queen or worker from the same starting material, he argued that similar influences must be at work in other cases. "Monstrosities" in man and other creatures might result from "weak health or misery." Chambers explicitly argued that "evolution" is not tantamount to "progress," but that good conditions could lead to progress and poor conditions to degeneration. Changes could be occurring, unnoticed by scientists, even today. Such changes might be rare but were not therefore impossible.

Although his theories seem quite similar to Lamarck's, Chambers tried to dissociate himself from his predecessor. While calling Lamarck a "naturalist of the highest character," his hypothesis was placed "with pity among the follies of the wise."[40] Going beyond Lamarck, Chambers was daring enough to include man in his naturalistic account of evolution. Even civilization was treated as a product of evolution. Speech was the only real novelty attending the creation of the human race. Chambers believed that just as the embryo passes through the stages of the lower animals during development *in utero,* so the human brain develops through the stages represented by the races of man. Because recent studies of language had shown that some primitives had a more complex language than Europeans, Chambers was in the awkward position of denigrating the difficulty of acquiring this unique human trait. Complexity of language, he rationalized, is not directly correlated with intelligence, because after all little children learn whatever language their parents speak.

Finally, Chambers defined man in a rather melancholy way: "Man is a piece of mechanism, which can never act so as to satisfy his own ideas of what he might be."[41]

While their own researches into the history of the earth inspired Chambers and Charles Darwin, geology was also a favorite study of Special Creationists.

Geology

In this section we will briefly describe some of the advances in geology which supplied important clues to the problem of species and evolution. Scientists thinking about changes in populations of living forms benefited from the development of theories of the changes that had occured in the history of the earth. Although applied knowledge of the earth is very old—involving mining, metallurgy, and surveying techniques—geology did not really emerge as a science until the nineteenth century. Even superficial observations make it obvious that floods, earthquakes, volcanoes, storms and lightening have changed the surface of the earth. Which factors were most impressive and seemed most important depended on local conditions and climate, but geologists wasted much of their energy in fruitless and pointless controversies about whether water or heat was the most important force in shaping the Earth. Abraham Gottlob Werner (1750-1817) emerged as leader of the Neptunists— who believed water was most important. James Hutton (1726-1797) was the guiding spirit of the Vulcanists. Progress in geology was impeded because many practitioners saw their subject not as a science but as the "handmaiden of theology."[42] Some geologists were so disturbed when their studies of the earth, particularly its fossils, seemed to contradict religious dogma that they discontinued their work.

When Charles Darwin was a student, the most popular geologists of the English school, William Buckland, Adam Sedgwick, and William Conybeare, often lectured about the marvelous way that geology supported theology. Such theologically guided geologists believed Neptunist theories were in harmony with the biblical stories of the great flood and the rather young age of the earth. But even in the eighteenth century James Hutton was laying the groundwork for a new geology.

James Hutton (1726-1797)

Hutton's theory of the earth, first communicated to the Royal Society of Edinburgh in 1785 may be regarded as the foundation of modern geology. In 1795 his ideas were published in two prolix volumes as *Theory of the Earth*. This work changed geology from the sport of amateurs into a true science. Wild speculations and religious dogma were to be abandoned for rigorous, coordinated rationalization based on observation and analysis of rock and earth strata. Hutton formulated the two major tenets of geology: the geological cycle and uniformitarianism; that is, that scientists must explain the history of the earth in terms of processes acting in the present. Supernatural agents or acts could not be invoked by geologists to extricate themselves from theoretical difficulties.

Since change is going on all the time, but very slowly, Hutton assumed un-limited time for the action of heat, pressure, and water in shaping the earth and its surface formations. Hutton viewed the earth as a self-renewing world machine operating in a law-bound fashion as a part of the great cosmic engine. Unfortunately for science, the bizarre theory of catastrophism, with its call for constant miracles and divine intervention captured scientific and popular opinion until the work of Charles Lyell. Hutton was a clear thinker and painted a convincing portrait of an earth with "no vestige of a beginning, no prospect of an end," but he was a tedious, obscure, and excessively boring writer. Until John Playfair (1748–1819) put Hutton's work into a more read-able form as *Illustrations of the Huttonian Theory* (1802), the new geology was in danger of being forgotten. Playfair's amplification of Hutton's work has been hailed as a major landmark in the history of British geology.

Practical men, concerned with mining and engineering, also contributed to understanding the earth, its strata and fossils. Chief among these was William Smith.

William Smith (1769–1839)

Although it was James Hutton who formed a completely naturalistic view of earth science, it was a largely self-educated drainage engineer and surveyor who was awarded the title "father of English geology" and, in 1831, the first Wollaston Medal from the Geological Society. Smith's great insight into the arrangement of rock strata was that beds of the same kind of stone, at differ-ent levels in the succession, could be recognized by the fossils they contained. Smith had noticed the relationship between fossils and geological strata by 1791, but he regarded himself as a practical field geologist rather than a scientist and did not publish his ideas until 1815.

While Smith seems to have admired the uniformitarian ideas of Lamarck and Hutton, he was pressured by his friends away from a purely naturalistic geology into more orthodox religious interpretations. He converted to "catastrophism" and in 1817 wrote that "each layer of fossils must be con-sidered as a separate creation or as an undiscovered part of an older creation." Accepting time on the great scale of Hutton, he could still claim that the use of fossils carried us back into "a region of supernatural events."[43] Smith's re-treat into supernatural explanations reflects the temper of the times quite well. Although there was great interest in science and advancing technology, there was no room for theories that would upset religious dogma. Natural theology flowered in the early nineteenth century. Many early structural geologists were clergymen eager to study the great "book of nature" to illu-minate the other book of God, the scriptures. The Reverend Adam Sedgwick (1785–1873), professor of geology at Trinity College, Cambridge, was one of these. Sedgwick admitted he knew nothing about geology in 1818 when he won the position of Woodwardian professor of science. Another candidate had been nominated for the position, but Sedgwick knew his rival had no chance—"for I knew absolutely nothing of geology, whereas he knew a good deal—but it was all wrong!"[44] Everywhere Sedgwick saw evidence to support diluvial geology and genesis. Since "truth must at all time be consistent with itself" the evidence drawn from observation of God's works must, when rightly interpreted, be consistent with conclusions drawn from God's words— the Scriptures. According to Sedgwick, "when science was well interpreted it must aid natural religion." An episode in Sedgwick's career that had peculiar consequences was his encouragement of an eager, but not terribly bright young student named Charles Darwin. A summer Darwin spent exploring the geology of North Wales with Sedgwick's guidance quickened Darwin's interest in natural science.

While the data became increasingly difficult to fit into the orthodox framework, pious geologists strived valiantly to do so. Not until the maturation of Charles Lyell's studies of geology were the chains that bound geology to theology finally strained to the breaking point.

Charles Lyell (1797–1875)

Young geologists used to say: "We collect the data, and Lyell teaches us to comprehend the meaning of them."[45] Lyell, the "great high priest of Uniformitarianism," reasserted Huttonian principles so persuasively that the concept of the historical and physical continuity of nature became inextricable linked with Lyell.

Although not a very enthusiastic student, Lyell was zealous in pursuit of his hobbies, such as lepidoptery and geology. While enrolled at Oxford he undertook several geological tours of England, Scotland, and the continent. A career in law was abandoned becaue of problems with his eyesight. This left Lyell free to pursue his real interest—geology. Written in an elegant style, with the logic of a legal brief, Lyell's *Principles of Geology* set a new standard for the profession. Lyell kept the three-volume work up to date with constant revisions. The twelfth edition was published posthumously in 1875. In addition to stimulating a new epoch of scientific geology, Lyell introduced Lamarck's ideas on species transmutation to the reading public. Charles Darwin called Lyell's *Principles of Geology* one of the major influences in the early development of his evolutionary theory. Later, Lyell became Darwin's friend and guide.

Although Lyell adopted a strict uniformitarian attitude towards geology, insisting that geological data be interpreted in accordance with forces known to be acting on the earth today, he did not extend this concept to the natural and historical succession of animate beings. For most of his life he resisted a uniformitarian system of organic evolution. Lyell believed that species could neither change nor adapt to altered conditions. Every species had somehow been created for its proper niche and spread out from that point to the limit of its powers, depending on climate, food supply, and competition with other life forms. Although there were obvious variations within a species, adaptation was limited so that a radical change in environment could only lead to death and replacement with already existing, better adapted species. The actual "creative force" that produced species was outside the scope of science for Lyell. Even when he finally accepted Darwinian transmutation of species, he refused to support the natural evolution of man.

Charles Darwin (1809–1882)

Although his family tree was a veritable Victorian *Who's Who,* Charles Darwin was not a very promising young man. He was the sixth of eight children born to Susannah (daughter of the eminent Josiah Wedgewood) and Robert Waring Darwin, physician. His son Francis speculated that Charles had inherited "sweetness of disposition from the Wedgewood side" and "the character of his genius . . . from the Darwin grandfather." Darwin was first sent to Shrewsbury School, a nearby boarding school, but gained little from the experience. Of his early schooling, Darwin complained: "Nothing could have been worse for the development of my mind than Dr. Butler's school, as it was strictly classical, nothing else being taught, except a little ancient geography and history." When he left school, Darwin felt his teachers and father rated him as "a very ordinary boy, rather below the common standard in intellect." He was deeply moritified when his father thundered at him, "You care for nothing but shooting, dogs, and rat-catching, and you will be a disgrace to yourself and all your family." Charles later reflected, "I think my father was a little unjust to me when I was young."

Charles was next dispatched to Edinburgh to study medicine in the tradition of his father and grandfather, but soon it was clear that his temperament was unsuited for a medical career. While anatomy and Latin merely bored him, the sight of blood made him ill. Seeing operations, then performed without anesthesia, convinced him that medicine was a "beastly profession." Besides his revulsion for medicine and his indifference to his education, Charles suspected that he would inherit enough money to subsist without working. This idea "checked any strenuous effort to learn medicine." During his time at Edinburgh from October 1825 to April 1827, he was more interested in amateurish studies of natural science than his school work. His hobbies did prove scientifically productive even at this stage for he was able to present his discoveries about the ciliated larva of Flustra to the Plinian Society in 1826. While Charles was at Edinburgh, an anonymous paper appeared in the Edinburgh *New Philosophical Journal* which vigorously supported the theory of the transmutation of species. The author was Darwin's friend Dr. Robert Grant. Once, when the two were out for a walk, Grant had astonished Darwin by bursting forth in high admiration of Lamarck and his views on evolution. Despite his association with Grant and his reading of *Zoonomia,* Darwin later claimed that in his youth he "did not then in the least doubt the strict and literal truth of every word in the Bible."

Above all else Charles Darwin loved hunting and what his father ridiculed as the "idle sporting life." Since Charles was not to be turned into a respectable physician, his father decided the next best profession was that of clergyman. Ironically, while Robert believed this was a suitable position for his son, he later admitted to Charles that he was himself a skeptic about religion. Charles began his studies toward the taking of holy orders at Christ's College, Cambridge in October of 1827, but his attitude towards his education was unchanged. "During the three years which I spent at Cambridge," he wrote, "my time was wasted, as far as the academic studies were concerned, as completely as at Edinburgh and at school." Dr. Darwin was not about to allow his "rat-catcher" of a son to pursue an idle life and disgrace the family, but he hardly knew what to make of this son who seemed destined for failure. The event that changed the course of Charles Darwin's life was the opportunity to serve as unpaid naturalist on the H.M.S. *Beagle.*[46]

The voyage of the *Beagle,* a voyage that lasted from December 27, 1831 to October 2, 1836, stands out as the great adventure and determining influence in Darwin's life. Indeed the first sentence of the *Origin of Species* reflects back on this experience and its intellectual outcome:

> When on board H.M.S. 'Beagle,' as naturalist, I was much struck with certain facts in the distribution of the organic beings inhabiting South America, and in the geological relations of the present to the past inhabitants of that continent. These facts . . . seemed to throw some light on the origin of species—that mystery of mysteries . . .[47]

Only through a peculiar chain of accidents and compromises did Darwin manage to secure a place on the *Beagle.* The offer was only extended to the young Darwin because the first two candidates had turned down the position. Although he was young and inexperienced, his friend Prof. Henslow had great confidence in him and recommended him for the appointment. Darwin's father objected to the project and his obedient 24-year old son was ready to comply with his father's wishes. Dr. Darwin said he would change his mind only if "any man of common sense" would support his son's strange idea.

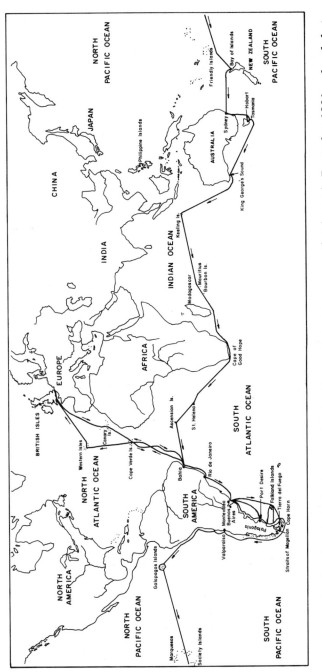

The route taken by the *Beagle* in circumnavigating the world during a voyage that began in December 1831 and ended at Plymouth, England in October 1836.

Uncle Josiah Wedgewood, unquestionably a man of common sense, pursuaded Robert to allow Charles to go. Overcoming the first set of obstacles, Charles almost missed the boat completely because of Captain Robert Fitz Roy's interest in phrenology. On meeting Darwin, Fitz Roy discovered that the shape of his nose suggested a certain lack of energy.

The *Beagle* itself, a 10-gun brig, refitted for the surveying mission as a three-masted bark, was a problem. With her rotting timbers and deck, she needed a thorough reconstruction. The accommodations, even for captain and naturalist, were far from luxurious. A corner of the chart table in the poop cabin at the stern of the ship was Darwin's entire "office." He slept in a hammock slung over the chart table. Charles soon discovered that he was very susceptible to seasickness and that his quarters were in the roughest-riding section of the ship. For a time there was an unwelcome suggestion that the mission of the *Beagle* be truncated. Fortunately for Darwin the full survey was carried out. The *Beagle* circled the earth and made stops at the Cape Verde Islands, South America, the Galapagos Islands, Tahiti, New Zealand, Australia, Mauritius, and South Africa.

As the *Beagle* proceeded southwards along the coast of South America, Darwin noticed how varieties differed with geographical location. Like Chambers, he must have worried over the uneconomical way all these slightly different creations would have taken up God's time and energy. These questions became most striking on the Galapagos Islands—a group of about 20 barren islands of volcanic origin on the equator, about 600 miles off the west coast of South America. The animals on these desolate, geologically young islands were similar to South American genera, but species existed that were unique to the islands. Beyond that, as Darwin finally realized when leaving the islands, different species of birds, tortoises, etc. could clearly be associated with specific islands. The Galapagos Islands served as a living laboratory of evolution where variation and reproductive isolation produced peculiar species living under essentially identical environmental conditions. Studying the creatures of these bleak islands was probably the single most decisive experience in Darwin's inexorable conversion from a naïvely orthodox future-clergyman into a disciplined scientist—naturalist, geologist, taxonomist, and above all, evolutionist.

By the time he returned to England, Darwin had undergone a remarkable change. He always regarded the voyage as "the first real training or education" of his mind. Above all, he acquired "the habit of energetic industry and of concentrated attention" which he applied to all his later scientific studies. All former interests gave way to a love of science as he discovered "that the pleasure of observing and reasoning was a much higher one than that of skill and sport." The voyage so improved his mind that on first seeing his son again Dr. Robert Darwin exclaimed: "Why, the shape of his head is quite altered."[48]

Having returned to England and a peaceful, well-ordered life, the bold adventurer of the *Beagle* was strangely transformed into a frail, nervous invalid. Like his father, Charles Darwin married a Wedgewood—his first cousin Emma, daughter of the uncle who had convinced Robert to let his son become a naturalist. The marriage in 1839 removed all pressure to choose a standard profession and earn a living. His wife brought a nice dowry of £5000 plus an allowance of £400 per year, while his father contributed £13,000. With a guarantee of well over £1000 per year, Charles was able to devote himself to his studies. The exact cause of Darwin's chronic ill health remains a mystery, since he did live to a ripe old age.[49] Despite his debilitating course of very Victorian-sounding maladies, he produced a prodigious amount of work, as well as a revolution in ideas about our species' place in nature—and, with his wife, Emma, some 10 children (only 7 survived to adulthood). This suggests that he had enjoyed the "iron constitution" of the true hypochondriac. Other more generous suggestions for his illnesses include hereditary weakness, arsenic poisoning as a result of medications then in vogue, or Chargas' Disease. Prof. S. Adler tried to rescue Darwin from the stigma of hypochondria by diagnosing his illness as Chargas' Disease. This is a prolonged, debilitating disorder caused by a South American relative of the parasite for African sleeping sickness. Darwin recalled being attacked by the "great black bug of the Pampas" in Mendoza, a province of Argentina. *Triatoma infestans* is now known to be the chief vector of Chargas' Disease.

From his diaries and notebooks, Darwin was able to write several volumes on geology and his *Voyage of the Beagle,* which became a popular success. Although the book did not bring him much money, Darwin enjoyed the status of literary celebrity. His reputation as a scientist was first built on his geological work, for his explanations of the formations of coral atolls were significant contributions to nineteenth century geology. During his years in London, Darwin served as secretary of the Geological Society and met many important leaders of the scientific and intellectual community. Charles Lyell, the geologist, and Joseph Dalton Hooker (1817-1911), the botanist, were his closest friends. They served as confidantes, critics, and later as agents in a narrowly avoided priority dispute with A. R. Wallace. Because of his poor health, Darwin chose to leave London and moved his family to Down—a village in Kent not too far from London—where they lived in almost total seclusion.

Darwin and Evolutionary Theory

Although Darwin's name is almost synonymous with the theory of evolution by natural selection, Darwin actually put more time and effort into topics not directly related to evolution. There are many questions concerning the way Darwin developed his theory of evolution by natural selection. For although he stated that his conversion from creationism to evolution occurred with his observations of the fossil mammals in Argentina and his reflections on the living birds and tortoises of the Galapagos Islands, 20 years were to elapse before he published his views on evolution. After his return to England, Darwin seems to have come to the conclusion that the remarkable fitness of animals to their environment was not a reflection of "design" or creation, but of adaptation. Adaptation was the result of evolution, which means descent with modification.

Darwin seems to have convinced himself of the fact of evolution as early as 1837 to 1839. By 1842 he had written up his ideas in a preliminary sketch. Yet he continued to ponder the question until he was virtually forced by A. R. Wallace to "rush" *Origin of Species* into print in 1859. One reason traditionally given for this delay is that he was afraid to openly voice such unorthodox views. Bruno and Galileo came to Darwin's mind in this context. But surely Darwin knew that the last burning for scientific heresy was in 1600. Being independently wealthy, Darwin had no worries about losing a job and source of income. Probably Darwin was less concerned with being labelled unorthodox than he was afraid that he might not be able to convince his peers of the soundness of his theories. The theory of evolution was not new, but it remained incomplete without an explanation of its mechanism. Darwin recognized that the question of whether evolution had occurred was separate from the question of the mechanism by which it occurred, although he and others sometimes confused the two. By 1859, besides finding himself about to be scooped by Alfred Russel Wallace, he finally began to believe that he had enough evidence to build a persuasive case, if not a compelling one. By the time Darwin roused himself to publish *Origin of Species* it could legitimately be said that "where Lamarck had one fact, Darwin had one hundred."

Darwin later said that the major literary influences on his thinking were Charles Lyell's *Principles of Geology* and Robert Malthus' *Essay on Population*. While reading *Essay on Population* for diversion in September of 1838 Darwin was struck by the way the Malthusian struggle could lead to "positive" as well as negative results. It is rather difficult today to imagine anyone reading a work so ponderous and essentially gloomy for amusement, but the outcome in Darwin's case was truly remarkable. (The essay also served Wallace as a source for the theory of natural selection.)

Robert Malthus (1766-1834): The Cheerful Prophet of Doom

When critics wanted to denigrate Charles Darwin's accomplishments they contemptuously charged that only a Victorian gentleman could have produced such an unoriginal book, a mere application of the doctrine of Malthus to the world of plants and animals. Like Darwin, Malthus created a revolution in the world view of his times. Both produced classic works which were hotly contested by many people who never troubled to read them. Both classics went through many editions and were revised so much in response to criticism and changing ideas that it became difficult to tell later just what the author really believed. Malthus wrote the *Essay on Population* after a dispute with his father concerning the nature of man and the prospects for the future. Daniel Malthus had been a friend of Rousseau and tried to inculcate into his son his liberal view of the perfectibility of man, but Robert rejected all such hopeful visions for the future and developed an awesome and apparently inexorable calculus of human misery.

The thesis propounded in the *Essay on Population* was that utopian visions were impossible because man's real nature was the source of all his problems, and man's defects were innate and inexpungable. The natural state of a large part of mankind was that of misery and vice, sickness, war, infanticide, and promiscuity. These vices either tended to prevent procreation or allowed it to proceed and then exterminated the excess. All of this misery was the inevitable outcome of man's two basic needs: food and sex. Misery was inevitable because population always and everywhere is pressing on the available resources. This was true throughout nature and applied to human, animal, and vegetable life.

Everywhere procreational capacity exceeded the food supply and led to violent competition for resources necessary to life. Malthus could prove this with the fascinating power of numbers. Population increases geometrically, so that the human population (if left unchecked) would tend to increase every 25 years in the manner: 1, 2, 4, 8, 16, 32, 64, etc. Food supply in contrast could only increase arithmetically: 1, 2, 3, 4, 5, 6, etc. As precise and unarguable as a multiplication table came the proof that in 200 years the ratio of population to food would be 256/9. In 300 years this would reach 4096/13. There is an inconsistency in Malthus' seemingly clear exposition of the future. The ratios he calculates are those that would obtain *if population growth continued unchecked.* Yet Malthus also argued that misery and vice provide constant checks on population growth. In later editions, Malthus compounded the confusion by adding another check on population, which he called "moral restraint."

It was obvious from reading Malthus that the poor would be affected by the horrors of competition to a greater extent than the rich. Thus, the misery of the poor classes could only increase.

Natural Selection

Malthus' vision of the terrible competition for resources and the ensuing struggle for survival provided Darwin with his mechanism of evolution by

natural selection. His proofs can be summarized as follows. First, Darwin
argued that variation and fertility exist in abundance. Next, it should seem
obvious that different varieties have different fitness in the struggle for existence. From this it must follow that selection will operate on all varieties and
allow the most fit to survive. The varieties that survive will leave the most
offspring. Only heritable variations are important in terms of evolution.
Darwin argued that natural selection, carried out by nature, must be superior
to artificial selection carried out by man. Therefore, progress in producing
better domesticated plants and animals through artificial selection over a
relatively brief time period proved the efficacy of selection. Natural selection
had two effects on population. First it led to divergence of character or
speciation. At the same time it led to extinction of the less fit varieties
and species.

Darwin hinted in rather melodramatic terms to his friend Hooker as early
as 1844 that he was working on the species problem from a very unorthodox
viewpoint. "I am almost convinced (quite contrary to the opinion I started
with" he wrote, "that species are not (it is like confessing a murder)
immutable."[50]

For systematists, the problem of defining "species" was central to biology.
This, in large part, was the legacy of Linnaeus, who believed that naming of
plants and animals was the great task of biological science. In describing and
cataloging species in such great detail, he had actually imposed a new standard
of fixity and rigidity on the term. For Darwin, once organic life was seen as
the product of long eons of change and diversification "species" became a
rather arbitrary term. It was merely used for the sake of convenience to refer
to groups of individuals which closely resembled each other. While it might
be necessary for cataloging and descriptive purposes, from an evolutionary
viewpoint "species" was merely a way of describing a group of organisms
which closely resembled each other at any given point in time. "Varieties"
which Linnaeus had dismissed as fluctuations of no real importance, now
took on an important role as "incipient species." Darwin believed that his
view of living entities in a dynamic state, rather than as static creations, provided the basis for a natural system of classification based on descent from
common ancestors and the divergence of character with time. The practical
problems of taxonomy remained, though, since in most cases all the information available to taxonomists is morphological. Descent has to be traced by
inference from morphology. However, many enthusiastic converts to evolution were happy to turn morphological and embryological relationships into
lines of descent.

The "origin" in Darwin's title clearly referred to the transmutation of one
species into another. Unlike the previous generation of evolutionists, he did
not start from some "primeval filament" or "living molecule" and derive all

of creation from it. Darwin suggested that evolution as "change" was almost self-evident. The real problem had been explaining the mechanism of change by natural selection. Speculation about the ultimate origin of life was outside of the scientific problem that Darwin chose to explicate. The starting point for his argument was the great abundance of variation which should be obvious to even the casual observer. This is documented with a wealth of detail in his first two chapters, "Variation under Domestication" and "Variation under Nature." Since Darwin did not know the cause of variation, his chapter on the "Laws of Variation" was one of the most vague and confusing in the book. By limiting his essential argument to measuring the effects of variation, without knowing the cause, Darwin attempted to remain on good scientific ground.

On the day of publication, all 1250 copies of the first printing of *Origin of Species* were sold out. After the impact of his first edition, Darwin revised and rewrote, often while struggling to answer his critics. The most difficult criticism came from a physicist, Lord Kelvin (William Thomson 1824–1907) and an engineer, Fleeming Jenkin. The former challenged Darwin's whole scheme of evolution through continuous small changes by declaring that the earth was not old enough to allow for such gradual transmutation.[51] Indeed some physicists granted a meager span of 10,000,000 years for geologists and biologists to maneuver in. Jenkin challenged Darwin on the basis of the mathematical effect of blending inheritance which would tend to damp out all novelties.[52] Given the theory of inheritance widely accepted in the 1860s, a rare new mutant obviously would be "swamped out" of the population by mating with normal individuals. Ironically, Mendel had supplied the key to the problem in 1865, but it was not applied until the next century. By the sixth edition of *Origin,* Darwin felt compelled to acknowledge Jenkin's argument and confess that he saw no simple way around it. Darwin was forced to fall back to a more Lamarckian position. In later editions of *Origin,* the effect of the environment and of use and disuse became more important, although natural selection was never abandoned.

Because of Darwin's difficulties in meeting the specious, but seemingly strong arguments of Jenkin, Kelvin, and others, the later editions of *Origin* are more contradictory and confusing than the first. Darwin felt he had been rushed into premature publication, for *Origin of Species* was only an "abstract" of the great book he had planned to write. He thought he might have crushed all opposition if he had been free to complete his "big book." It is also possible that friend and foe alike would instead have choked on the undigestible mass of data. Nevertheless, the problems with inheritance and the age of the earth would have remained. The "abstract" published in 1859 was a clear statement of the fact and mechanism of evolution and was quite convincing to readers with open minds. Indeed, when T. H. Huxley read the book, he made the classic comment of a scientist who realizes he has missed an obvious point: "How extremely stupid not to have thought of that onself."

Descent of Man and Selection in Relation to Sex

If Darwin seemed reluctant to publish *Origin of Species,* he found discussion of the evolution of man an even more unpleasant task. Darwin merely hinted in the *Origin* that his ideas might throw some light on the origin of man. Probably he hoped that someone else would write the book which would elaborate on this theme. Although evolutionists like T. H. Huxley (1825-1895), Ernest Heinrich Haeckel (1834-1919), and Herbert Spencer (1820-1903) were busy explaining man's place in nature or the application of the mechanism of survival of the fittest to human society, this was not the same sober, rational, and scientific book that demanded to be written. Darwin even encouraged Wallace to write the book on human evolution, volunteering his notes on the subject. But Wallace was dabbling in spiritualism, and could not be relied on to keep "divine intervention" out of the picture. Indeed, Darwin realized that many supporters of evolution still attempted to retain some role for divine intervention where human intelligence and soul were concerned. Perhaps, they argued hopefully, the intellectual and spiritual faculties of man constituted an unbridgeable chasm separating human beings from the animals.

But like a machine, Darwin inexorably ground out the makings of a uni-formitarian continuity between man and other animals. Darwin pledged at the outset that the sole object of his *Descent of Man* was "to consider, firstly, whether man, like every other species, is descended from some pre-existing form; secondly, the manner of his development; and thirdly, the value of the differences between the so-called races of man."[53] Arguments that related to physical similarity and common descent of species in the *Origin* were extended in the *Descent of Man* to provide continuity among all species—even man. Through the analysis of qualities shown by animals, such as curiosity, mem-ory, imagination, reflection, loyalty, and the tendency to imitate, Darwin proceeded to the evolution of human qualities. Every anecdote he could muster was used to bolster his basic premise. Thus, the feeling of his dog, Polly, for her human master, is cited as probably homologous to the religious awe man feels for God.

Since he was very cautious and vague about the time and place, as well as the racial make up of the first "humans," it is difficult to be sure where Darwin stood on many issues. He suggested so many possibilities and ranged so widely that adherents of all kinds of ideologies were able to cite Darwin as "scientific evidence" for their cause. "Darwinism" has been used by racists and sexists as well as by proponents of equality, and by warmongers as well as pacifists.

As the title indicates, a major portion of the book was devoted to the subject of sexual selection and was only indirectly related to man. The original theory of natural selection could not be directly applied to the evolution of the secondary sexual characteristics. Darwin may have been more heavily indebted to his grandfather for the idea of sexual selection than he wished to admit. Traits involved in competition between males for attracting the females of their own species are not strictly speaking, useful in the struggle for existence carried out between species. Indeed, some secondary sexual traits might even be dangerous or cumbersome to the animal, such as the bright plummage of male birds or the large decorative antlers of certain deer. Without knowledge of hormonal regulation of development and of the laws of heredity, it was impossible to explain how males could transmit ancestral female characteristics to their daughters, and vice versa.

Darwin's ideas about the traits belonging to each sex were typical of his time. Like Aristotle before him, he saw male traits as strength, bravery, and intelligence: "Man is more courageous, pugnacious and energetic than woman," wrote Darwin, "and has a more inventive genius." Females, in contrast, were described as passive, weak in body, and deficient in brains. Even here Darwin was sufficiently ambiguous to leave room for other interpretations. Darwin noted that the male brain was larger, "but whether or not proportionately to his body, has not, I believe, been fully ascertained."[54] Since his theory of sexual selection depended on intrasexual competition for mates, Darwin overemphasized male traits and generally neglected the very demanding female functions of care, feeding, and training of the young. Needless to say, this further "scientific evidence" for male superiority won him many adherents. Indeed the one hundredth anniversary tribute to Darwin, *Sexual Selection and the Descent of Man 1871-1971,* retains essentially the same emphasis and perspective.[55] Animals in which the average female is larger than the average male have been less studied, although the average female is larger in about 30 of the 122 families of living mammals.[56] Among these creatures are rabbits and hares, one family of bats, three families of baleen whales, a subfamily of seals, two tribes of antelopes, the aardwolf, klipspringer, short-snouted elephant shrew, and the hairy-nosed wombat. The females may have become larger because of competition for some limiting resource. Darwin's theory of sexual selection dominated research on the question for more than a century, but researchers are now looking at these ideas more critically. No single theory has so far been able to account for sexual dimorphism with respect to size, the diversity of mating systems, and intrasexual competition. These questions are obviously quite complicated, since similar species may differ as to relative size of the two sexes even though habitat, food preferences, and behavior seem similar.

It is remarkable just how prescient Darwin's views on the descent of man really were considering that human paleontology and cultural anthropology were disciplines in their infancy at the time. Old, plagued by chronic illness, Darwin could not carry out his own fieldwork in these areas. Without authentic or even dubious evidence of "missing links," he could rely only on evidence from comparative anatomy and psychology. From such sparse data, Darwin surmised that man's ancestor was related to the gorilla and chimpanzee; that he originated in a hot climate, probably in Africa; and that the ancestor of man appeared between the Eocene and Miocene periods. By the time Darwin was ready to publish *Descent,* several other naturalists had already concluded that "man is the co-descendant with other species of some ancient, lower, and extinct form," but many others still rejected what Darwin considered good documentation of the antiquity of man. Embittered by the venomous attacks on his work and reputation, Darwin remained confident that *all* life forms had indeed evolved, but began to doubt "whether humanity is a natural or innate quality."[57] While many scientists, philosophers, and clergymen asserted that man's origin would never be discovered through science, Darwin admonished them to remember that "ignorance more frequently begets confidence than does knowledge: It is those who know little, and not those who know much, who so positively assert that this or that problem will never be solved by science."[58]

During the 20 years in which he secretly worked on the theory of evolution, Darwin established his reputation as a naturalist with less controversial subjects. For eight long years he studied the taxonomy and physiology of living and fossil barnacles. This was good training for understanding the minute distinctions that separate some species, but later even Darwin wondered whether the project had been worthwhile. In 1868 he published *Variation in Animals and Plants under Domestication,* in which he described his very fuzzy ideas about the mechanism of inheritance. A year after the appearance of *Descent of Man* the companion volume *The Expression of the Emotions in Man and Animals* laid down the fundamentals of modern research into ethology. The methods he used are still basic to this field: studies of infants, the insane, paintings and sculptures, photographs of expression evaluated by different judges, and comparative study of expression among different peoples. As Konrad Lorenz points out in his introduction to *Expression,*

> behaviour patterns are just as conservatively and reliably characters of species as are the forms of bones, teeth, or any other bodily structures. Similarities in inherited behavior unite the members of a species, of a genus, and even the largest taxonomic units in exactly the same way in which bodily characters do so.[59]

Darwin showed that the evolution of behavior is like the evolution of organs. That is, behavior patterns may outlive their original function and be-come as "vestigial" or "rudimentary" as physical characteristics. Thus, just as the first gill slit became an ear opening when a change from an aquatic to a terrestrial life occurred, so human "snarling" developed out of biting (a means of aggression which is practically extinct in adult humans). The book is relatively unknown, but has a lively and imaginative flavor that suggests Darwin has often been underrated as an original thinker.

The amount of truly fundamental work carried out by this gentle, modest, semi-invalid is prodigious in quantity and astonishing in scope. Darwin asked direct, even child-like questions and provided answers so ingenious and ele-gant in their simplicity that they often served to undercut judgments of his real worth.

Darwin and Wallace

One reason Darwin's originality and genius have been called into question is that he shared the discovery of natural selection with another man. Darwin wrote two trial essays on evolution for his own use in 1842 and 1844, but it was not until about 1854 that he began trying to bring order to his mountains of data. In 1856, Lyell, although still not converted to evolution, urged Darwin to publish his ideas before someone else anticipated him. In his slow, methodical way, Darwin began to write out his ideas in such monstrous detail that many volumes would have been required to present his case. When about halfway into his project, Darwin was shocked to receive a letter from Alfred Russel Wallace, a naturalist working in Malaya, along with an essay, "On the Tendency of Varieties to Depart Indefinitely from the Original Type." Wallace wanted Darwin's opinion of this essay. Should it have any merit, Wallace re-quested that it be sent on to Lyell. Darwin was thunderstruck—he found that Wallace had exactly the same theory as his own. Unlike his "badly written" essay of 1844, Wallace's essay was "admirably expressed and quite clear." Left to himself in this crisis, Darwin might have withdrawn from the potential priority battle. Luckily for him, Lyell and Hooker took over as mediators and champions of fair play. A joint presentation of papers by Wallace and Darwin was quickly arranged before the Linnaean Society. The papers ap-peared in the *Proceedings of the Linnaean Society* for August of 1858. After this unsettling experience, Darwin polished off his famous *Origin of Species* after "thirteen months and ten days' hard labour."

In his autobiography Darwin declared, "I cared very little whether men attributed most originality to me or Wallace." However, in his letters to his friends Lyell and Hooker, he admitted how hard it was to think of losing his "priority of many years' standing" and see his originality "smashed."[60]

Alfred Russel Wallace (1823–1913)

Although it may seem that Wallace was cheated of much of the glory for contributions to evolutionary science, it is unlikely that his essay alone would have created much impact. Wallace seems to have accepted his role as minor, co-discoverer rather philosophically as another disappointment in a life full of hard luck. Generously, he coined the term "Darwinism" to distinguish the new theory of evolution from its predecessors.[61]

Wallace was an enthusiastic if amateurish naturalist by 1844 when he became a master at the Collegiate School at Leicester. His own education had been "rudimentary" and he had held a variety of jobs. In 1848 he was able to persuade the entomologist Henry Walter Bates (1825–1892) (famous for the discovery of "mimicry" in animal coloration) to accompany him on an expedition to the Amazon. The two planned to study natural history in the tropics and pay for their trip by selling their collections. Wallace was already interested in the question of the origin of species when he set out, having been inspired by reading the work of Alexander von Humboldt (1769–1859) and Charles Darwin. In the tropics Wallace was sure he would find the materials to answer the "question of questions." He was not to have much success on this venture—the collections, obtained with such difficulty, were lost when the ship caught fire on the return voyage in 1852.

Recovering from this disaster, Wallace began a new phase of his research in the Malay Archipelago, observing and collecting for more than eight years (1854–1862). In 1855 he published an essay, "On the Law which has Regulated the Introduction of New Species," which was largely ignored, although it is now regarded as an important pre-Darwinian contribution to evolutionary theory. Wallace came closer to the solution of the problem of the origin of species during a bout of illness when he remembered the arguments of Malthus and in a flash of insight saw natural selection as the mechanism of evolution. In about two days he had written a paper for publication entitled "On the Tendency of Varieties to Depart Indefinitely from the Original Type," and sent it to Darwin. When Darwin read this paper in 1858 he was shocked. "I never saw such a striking coincidence," he wrote, "if Wallace had my MS. sketch written out in 1842, he could not have made a better short abstract!"[62]

Darwin's friends settled the priority issue in a quiet, efficient manner. Wallace seems to have accepted his subordinate position gracefully. Even Darwin felt that Wallace was overly modest: "You are the only man I ever heard of who personally does himself an injustice, and never demands justice." Wallace published several important works and helped to spread the Gospel of evolutionary theory as a popular scientific lecturer. His *Geographical Distribution of Animals* (1876) is a landmark in the science of zoogeography. His *Island Life* and *Malay Archipelago* (1869) were fine descriptions of the flora and fauna of the islands and great travel books.

Wallace has been described as "half genius, half crank." In his later years, he not only dabbled in spiritualism but published his ideas on the subject, which embarrassed some of his friends. His support of women's suffrage, land nationalization, and socialism were so advanced as to be quite unorthodox in his milieu.

Reactions to Darwinism

Darwin and Wallace created a revolution in thought by challenging the long-cherished picture of nature as the result of God's design, intention, and direct intervention; and the earth as a place created especially for the comfort of man. Darwin had turned the familiar arguments of natural theology inside out and created a world of materialism and chance in which the fit between creatures and their environment was not proof of the wise design of the Creator, but the result of natural selection acting impersonally on fortuitous variations over vast periods of time. Some clergymen rushed to put together a new concept of nature and the relationship of God to the world. As Chambers had suggested, it was not impossible to see evolution as consistent with a more glorious, if more distant and abstract, concept of God. Others had relied so heavily on the argument from design that they could not accept any alternative.

Darwin's chronic ill health and strict resolution to avoid all controversy might have damaged the cause of evolutionary science. Fortunately for Darwin, the battle was taken up by some extremely pugnacious, determined, and clever scientists. Chief among them were Thomas Henry Huxley and Ernst Haeckel. The application of Darwinian ideas to society was actively carried out by Herbert Spencer and others. The effect of Haeckel's writings on a young mind is best described by Richard Goldschmidt in *Portraits from Memory*. On reading Haeckel's history of creation, the young man felt "all problems of heaven and earth were solved simply and convincingly; there was an answer to every question which troubled the young mind. Evolution was the key to everything and could replace all the beliefs and creeds which one was discarding."[63]

Thomas Henry Huxley (1825–1895)

Although his advocacy of evolutionary theory won him the title "Darwin's Bulldog," the irrepressible Thomas Henry Huxley was an eminent scientist in his own right. In his autobiography, Darwin described Huxley as a clever man, remarkable for his quickness of apprehension and wit. Unlike the well-to-do Darwin, who had had access to the best available (though dull) education,

Huxley had only two years of formal schooling between the ages of eight and ten. He had the misfortune to be the seventh of eight children born to Rachel and George Huxley, an unsuccessful schoolmaster. Huxley was self-educated but widely read. He taught himself French, Latin, German, Italian, and Greek.

Like Darwin and Wallace, Huxley developed his talents as a naturalist through long voyages to exotic places. A prodigious worker, Huxley turned out over 150 research papers covering fields as diverse as zoology, paleontology, geology, anthropology, and botany. He also produced 10 scientific textbooks, books of essays, and numerous controversial articles on topics from education to religion and ethics. Huxley coined several words still used in and out of science, such as "biogenesis" and "agnostic"—and "bishipophagous" (to refer to his ability to eat up clergymen in his famous debates). Part of Huxley's success rested on his wonderful ability with language, either spoken or written. (His grandsons, Julian and Aldous Huxley carried on the tradition.)

One of Huxley's most famous debates was with Bishop Wilberforce, widely known then as "Soapy Sam." The occasion was the June 1860 meeting of the British Association at Oxford. Huxley did not like the medieval atmosphere at Oxford and did not want to be "episcopally pounded," but was talked into attending by Robert Chambers, the still anonymous author of *Vestiges.* Although the focal point of controversy was the *Origin of Species,* Darwin was conspicuously absent. The topic was obviously of great interest for some 700 people were packed into the lecture room to hear the Bishop denounce the dreadful "monkey theory." The crowd was quite hostile to Darwinism and warmly cheered the Bishop's lecture, although it was obvious to Huxley he knew nothing whatever about evolutionary science as science. Finally, the Bishop turned to Huxley and inquired politely "was it through his grandfather or his grandmother that he claimed his descent from a monkey?" Huxley knew then that he could destroy his adversary. After succinctly explaining Darwin's ideas and exposing the Bishop's ignorance, Huxley closed by saying that he would "rather have a miserable ape for a grandfather" than a man who used his great gifts and influence "to introduce ridicule into a grave scientific discussion." Huxley had made a monkey of the Bishop.[64]

Impact of Darwinism

Although other triumphs of nineteenth century biology, such as bacteriology and physiology, have had a greater effect on life and health than evolution, no other science has had more impact on human thought. Evolutionary theory was extended to many fields far removed from the biological limits to which Darwin generally confined himself. Darwin's writings were cited as scientific "proof" for movements and philosophies as divergent as socialism, fascism, communism, capitalism, atheism, and evolutionary theology. In America, Darwinism found a sympathetic reception among scientists, intellectuals, businessmen, and politicians promoting "manifest destiny," even if evolution was not granted a place in American high school biology texts.

In evolutionary science, many saw a political and economic message: Since the status quo is the result of eons of natural evolution, things *must be* the way they are. Therefore, present hardships of industrial society must be accepted as the result of the workings of inexorable natural laws. "Reform" was an attempt to tamper with evolutionary law. Such reforms would be a futile, misguided, even dangerous interference with natural processes and the operation of natural law. Yet Marx and Engels tried to use the very same evolutionary "laws," to justify the inevitability of communism. Darwin had written so much, revised so often, and been so vague that all sides could claim that proper interpretation of the Darwinian "Bible" favored their position and only theirs. Remaining quietly in his retreat at Down, Darwin neither encouraged nor discouraged the development of various forms of "social Darwinism."

Social Darwinism

The elaboration of the doctrine of social Darwinism—the application of "evolutionary laws" to human society—was most closely associated with Herbert Spencer (1820-1903), a railway engineer and inventor, who later turned to a career in journalism. Among Spencer's many books on philosophy, sociology, education, and science, *Principles of Biology* was particularly important as a link between the general reader and the scientist. Although he and Hooker checked much of this work before publication, Huxley characterized Spencer's idea of a tragedy as a theory destroyed by a fact.

According to Spencer, society was analogous to an organism; changes in the social organism must be measured on the scale of evolutionary progress. In the very remote future, society would evolve inevitably towards a better state. Some social Darwinists adopted a pessimistic view of social evolution as a Malthusian struggle, while others were quite optimistic about the inevitability of progress. But all social Darwinists agreed on one thing—nature selected the best and the fittest and to them she gave her rewards. The best were the ones who were most fit as demonstrated by their riches and their position in society. The unfit were the "negligent, shiftless, inefficient, silly, and imprudent." The workings of evolutionary law were inexorable: The fit would be rewarded and the unfit would be destroyed. This led some observers to remark that the whole "scientific" package was nothing but "biological Calvinism." The fit were urged to reproduce their own kind while they worried about preventing the multiplication of the unfit.

It is in this concern with the differential fertility of rich and poor that the analogy between biological and social Darwinism most obviously breaks down. The irony of the situation should be obvious to anyone who critically analyzes this false analogy. In biology, fitness and successful competition are measured by survival and reproductive success, that is leaving a larger number of off-spring. In society, the measure of success was the accumulation of wealth, not the number of offspring. Indeed, Americans feared that the country was committing "race suicide" because the fit, so worn out in the struggle for wealth, were not doing their share in the struggle to leave offspring. The un-fit, in contrast, were doing nothing but reproducing their own unfit kind. Darwin's cousin, Francis Galton (1822-1911), did much to stimulate this fear through his development of the "science" of eugenics.

Opponents of the "dog-eat-dog" version of social Darwinism eventually be-gan to fight back with a different vision of human society. Still committed to a belief in Darwinism, Lester Ward, John Dewey, and the pragmatists chal-lenged the view of society founded on struggle and competition. Peter Kropotkin's (1842-1921) *Mutual Aid* was one of the earliest reinterpretations of Darwinism based on cooperation instead of competition. The stimulus to this work was Kropotkin's desire to answer T. H. Huxley's harsh description of the role of competition in evolution.

Kropotkin was a Russian prince who became an anarchist-nihilist activist. Many of his biological ideas developed as a result of surveys of the flora and fauna of Manchuria and Siberia. Because these areas were certainly not like the Garden of Eden, the severity of the struggle for existence should have been exceptionally harsh. However, Kropotkin found much competition be-tween species, but little evidence of intraspecific struggle. Thus, he proposed that in addition to the "law of struggle," there is a "law of mutual aid and support" which is a primary factor in the evolution of many species. He ar-gued that this law could be found, at least implicitly, in Darwin's *Descent of Man*. Consciousness of species solidarity produced an instinct that drove the social insects and led wolves to hunt in packs. Obviously nature was neither all pitiless struggle nor all peace and harmony. Recent studies of "socio-biology" tend to support Kropotkin's vision of "mutual aid."

Those who praised Darwin and those who condemned him often failed to understand him. Even those who spoke glibly of "manifest destiny" and "racial purity" condemned the introduction of evolutionary theory into the high school curriculum. The best known example of the conflict over the teaching of evolution is the Scopes trial. This event was commemorated by the Tennessee Historical Society with a marker on the courthouse grounds in Dayton which reads:

Here, from July 10 to 21, 1925, John Thomas Scopes, a County High School teacher, was tried for expounding the theory of the simian descent of man, in violation of a lately passed state law. William Jennings Bryan assisted the prosecution; Clarence Darrow, Arthur Garfield Hayes and Dudley Field Malone the defense. Scopes was convicted.

In 1972, a half century after the famous "monkey trial," a survey of Dayton high school students revealed that three-quarters still believed the King James Bible's version of the creation rather than Charles Darwin's. While the trial of John Thomas Scopes created a national sensation the outcome was far from clear. Although Scopes was found guilty, the decision was later reversed on a technicality and evolutionists viewed the case as a triumph for modern science. However, a recent investigation of science teaching at the elementary and high school levels indicates that the teaching of evolution actually declined after the Scopes trial.[65] Indeed pressure against school boards still affects the teaching of evolution today. Recent drives by antievolutionists have tried to either ban the teaching of evolution or have demanded "equal time" for "Special Creation" as described in Genesis. This has raised many questions about the separation of Church and state, the teaching of controversial subjects in public schools, and the ability of scientists to communicate with the public.[66]

The controversy over evolution proves more clearly than any other episode in the history of science that science is not limited to the laboratory nor is it a catalog of facts. Science has a profound impact on society and is an integral part of the history of ideas and cultures. In no other case have the offshoots of a science stimulated more social challenges than evolution. Nor has any other scientific theory been so thoroughly exploited and corrupted by partisans seeking a scientific prop for their dogmas.

Scientifically, the weakest part of Darwin's work was his theory of pangenesis—an excessively vague and unfounded hypothesis by which he attempted to explain the mechanism of inheritance. Although Darwin was totally unaware of it, an obscure monk, called Gregor Mendel, was conducting eccentric experiments on the common garden pea which unravelled the fundamental laws of genetics, as we shall see in Chapter 13.

Notes

1. W. Coleman (1971). *Biology in the Nineteenth Century*. New York: Wiley, p. 2.

2. J. Fruton (1972). *Molecules and Life*. New York: Wiley-Interscience, p. 181.

3. A. O. Lovejoy (1964). *The Great Chain of Being*. Cambridge: Harvard University Press.

4. E. Topsell (1658). *The History of Four-Footed Beasts and Serpents and Insects: Taken principally from the "Historiae Animaluim" of Conrad Gesner.* 3 vols. Reprinted 1967. New York: Da Capo. (Facsimile of 1658 edition.)

5. E. Mayr (1963). *Animal Species and Evolution.* Cambridge, Mass.: Harvard University Press, p. 14.

6. E. Nordenskiold (1936). *The History of Biology.* New York: Tudor, p. 202.

7. C. Raven (1942). *John Ray Naturalist.* Cambridge: Cambridge University Press.

8. N. Gourlie (1953). *Carl Linnaeus: The Prince of Botanists.* London: Witherby.

9. C. Linnaeus (1964). *Systema Naturae, 1735.* Nieuwkoop, Holland: De Graaf, p. 23.

10. Linnaeus, pp. 7, 11.

11. E. L. Green (1907). Linnaean Memorial Address: *Proc. Wash. Acad. Sciences 9:* 241–272, p. 262.

12. Linnaeus, p. 26.

13. A. R. Hall (1966). *The Scientific Revolution 1500–1800.* Boston: Beacon, p. 295.

14. Linnaeus, p. 61.

15. C. L. Porter (1967). *The Taxonomy of Flowering Plants.* San Francisco: Freeman, pp. 65–66.

16. E. L. Greene (1909). Linnaeus as an Evolutionist. *Proc. Wash. Acad. Sci. 11:* 17–26; L. Eisley (1961). *Darwin's Century.* New York: Anchor.

17. B. De Maillet (1968). *Telliamed.* Translated by A. V. Carozzi. Urbana: University of Illinois Press, p. 230.

18. P. L. Maupertuis (1966). *The Earthly Venus.* Translated by S. B. Boas. New York: Johnson Reprint, p. xxi.

19. Maupertuis, p. 4.

20. C. Darwin (n.d.). *The Origin of Species and the Descent of Man.* New York: Modern Library, p. 3.

21. Lovejoy, p. 230; P. Farber (1972). Buffon and the Concept of Species. *J. Hist. Biol. 5:* 259–284; D. L. Hall (1965). The Effect of Essentialism on Taxonomy. *Brit. J. Phil. Sci. 15:* 314–326; *16:* 1–18.

22. Lovejoy, p. 230.

23. F. M. Broodie (1974). *Thomas Jefferson.* New York: Norton, pp. 155, 34, 194.

24. A. S. Packard (1901). *Lamarck, The Founder of Evolution.* New York: Longmans, Green.

25. Darwin, pp. 3–4.

26. J. B. Lamarck (1809). *Zoological Philosophy.* Translated by H. Elliot. (1914). London: MacMillan.

27. Lovejoy, p. 251.

28. Lamarck, pp. 169–170.

29. Lamarck, p. 173.

30. E. S. Russell (1916). *Form and Function.* London: Murray, p. 55.

31. D. King-Hele (1968). *Erasmus Darwin.* New York: Scribner.

32. F. Darwin, ed. (1958). *The Autobiography of Charles Darwin and Selected Letters.* New York: Dover, p. 13.

33. E. Darwin (1794–96). *Zoonomia.* 2 vols. London: Johnson, Vol. 1, p. 572; E. Darwin (1803). *The Temple of Nature.* London: Johnson, p. 28.

34. G. Cuvier (1837). *The Animal Kingdom.* Translated from the French. London: Henderson, p. xiii.

35. Cuvier, p. 8.

36. R. Chambers (1969). *Vestiges of the Natural History of Creation* (1844). New York: Humanities.

37. M. Millhauser (1959). *Just Before Darwin: Robert Chambers and Vestiges.* Middletown: Wesleyan University Press.

38. Chambers, p. 152ff.

39. Chambers, p. 204.

40. Chambers, p. 230.

41. Chambers, p. 344.

42. C. C. Gillispie (1959). *Genesis and Geology.* New York: Harper Torchbook.

43. Eiseley, p. 79; A. Geike (1962). *The Founders of Geology.* 2nd Ed. New York: Dover.

44. Gillespie, p. 112.

45. Geike, p. 45.

46. F. Darwin, Chapter 1, "The Darwins"; Chapter 2, "Autobiography."

47. C. Darwin, p. 11.

48. F. Darwin, pp. 29–30.

49. R. Colp (1977). *To Be an Invalid: The Illness of Charles Darwin.* Chicago: University of Chicago Press.

50. F. Darwin, p. 184.

51. J. D. Burchfield (1975). *Lord Kelvin and the Age of the Earth.* New York: Science History Publications.

52. Eiseley, pp. 209–11, 214–216.

53. Darwin, p. 390.

54. Darwin, p. 867.

55. B. Campbell (1972). *Sexual Selection and the Descent of Man, 1871–1971.* Chicago: Aldine Publishing.

56. G. B. Kolata (1977). Sexual Dimorphism and Mating Systems: How Did they Evolve? *Science 195:* 382–383.

57. F. Darwin, p. 7.

58. Darwin, p. 390.

59. C. Darwin (1965). *The Expression of the Emotions in Man and Animals.* Chicago: University of Chicago Press, Introduction.

60. F. Darwin, p. 45, 196–198.

61. A. R. Wallace (1891). *Darwinism, an Exposition of the Theory of Natural Selection With Some of Its Applications.* London: MacMillan.

62. F. Darwin, p. 196.

63. R. Goldschmidt (1956). *Portraits from Memory.* Seattle: University of Washington Press, p. 35.

64. T. H. Huxley (1959). *Man's Place in Nature.* Ann Arbor: University of Michigan Press, Introduction by A. Montague, pp. 2–3.

65. J. V. Grabiner and P. D. Miller (1974). Effects of the Scopes Trial. *Science 185:* 832–837.

66. D. Nelkin (1976). The Science Textbook Controversies. *Sci. Amer. 234:* 33–39.

References

Adams, F. D. (1938). *The Birth and Development of the Geological Sciences.* New York: Dover.

Appleman, P., ed. (1976). *Thomas Robert Malthus: An Essay on the Principle of Population, Text, Sources and Background Criticism.* New York: Norton.

Appleman, P., ed. (1970). *Darwin.* New York: Norton.

Bailey, E. B. (1967). *James Hutton—the Founder of Modern Geology.* New York: Elsevier.

Bailey, E. B. (1962). *Charles Lyell.* Garden City: Doubleday.

Barzun, J. (1958). *Darwin, Marx, Wagner.* Garden City: Doubleday.

Bennet, A. W. (1870). The Theory of Selection from a Mathematical Point of View. *Nature 3:* 30–31.

Bibby, H. C. (1959). *T. H. Huxley: Scientist, Humanist and Educator.* London: Watts.

Bibby, H. C., ed. (1967). *The Essence of T. H. Huxley.* London: MacMillan.

Blum, H. F. (1968). *Time's Arrow and Evolution.* Princeton: Princeton University Press.

Bonney, T. G. (1895). *Charles Lyell and Modern Geology.* New York: MacMillan.

Brunet, P. (1929). *Maupertuis.* Paris: A. Blanchard.

Burchfield, J. D. (1975). *Lord Kelvin and the Age of the Earth.* New York: Science History Publications.

Campbell, B. (1972). *Sexual Selection and the Descent of Man, 1871–1971.* Chicago: Aldine.

Cannon, H. G. (1959). *Lamarck and Modern Genetics.* Springfield, Ill.: Thomas.

Chambers, Robert (1844). *Vestiges of the Natural History of Creation.* London: Churchill.

Coleman, W. (1964). *Georges Cuvier, Zoologist.* Cambridge, Mass.: Harvard University Press.

Colp, R. (1977). *To Be an Invalid: The Illness of Charles Darwin.* Chicago: The University of Chicago Press.

Cox, L. R. (1948). William Smith and the Birth of Stratigraphy. *International Geological Congress:* 18th Session.

Cuvier, G. (1837). *The Animal Kingdom, arranged according to its organization, serving as a foundation for the natural history of animals.* Translated from the French. London: Henderson.

Darwin, C. (1962). *The Voyage of the Beagle.* Garden City: Doubleday.

Darwin, C. (1859). *The Origin of Species by Means of Natural Selection or The Preservation of Favoured Races in the Struggle for Life.* Edited and Introduced by J. W. Burrow (1968); from Darwin's 1859 edition. Maryland: Penguin.

Darwin, C. (n.d.). *The Origin of Species and the Descent of Man.* New York: Modern Library.

Darwin, C. (1872). *The Expression of the Emotions in Man and Animals.* Republished in 1965. Chicago: The University of Chicago Press.

Darwin, C. (1868). *The Variation of Animals and Plants under Domestication.* 2 vols. London: Murray.

Darwin, C. (1868). *On the Various Contrivances by which British and Foreign Orchids are Fertilized by Insects, and on the Good Effects of Intercrossing.* London: Murray.

Darwin, C. (1875). *The Movements and Habits of Climbing Plants.* London: Murray.

Darwin, C. (1876). *The Effects of Cross and Self-Fertilization in the Vegetable Kingdom.* London: Murray.

Darwin, C. (1880). *The Power of Movement in Plants.* London: Murray.

Darwin, C. (1881). *The Formation of Vegetable Mould, Through the Action of Worms, with observations of their Habits.* London: Murray.

Darwin, C. and Wallace, A. R. (1858). On the tendency of species to form varieties; and on the perpetuation of varieties and species by natural means of selection. *Proc. Linnean Soc.*

Darwin, F., ed. (1958). *The Autobiography of Charles Darwin and Selected Letters.* New York: Dover.

de Beer, G. (1963). *Charles Darwin; Evolution by Natural Selection.* New York: Nelson.

de Beer, G. (1964). *Atlas of Evolution.* London: Nelson.

Dobzhansky, T. (1951). *Genetics and the Origin of Species.* 3rd Ed. New York: Columbia University Press.

Dobzhansky, T. (1955). *Evolution, Genetics and Man.* New York: Wiley.

Drachman, J. M. (1930). *Studies in the Literature of Natural Science.* New York: MacMillan.

Dupree, A. H. (1959). *Asa Gray 1810–1888.* Cambridge, Mass.: Harvard University Press.

Eiseley, L. (1958). *Darwin's Century: Evolution and the Men Who Discovered It.* Garden City: Doubleday.

Eiseley, L. (1959). Charles Lyell. *Sci. Amer. 201:* 98–106.

Eiseley, L. (1960). *The Firmament of Time.* New York: Atheneum.

Flourens, P. M. J. (1858). *Histoire des Travaus de Georges Cuvier.* Paris: Garnier.

Fothergill, P. G. (1952). *Historical Aspects of Organic Evolution.* London: Hollis and Carter.

Garn, S. M. (1961). *Human Races.* Springfield, Ill.: Thomas.

Geige, Sir Andrew (1962). *The Founders of Geology.* New York: Dover.

George, W. (1964). *Biologist Philosopher, A Study of the Life and Writings of Alfred Russel Wallace.* New York: Abelard-Schuman.

Gillispie, C. C. (1959). *Genesis and Geology.* New York: Harper Torchbook.

Gillispie, C. C. (1958). Lamarck and Darwin in the history of science. *Amer. Sci. 46:* 388–409.

Glass, B., ed. (1959). *Forerunners of Darwin: 1745–1859.* Baltimore: Johns Hopkins University Press.

Gourlie, N. (1953). *Carl Linnaeus: The Prince of Botanists.* London: Witherby.

Grabiner, J. V., and Miller, P. D. (1974). Effects of the Scopes Trial: Was it a victory for evolutionists? *Science 185:* 832–837.

Grant, S. (1884). *Story of the University of Edinburgh.* London: Longmans, Green.

Gray, A. (1963). *Darwinia.* Edited by A. H. Dupree. Cambridge, Mass.: Harvard University Press.

Green, J. C. (1959). *The Death of Adam: Evolution and Its Impact on Western Thought.* Iowa: The Iowa State University Press.

Gruber, H. E. (1974). *Darwin on Man. A Psychological Study of Scientific Creativity. Together with Darwin's Early and Unpublished Notebooks.* New York: Dutton.

Hallam, A. (1973). *A Revolution in the Earth Sciences: From Continental Drift to Plate Tectonics.* Oxford: Clarendon.

Himmelfarb, G., ed. (1960). *Thomas Robert Malthus, On Population.* New York: Modern Library.

Himmelfarb, G. (1968). *Darwin and the Darwinian Revolution.* New York: Norton.

Hofstadter, R. (1955). *Social Darwinism in American Thought.* Revised Edition. Boston: Beacon.

Hooton, E. A. (1946). *Up From the Ape.* New York: MacMillan.

Huxley, F. (1959). Reappraisals of Charles Darwin: Life and Habit. *Amer. Scholar* Autumn, pp. 489–499; Winter, pp. 85–93.

Huxley, J. (1943). *Evolution, the Modern Synthesis.* New York: Harper.

Huxley, J. (1953). *Evolution in Action.* New York: Harper.

Huxley, J. (1959). *The Living Thoughts of Darwin.* Grennwich: Fawcett.

Huxley, L. (1900). *The Life and Letters of Thomas H. Huxley.* 2 vols. New York: Appleton.

Huxley, T. H. (1909). *Autobiography and Selected Essays.* Boston: Houghton Mifflin.

Huxley, T. H. (1959). *Man's Place in Nature.* Michigan: The University of Michigan Press.

Irvine, W. (1972). *Apes, Angels, and Victorians. The Story of Darwin, Huxley, and Evolution.* New York: McGraw-Hill.

Jenkin, Fleeming (1867). "The Origin of Species." *North British Review 46:* 149–171.

Judd, J. W. (1896). *The Student's Lyell*; a manual of Elementary Geology. New York: Harper.

Kettlewell, H. B. D. (1959). Darwin's Missing Evidence. *Sci. Amer. 200:* 28–53.

Keynes, J. M. (1933). *Essays in Biography.* New York: Harcourt, Brace.

King-Hele, D. (1963). *Erasmus Darwin: Doctor, Scientist, Poet, Inventor and Talker.* New York: Scribner.

King-Hele, D., Ed. (1968). *The Essential Writings of Erasmus Darwin.* London: MacGibbon and Kee.

Koestler, A. (1971). *The Case of the Midwife Toad.* New York: Random House.

Kolata, G. B. (1977). Sexual Dimorphism and Mating Systems: How Did they Evolve? *Science 195:* 382–383.

Krause, E. (1880). *Life of Erasmus Darwin.* New York: Appleton.

Kropotkin, P. (n.d.). *Mutual Aid, A Factor in Evolution* and T. H. Huxley *"The Struggle for Existence."* Boston: Extending Horizons.

Kurten, B. (1972). *Not From the Apes.* New York: Vintage.

Lamarck, J. B. (1809). *Zoological Philosophy, an Exposition with Regard to the Natural History of Animals.* Translated by Hugh Elliot. London: MacMillan.

Larson, J. L. (1971). *Reason and Experience. The Representation of Natural Order in the Work of Carl von Linné.* Berkeley: University of California Press.

Loewenberg, B. J. (1959). *Darwin, Wallace, and the Theory of Natural Selection.* Cambridge: Arlington.

Lovejoy, A. O. (1964). *The Great Chain of Being.* Cambridge, Mass.: Harvard University Press.

Lyell, C. (1892). *Principles of Geology; or the Modern Changes of the Earth and Its Inhabitants Considered as Illustrative of Geology.* London: Murray.

Lyell, K., ed. (1881). *Life, Letters, and Journals of Sir Charles Lyell.* London: Murray.

Macbeth, N. (1971). *Darwin Retired.* New York: Dell.

Marchant, J. (1916). *Alfred Russel Wallace; Letters and Reminiscences.* New York: Harper.

Maupertuis, P. L. (1966). *The Earthly Venus.* Translated by S. B. Boas. Sources of Science, No. 29. New York: Johnson Reprint.

Mayr, E. (1972). The Nature of the Darwinian Revolution. *Science 176:* 981–989.

McKinney, L. L. (1972). *Wallace and Natural Selection.* New Haven: Yale University Press.

Millhauser, M. (1959). *Just Before Darwin: Robert Chambers and Vestiges.* Middletown: Wesleyan University Press.

Nelkin, D. (1976). The Science Textbook Controversies. *Sci. Amer. 234:* 33–39.

Newman, H. H. (1932). *Evolution Yesterday and Today.* Baltimore: Williams and Wilkins.

Olby, R. C. (1967). *Charles Darwin.* London: Oxford University Press.

Oparin, A. I. (1953). *The Origin of Life.* 2nd Ed. Translated by S. Morgulis. New York: Dover.

Osborn, H. F. (1894). *From the Greeks to Darwin.* New York: MacMillan.

Packard, A. S. (1901). *Lamarck, the Founder of Evolution.* New York: Longmans, Green.

Peattie, D. C. (1937). *Green Laurels.* London: Harrap.

Peterson, K. (1961). *Prehistoric Life on Earth.* New York: Dutton.

Phillips, J. (1844). *Memoirs of William Smith, LL.D.* London: Murray.

Playfair, J. G. (1822). *The Works of John Playfair.* Edinburgh: Constable.

Playfair, J. G. (1805). Biographical Account of the late Dr. James Hutton. *Transactions of the Royal Society of Edinburgh 5:* 39.

Poulton, E. B. (1909). *Charles Darwin and the Origin of Species.* London: Longmans, Green.

Radl, E. (1930). *The History of Biological Theories.* Translated by E. J. Hatfield. London: Oxford University Press.

Raven, C. E. (1942). *John Ray Naturalist.* Cambridge: Cambridge University Press.

Rudwick, Martin J. S. (1972). *The Meaning of Fossils: Episodes in the History of Palaeontology.* New York: American Elsevier.

Russell, E. S. (1916). *Form and Function: A Contribution to the History of Animal Morphology.* Reprinted 1972. London: Gregg International.

Russett, C. E. (1976). *Darwin in America: The Intellectual Response 1865–1912.* San Francisco: Freeman.

Schmidt, O. (1895). *The Doctrine of Descent and Darwinism.* New York: Appleton.

Schofield, R. E. (1963). *The Lunar Society of Birmingham.* Oxford: Clarendon.

Sears, P. B. (1950). *Charles Darwin, the Naturalist as a Cultural Force.* New York: Scribner.

Sheppard, T. (1917). William Smith: His Maps and Memoirs. *Proceedings of the Yorkshire Geological Society 19:* 75.

Simpson, G. G. (1967). *The Meaning of Evolution.* New York: Bantam.

Smith, D. E. (1923). *History of Mathematics.* Vol. I. Boston: Ginn.

Smith, H. W. (1953). *From Fish to Philosopher.* Boston: Little, Brown.

Spencer, H. (1893). The Inadequacy of Natural Selection. *Popular Science Monthly 42:* 807.

Stauffer, R. C., ed. (1975). *Charles Darwin's Natural Selection. Being the Second Part of His Species Book Written from 1856 to 1858.* New York: Cambridge University Press.

Taton, R., ed. (1963–66). *History of Science.* 4 vols. Translated by A. J. Pomerans. New York: Basic Books.

Tax, S., ed. (1960). *Evolution after Darwin.* 3 vols. Chicago: University of Chicago Press.

Thompson, S. P. (1910). *The Life of William Thomson.* London: MacMillan.

Wallace, A. R. (1899). *The Wonderful Century.* New York: Dodd, Mead.

Wallace, A. R. (1891). *Darwinism.* London: MacMillan.

Wallace, A. R. (1905). *My Life.* New York: Dodd, Mead.

Wendt, H. (1955). *In Search of Adam.* Boston: Houghton Mifflin.

Williams, E. A. (1966). *Darwin's Moon: A Biography of Alfred Russell Wallace.* London: Blackie.

Wilson, E. O. (1975). *Sociobiology: The New Synthesis.* Cambridge: Belknap Press of Harvard University Press.

Wilson, L. G. (1970). *The Species Notebooks of Charles Lyell*. New Haven: Yale University Press.

Woodward, H. B. (1911). *History of Geology*. London: Watts.

13

GENETICS

A New Science—An Old Problem

Genetics is a strange combination of the newest and oldest fragments of human knowledge. Just as the term "biology" was coined at the beginning of the nineteenth century to replace the outmoded "natural philosophy," so "genetics" was coined at the beginning of the twentieth century to replace eons of speculation on the nature of "generation," or "inheritance," or "heredity." In 1923 Dr. William Morton Wheeler, Professor of Biology at Harvard University, saw genetics as a new offshoot of natural history "so promising, so self-conscious, but alas, so constricted at the base."[1] Dr. Wheeler greatly feared that the bud would be abortive. Yet in 1959 when the great biologist Sewall Wright reappraised the situation, he could say that genetics, instead of aborting, had become the new rootstock of all biological science. He saw the whole field of biology transformed by genetics into a unified discipline that would eventually rival the physical sciences.[2] In the nineteenth century it would have been impossible to visualize fundamental genetic principles that would provide the unity that makes physics so elegant.

The diversity of means of reproduction and the differences in the development of the young were long a source of confusion to natural philosophers. Reproduction could be internal or external, the young could be "born" as eggs, worms, larvae, miniature versions of their parents, or totally different in form from the adult. Scientists can either focus on similarities or differences between phenomena. With such a baroque profusion of reproductive patterns to study, and no theoretical framework to link their observations, it is no wonder scientists were obsessed by peculiarities and failed to comprehend the underlying laws of inheritance formulated by Mendel in the 1860s. Once the laws of genetics are known, a simple elegance emerges in the underlying mechanism of information transfer. Why then was the development of genetic science delayed until the twentieth century?

Ancient Beliefs and Myths

Perhaps the greatest obstacle to the advance of genetics as a science was not
that scientists knew too little about the subject, but rather that everyone had
too much information about heredity—most of it wrong. In no other area of
biology is there a greater heritage of error, compounded with confusion between religious and scientific concepts than the subject of heredity. Stone
age beliefs still exist in the space age world. The importance of the two sexes
and the relation of the act of intercourse to reproduction was of course known
in ancient times. This information was put to practical uses such as castration
of animals and men to make them more useful and tractable for special purposes. Although scientists did not discover sexuality in plants until the seventeenth century, since very ancient times in Assyria the date palm had been
carefully cross-pollinated and the ratio of male to female trees regulated to
secure good crops.

Stories of strange hybrids and monstrous births added color and confusion
to popular ideas about reproductive possibilities. Although reproduction
normally involved different sexes of the same species, myth, legend, and tales
of strange creatures encouraged the belief that unusual matings could produce
bizarre new products. The common observation that the intraspecific difference between the sexes is sometimes greater than between the sexes of related
species may have stimulated such ideas. Mythological creatures like the
centaur, a combination of man and horse, or the minitaur, a combination of
bull and man, were obviously the result of crosses between animals and humans. Strange plants and animals implied peculiar parentage.

In ancient times the practice of inbreeding was regarded as beneficial. The
Paraohs of Egypt often married their sisters, or half-sisters. Uncle-niece marriages were regarded as very good by the Greeks. Although mother-son
matings were forbidden, the offspring of Oedipus and Jocasta were exceptionally fine specimens. In Leviticus the practice of hybridization was actually
forbidden: "Thou shalt not let thy cattle gender with a diverse kind; thou
shalt not sow thy field with mingled seed. . . ." Despite the encouragement
of inbreeding, the mule is mentioned in Genesis. A Sumerian proverb asks:
"O mule, will your sire recognize you, or will your dam recognize you?"[3]
This hybrid was mentioned also by Homer, Hesiod, and Plato and the sterility
of the mule was discussed by Empedocles, Democritus, Aristotle, and Maupertuis. Important scientists such as Aristotle and Conrad Gesner gave re-
spectability to tales of more bizarre matings. Aristotle wrote that in Libya
different species would meet at the water holes and mate, despite their differences, as long as they were of similar size and gestational periods. This belief was captured in the proverb, "Something new is always coming from Libya."[4]

Camels seem to have been particularly likely to engage in experimental
matings. The giraffe was supposed to be the product of a cross between a

camel and a leopard. The two-humped camel was itself the product of a
normal female camel and a wild boar. An even more unusual mating of the
camel was with the sparrow, producing the ostrich. Dogs were also quite
promiscuous and likely to attempt matings with wolves, foxes, and goats (to
produce the wild boar). Risking their lives, dogs would even attempt matings
with lions and tigers. Human beings may have encouraged some of these
crosses since Aristotle says that female dogs were sometimes taken to lonely
spots and tied up. "If the tiger be in an amorous mood he will pair with her,"
said Aristotle, "if not he will eat her up, and this casualty is of frequent
occurrence."[5]

Pliny told of the mating of eels and vipers which occurred on beaches. The
viper would thoughtfully leave his poison in a hollow rock and meet the eel.
The same story was still being told by Renaissance authors. Some bizarre hybrids seem to have originated relatively late. The jumar was a cross between
the horse and the cow, apparently first reported in the sixteenth century and
described by Gesner, della Porta, Aldrovandus, Reaumur, Voltaire, Locke,
Buffon, Bonnet, van Haller, and Spallanzani. John Locke, the great seventeenth century philosopher, reported that he himself had seen a chimera
which was plainly the product of a mouse and a cat. He was also sure that
monsters had originated from matings of women with apes. Certainly there
were many reports of peculiar hybrids produced by crosses between human
beings and animals. Apes were explained as the product of humans and some
unknown quadruped. In the tenth century, the manatee was explained as a
hybrid of a fish and an Arab.

Apparently there was much actual experimentation on the subject, not of a
purely scientific kind. The practice of bestiality was punishable by death
among the ancient Hebrews, but among some other peoples at the time intercourse with animals was a part of religious ceremonies. Pindar, Herodotus,
and Strabo all wrote about a city in Egypt where women had intercourse with
goats for the worship of Pan. Pliny cited cases of women who gave birth to a
hippocentaur, elephant, and serpent. In some cases animals gave birth to human beings. Plutarch reported the case of a girl born to a mare and another
to an ass. A boy born to a goat was quite human except for having goat-like
legs. In Scandinavia and Russia tales of women who were kidnapped by bears
and returned to give birth to men of great strength were not uncommon.

Plant hybrids were also imaginatively explained. The orange was produced
by grafting the citron on the pomegranate, while kumquats could be produced
by a graft of orange onto the olive. The banana was the hybrid obtained by
putting the seed on the date palm into the corm of the colocasia.

The Bible is a key to many ancient ideas about theories of heredity and
the realities of animal and plant breeding. Here too is the source of the confusion which stems from using "inheritance" to mean all that we receive from

our ancestors, in terms of legal property, blessings, curses, and our physical characteristics. For example, Jacob put to good use the magical belief that offspring are influenced by what their mother experiences during pregnancy. While tending sheep for his uncle, Jacob was allowed to keep all the striped and spotted lambs. Since these were very rare, Jacob increased his percentage by peeling hazel rods into designs of stripes and showing them to the ewes. Not only did he increase the number of striped lambs, but when he discovered how well the trick worked he began to apply the principle of artificial selection. Jacob showed the design only to the best of the ewes so that his herd grew in vigor as well as in numbers. The idea that maternal impression could influence the physical and mental traits of the offspring was still accepted at the beginning of the twentieth century. The interesting thing about twentieth century "prenatal culture" was that some authors claimed to have accepted the germ plasm theory of Weismann, but still managed to believe that acquired characteristics were inherited.[6]

Many myths provided the ancients with examples of virgin birth, parthenogenesis, and uniparental reproduction. Athena sprang forth, complete with armor, from the head of Zeus. Not wishing to be outdone, his wife, Hera, brought forth Hephaestus (Vulcan) by herself. Since she was not as powerful as Zeus, a defective birth occurred; Hephaestus was a cripple. Not only the gods engaged in parthenogenesis, for "wind eggs" developed without the aid of the male, being fertilized by the wind.

Hippocrates believed that both parents contributed "pangenes" from their blood through their sexual products. Aristotle rejected this theory and supported the idea that the male provided the "form" or blueprint for the embryo while the mother's contribution was merely "matter" and service as an incubator. Although Aristotle was a keen observer and usually fairly critical in evaluating the reports of others, he did perpetuate many errors concerning inheritance and generation. He accepted the spontaneous generation of lower creatures and stories of the inheritance of acquired characteristics. A man branded on the arm, he reported, fathered a son with a similar mark on his arm.

The idea that the sex of the offspring is determined by the heat of the womb was held by Empedocles. Aristotle explained female offspring as a defect due to a deficiency of heat. (Explaining opposite sexed twins was very difficult with this theory.) Shepherds attempted to put the "heat determination of sex" to practical advantage by encouraging matings when the wind was from the north or south according to whether they wanted more rams or ewes. Other natural philosophers believed that the sex of the offspring was already determined in the seed of the father. The right testes produced males and the left produced females. Lucretius and Plato held the remarkable idea that the relative contributions of the parents depended on the level of emotional interest at the time of conception. The degree of enthusiasm would determine whether the child was more like the father or mother.

Theories of Heredity

Two lines of experiment and observation eventually led to modern genetics. These were studies of plant hybridization and evolutionary views of variation. The latter so overshadowed the former that the work of Gregor Mendel was ignored and forgotten during "Darwin's century." But Darwin's attempts to deal with inheritance were totally unsuccessful. He and his followers attempted to deduce the rules of inheritance from observations of variation and change. Mendel realized that it was first necessary to understand the rules that make inheritance constant from generation to generation.

Darwin's Theory of Inheritance

Darwin regarded the study of heredity as the study of the direct transmission of qualities from parents to offspring, as influenced by external conditions. The cobwebs of confusion hung heavily over his writings on the "law of variation" and his discussions of the mechanism of inheritance. The worst aspect of his theory of heredity was acceptance of "blending" inheritance. Bold enough to break with the prevailing natural theology of his day, he could not overcome the universal prejudice in favor of a blending of characters during inheritance. This meant that a unique variant individual would have to mate with an average member of the species. Variation would tend to be damped out before natural selection could stabilize or act in favor of the new variety. The first generation should have only 50 percent of the variation and the next generation only 25 percent. Darwin based his assumption of blending inheritance on Kölreuter's "law of intermediacy of true hybrids."

Although Darwin wrote extensively on sexual selection as a means of perpetuating traits not of immediate use in the struggle for existence, he did not appreciate the importance of sexual reproduction in leading to reassortment and recombination. Quite incorrectly, he regarded sexual reproduction not as a means of providing diversity in the population, but uniformity. Sexual reproduction certainly could not be the cause of variation. Pressed by his critics, Darwin increasingly came to rely on a "soft" explanation of heredity. That is, he increasingly allowed for an environmental influence on organisms, especially on their reproduction systems as a means of introducing and perpetuating variation. Collecting data from animal breeders and gardeners, Darwin was struck with the great variety of odd and frivolous traits that had been selected in certain domestic breeds. Such evidence convinced him that domestication might actually stimulate variation.

In a valiant attempt to accommodate heredity, variation, and evolution within a general framework, Darwin revived the Hippocratic theory of inheritance and renamed it the "provisional hypothesis of pangenesis." *Variation in Animals and Plants,* published in 1868, explained how every organ, tissue, and cell gives off minute "units" which Darwin called "gemmules." In the reproductive organs, the gemmules were assembled and incorporated into the sexual products. At conception, gemmules from both parents combined to form the embryo. The characteristics of the new individual depended on whether the gemmules for particular traits came from the maternal or paternal line. Not all transmitted gemmules need be expressed. Therefore, unused gemmules might appear in later generations. Such "hidden" gemmules could be invoked to explain obscure aspects of inheritance.

Among the problems of heredity which Darwin examined were atavism, that is, the recurrence of qualities from past rather than the immediate ancestors. Aristotle used this phenomenon to argue against Hippocrates' theory of pangenesis, since Hippocrates seemed to believe that the heritary particles were given off during intercourse and passed directly into the semen. Darwin's claim that not all the "gemmules" were expressed met this objection. However, sexual dimorphism and the way that the secondary sexual characters of one sex could be transmitted through the other sex were still very difficult to explain. Darwin was also aware of the "prepotency of transmission" (dominant) and "latency" (recessive) of some traits. Pangenesis could explain some of these difficulties, as well as regeneration and malformations. If the limb of a frog or salamander were cut off, limb gemmules circulating in the blood could go to the site and express themselves in the formation of a new limb. Malformations could result from the wrong gemmules expressing themselves at some site. Ambiguous and unverified, at least this theory had the virtue of suggesting some particulate basis for heredity, no matter how vaguely defined this might be.

Darwin's provisional hypothesis may have stimulated the search for better explanations and various experimental tests. Eventually pangenesis fell into well deserved obscurity. It is interesting to note that after the theory of evolution had obscured genetics so totally in the nineteenth century, when genetics established itself as a science in the twentieth century, Darwinian science fell for a time into the shadows. Darwin's theory of evolution suffered through many vicissitudes in popularity before neo-Darwinism was integrated with the new genetics.

Plant Hybridization

The line of research which led most directly to Mendelian genetics was work on plant hybridization. Unfortunately for progress in science, botany and zoology have tended to develop as quite separate fields, and communication between them was problematic. Darwin was exceptional in his range of interests, as was apparent in his great knowledge of the plant and animal world. Botanists were interested in plant hybridization as part of the proof for sexual

reproduction in plants. Of course sexual reproduction in animals did not require scientific demonstrations. Nehemiah Grew suggested the sexual nature
of plant reproduction as early as 1676, but it was not until the 1690s that
Rudolf Camerarius (1665-1721) provided sound experimental proof. Observations on natural or accidental hybrids appeared sporadically, but the first
botanist to systematically make and test hybrids was Joseph Gottlieb Koelreuter.

Joseph Gottlieb Koelreuter (1733-1806)

While experimenting with tobacco plants, the German botanist Joseph Gottlieb
Koelreuter succeeded in producing his first hybrids. He first discussed these
experiments in his "Preliminary report of experiments and observations concerning some aspects of the sexuality of plants." One of the major effects of
this work was to challenge the peculiar notions that Linnaeus held concerning
plant hybrids. Linnaeus simply assumed that any plant which showed characters intermediate between two known species must be a hybrid. He had no
scruples about conducting breeding tests to determine purity of type or reconstructing such hybrids from the presumptive parents. His students produced lists of more than one hundred "hybrids," among which perhaps six
were truly hybrids and the remainder were impossible combinations.

According to Linnaeus, the contributions of male and female were quite
different. In animal hybrids the outer layer and vascular system were derived
from the male and the inner layer, including the nervous system, derived from
the female parent. The two-layer system was also invoked in plant hybrids
where the leaves and the rind of the stem were supposedly the paternal contribution, while the inner portions (central part of the flower, the "fructifica-
tion," and the pith of the stem) were the maternal contribution. In contrast
to Linnaeus, Koelreuter meticulously carried out many experiments in hybridization as well as studies of fertilization and pollination. The scope of
this work can best be appreciated by noting that he carried out more than 500
different hybridization tests with 138 species, and studied the shape, size, and
color of pollen grains from more than 1000 plant species.

To avoid the errors that stem from self-pollination, Koelreuter developed
techniques for artificial hybridization: deliberately removing the anthers, pol-
linating by hand, and then covering the flower to prevent extraneous pollen
from confusing the results. He realized that the hybrids were usually intermediate between both parents, although in some cases they resembled one
parent more than the other. Reciprocal crosses were found to produce identical products. Hybrids formed from very different parents were generally
sterile, but some hybrids exhibited greater vigor than the parents, while others
produced very variable offspring. These experiments also proved the naïveté
of the preformationist theory of generation. Koelreuter thought that the

"virtue" of the pollen was somehow transmitted by a "fluid." In some hybrids the pollen was sterile, but Koelreuter found that the ovary could be back-crossed with pollen of the parent species.

Because of his pious conviction that new species could not be produced, Koelreuter was concerned with how hybridization was prevented in nature. He thought it remarkable that the initial hybrids (F_1 generation) were uniformly intermediate between the parents, while the next generation (F_2) was extremely diverse. Koelreuter explained his findings in a manner described as "theological and alchemical." For Koelreuter, the peculiar diversity of the F_2 generation was the result of his own interference with nature, that is, of forcing matings between species which God had not intended to occur. Although Koelreuter planned to show that hybridization could also be scientifically studied in animals, he was never able to carry out such crosses. Even his botanical experiments were often terminated prematurely for lack of funds, facilities, equipment, and competent assistants.

Generally ignored by his contemporaries, Koelreuter died embittered by his own obscurity and the glory accorded to Linnaeus. However, both Mendel and Darwin knew and used Koelreuter's work. Although he was unable to carry out the ambitious experimental program he had planned, Koelreuter was at least satisfied that he had unequivocally settled the question of plant sexuality. In 1761 he wrote that "even the most stubborn of all doubters of the sexuality of plants would be completely convinced" as a result of his experiments.[7] Here he failed to appreciate the strong grip of tradition and conservatism among botanists. Friedrich Schelver (1778-1832), a professor at Heidlberg, and August Henschel (1790-1856), a medical practitioner and university tutor in Breslau, upheld the ancient doctrine of asexuality in plants. Experiments were to be distrusted when they conflicted with venerable traditions. Henschel, a disciple of nature philosophy, argued that Koelreuter had mutilated his plants by castrating the flowers, forcibly dusting them with foreign pollen, and growing them in pots—not open fields. Such unnatural conditions inevitably reduced fertility and produced monstrosities, varieties, and degenerate types from the parental stock, rather than true hybrids. Because of the popularity of nature philosophy and the lack of experimental support for Koelreuter's pioneering work, many naturalists were ready to accept such flimsy ideological criticism.

Carl Friedrich von Gaertner (1772-1850)

Although Carl had earned the degree of Doctor of Medicine, he took up botanical experiments to complete the work of his father, Joseph Gaertner (1732-1791), a famous botanist who had been a friend of Koelreuter. As a dutiful son, Carl first dedicated himself to publishing his father's notes for his *De Fructibus et Seminibus Plantarum.* The man so appropriately named

"Gaertner," performed some 10,000 separate experiments with 700 species, representing some 80 different genera of plants, from which he produced 250 different hybrid forms. Independently wealthy, Gaertner could carry out this ambitious program of experimental hybridizations in his own gardens. His work was published in 1849 as *Experiments and Observations on Hybridization in the Plant Kingdom.* Although dull and repetitive, the book was encyclopedic in scope. Darwin and Mendel studied it carefully.

While Gaertner's work contained "very valuable observations," as Mendel said, he added little to the theoretical aspects of the debate. While he emphasized the great variability of the F_2 plants compared to the F_1, he treated all offspring as whole organisms, rather than analyzing the separate traits found in his hybrids.

Charles Naudin (1815–1899)

Many botanists observed what we now recognize as dominance and segregation in their hybrids, but because they were accustomed to treating the plant as a whole, none of them achieved Mendel's clear insight into the laws on inheritance. Other botanists failed to recognize that some traits were recessive (or latent) and were masked by other traits, but not lost. Failing to deal with individual traits, botanists could not see that some segregation mechanism acted in the process of reproduction. Nevertheless, Naudin deserves a place in the history of genetics as a forerunner of Mendel, since he suggested that segregation occurred in reproduction.

Attempting to clarify the taxonomic relationships between the genera and species of the potato and cucumber families, Naudin tried to use hybridization in his taxonomic work. In the 1850s he became intrigued with the evolutionary significance of hybridization although he clung to the old belief that hybrids were unnatural entities. Thus, his observation that offspring of a hybrid *Primula* had nearly reverted to the parental species led him to suggest that segregation of the two species occurred in the hybrid. He argued that nature was "eager to dissolve hybrid forms" and did so by "separation of the two specific essences."[8]   In addition to recognizing the need to use many plants from each hybrid type in such studies, Naudin also carried some hybridization experiments to the third or even fifth generation. But since he continued to see the species as a whole, his analyses never approached the sophistication of Mendel's work.

Naudin's work was quite well known to Charles Darwin, with whom he corresponded between 1862 and 1882. Regretably, Darwin did not appreciate the implications of Naudin's hypothesis, but tended to dismiss it because he could not see how it would explain reversion to distant ancestral traits. This suggests that Darwin would not have appreciated Mendel's work either, even if he had had direct knowledge of it.

Gregor Mendel (1822–1884)

Work on plant hybridization before Mendel bears no comparison with his elegant statistical and experimental approach. Although many scientists were interested in the problems of heredity and plant hybridization, until the twentieth century no one came close to Mendel's achievement. Mendel's story has been called both a mystery and a minor tragedy, since his work remained in obscurity for over 30 years. The major questions surrounding Mendel are, how did he solve the problem that no other nineteenth century mind could cope with and why was his work so long ignored?

Mendel was the only son among five children born into a peasant family in a small Silesian village. Even the local schoolmaster appreciated Mendel's exceptional ability and encouraged his parents to continue his education. Later, as a University student, Mendel constantly struggled for funds, but as he wrote in his autobiography, "all his efforts remained unsuccessful because of lack of friends and recommendations." This losing battle left him constantly anxious and finally caused serious illness. Eventually Mendel joined the Augustinian Monastery at Brünn, not because he felt called to the church but because he "felt compelled to step into a station of life, which would free him from the bitter struggle for existence." Thus, for Mendel, "circumstances decided his vocational choice."[9] As a monk Mendel was quite conscientious, but was unable to carry out certain pastoral duties, such as visiting the sick. Fortunately, he proved to be very successful as a teacher. However, his attempt to pass the examination for gymnasium teachers was unsuccessful, apparently due to inadequate preparation. To obtain better training in natural science, Mendel was allowed to spend four terms, from 1851 to 1853, at the University of Vienna. Here he studied physics, chemistry, zoology, entomology, botany, paleontology, and mathematics. At the University, he was influenced by outstanding scientists such as Christian Doppler (1805–1853) for whom he served as "assistant demonstrator" in physics; Andreas von Ettinghausen, mathematician and physicist; and Franz Unger. Unger was an important figure in the development of cell theory, but he had been attacked by the clergy for denying the fixity of plant species. From Unger, Mendel would have learned to regard the cell as the focal point of organization of plants and animals.

After returning to Brünn, Mendel began to explore a variety of biological problems. He studied the pea weevil, began growing 34 strains of peas, became a beekeeper with about 50 hives, attempted crosses between American, Egyptian, and European bees, and also kept mice. Probably he did some genetic studies on mice, but felt it was inappropriate for a priest to report on such activities. Like Dalton and Lamarck he became interested in meteorology and served as the Brünn correspondent for Austrian regional weather reports, making daily records of temperature, humidity, rainfall, and barometric pressure.

Unfortunately for science, Mendel was elected abbot of the monastery in 1868 and thereafter administrative work curtailed his scientific research as his Order became embroiled in a controversy with the government over the payment of new ecclesiastical taxes. Although he suffered from uremia and dropsy, and his pulse rate measured 120 beats per minute, Mendel persisted in smoking 20 cigars a day. He died from chronic kidney disease on June 6, 1884. Because his successor at the monastery, Abbot Rambousek, burnt Mendel's private papers, we have almost no direct information on Mendel's sources or inspiration. Mendel's contemporaries seems to have regarded him as a nice old cleric with some silly but harmless pastimes. Surely an educated man who spent his time counting thousands of wrinkled peas must be a bit queer.

In the summer of 1854 Mendel began working on 34 strains of peas, testing them for constancy in the transmission of selected traits in 1855. The next year he began the famous series of experiments that led to the paper read to the Brünn Society for Natural History in 1865 and published in the proceedings of the Society in 1866. Mendel's choice of experimental system was excellent, but his choice of scientific correspondent was disasterous. Mendel sent a copy of his paper to Carl Nägeli (1817–1891), professor of botany at Munich. Although Nägeli was interested in plant heredity, he completely failed to appreciate what Mendel had done.

Nägeli thought that all hybrids produce variable offspring. The mathematical relationships among unit characters which Mendel had studied meant nothing to him. Nägeli dismissed Mendel's work as "merely empirical, not rational" and never referred to Mendel's work on peas in any of his own publications. However, since Mendel had mentioned further experiments to be done on other plant species, Nägeli kept up the correspondence and encouraged Mendel to switch his experiments from peas to *Hieracium* (wild hawkweed). Nägeli believed that Mendel's "pure" forms in the F_2 generation must actually be mixtures and that further inbreeding should reveal the presence of the hidden characters. Experiments on *Hieracium* seemed to prove this because perpetual hybrids were common in this genus. For five years Mendel tried to hybridize *Hieracium* and reproduce the results he had obtained so neatly for peas, but he was totally frustrated in these attempts. It is now clear that in *Pisum* Mendel was observing the typical case, but in *Hieracium* the peculiar results were due to apomixis (parthenogenesis by pseudogamy which results in seeds of purely maternal origin—meiosis and fertilization do not occur). Such work on a peculiar system may have led Mendel to question the general validity of his work on peas. Studying many other species, Mendel found some that behaved like peas. This suggests he was trying to extend the range of his work and provide confirmation of the laws derived from the experiments on peas, but he did not publish these experiments. Could he have convinced his contemporaries of the value of his work if he had offered it all to them? This is impossible to judge, but it is apparent he never convinced Nägeli, the one well-established botanist who did know his work. Indeed, Mendel seemed to assume that other scientists would not appreciate his work, as indicated by a passage in one letter to Nägeli:

> I am not surprised to hear your honour speak of my experiments with mistrustful caution. I would not do otherwise in a similar case.[10]

Peas on Earth

There were few scientists in the nineteenth century who could have appreciated Mendel's experimental and mathematical approach to heredity. Although plant hybridization experiments were common, no botanists at the time did mathematical studies of *all* the offspring of their hybrids. Mendel's work was unique for its statistical orientation and because he counted all the progeny from his crosses, rather than just the "interesting" ones. While others treated species characters as a unit, Mendel worked with pure lines and studied simple, separable characters. In this he has been called the "Lavoisier of botany," since his factors can be compared with Lavoisier's elements.

Apparently Mendel had a clear goal in mind at the outset of his tests of 34 varieties of garden peas for their purity of type and suitability as an experimental system. "The value and utility of any experiment," he wrote, "are determined by the fitness of the material to the purpose for which it is used." He realized that his approach was novel in three fundamental ways. Many before had done experiments on plant hybrids, but

> not one has been carried out to such an extent and in such a way as to make it possible to determine the number of different forms under which the offspring of hybrids appear, or to arrange these forms with certainty according to their separate generations, or definitely to ascertain their statistical relations.[11]

The experimental design reported in Mendel's classic paper is so elegant that Sir Ronald Fisher suggested that "the experiments were in reality a confirmation, or demonstration, of a theory at which he had already arrived."[12] Of all the varieties tested, 22 kinds of peas were selected for further experiments. Mendel carried out seven series of crosses which varied by one factor. From these experiments he established the famous 3:1 ratio. He also carried out two bifactorial crosses and one trifactorial cross to demonstrate the independent segregation of traits. Other bi- and trifactorial crosses were used to test predictions based on his theory of independent factors acting in inheritance.

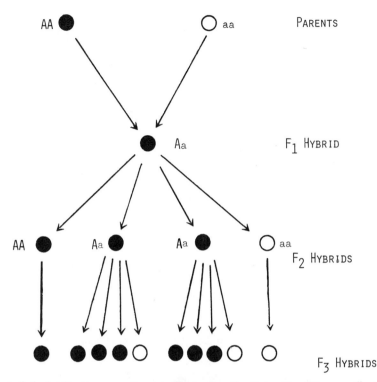

Mendel's hybridization experiments illustrating the 3:1 ratio. Diagram illustrates the results of crossing two strains of pea differing in a particular trait. The dominant trait appears in individuals with either genotype AA or Aa (●). The Recessive trait appears only in individuals with the genotype aa (○).

It is indeed possible that Mendel first hypothesized that one could quite simply predict the number of different types produced by random combinations of two kinds of female sex cells with two kinds of male sex cells. This supposes that in the formation of the sex cells factors which determine particular traits separate from each other. Thus, a hybrid—"Aa"—with factors for traits "A" and "a" would produce sex cell with either factor A or a. Combinations among such sex cells would produce one pure A type, two hybrid Aa, and one pure a. Where A is dominant to a, the ratio of types that can be observed would be 3:1. Mendel did not stop with the F_2 generation but continued some of his experiments for five or six generations. In all generations the hybrids gave the 3:1 ratio.

Carl Correns later summarized Mendel's discoveries in terms of two Mendelian "laws of heredity," which were called the "law of segregation" and the "law of independent recombination."

Did Mendel Cheat?

A close examination of Mendel's statistics convinced Ronald A. Fisher that Mendel's ratios are closer to the theoretical expectation than sampling theory would predict, and such results could not be obtained without an "absolute miracle of chance." Several explanations have been offered for Mendel's statistics. It is possible that Mendel stopped counting at the point where the numbers came close to the expected ratio. He may have unconsciously tended to classify any doubtful individuals to fit the expectation. Another very generous explanation claims that Mendel's assistants understood what he wanted and tried to please him, while saving themselves from some tedious work by giving him the expected results. Sturtevant reviewed Fisher's analysis and found all such explanations unsatisfactory and out of character with the tone of Mendel's paper.[13] (Perhaps the prayers of this worthy monk were answered by ratios that came closer to theoretical expectation than statisticians would predict.)

Mendel was fortunate not only in his statistical results. His choice of the pea plant and the seven particular characters he studied was a near miracle. In all cases clear patterns of dominance and recessiveness appeared. The traits in each pair were distinct or discountinuous variations, having no intermediate grades. Furthermore, the genes for each of his traits turned out to be on different chromosomes, so that his studies were not complicated by linkage.

TABLE 1. A Summary of Mendel's Results.*

	Number of plants		Ratio
Trait studied	Dominants	Recessives	Dominants/Recessives
Length of stem: tall vs. short.	787	277	2.84:1
Position of flower: axial vs. terminal.	651	207	3.14:1
Shape of pod: inflated vs. constricted.	882	299	2.95:1
Color of pod: green vs. yellow.	428	152	2.82:1
Shape of seed: round vs. wrinkled.	5474	1850	2.96:1
Color of cotyledons: yellow vs. green.	6022	2001	3.01:1
Color of seed-coat: gray vs. white.	705	224	3.15:1

*Results are tabulated from Mendel's results for the F_2 generation in his seven major series of experiments. For each trait studied, the characteristic listed first is the dominant one. (After Iltis, 1932.)

Why Was Mendel's Work Ignored?

A common explanation for the neglect of Mendel's work could be called the "obscure journal theory." Yet the *Proceedings* of the Brünn Society were not really that unknown. Indeed, Conway Zirkle argued that no paper published in the less frantic days of the 1860s was as obscure as any paper published today, because of the great difference in the number of papers published.[14] The *Proceedings* were part of the holdings of many libraries in universities and learned societies such as the Royal and the Linnean societies of London; four copies were sent to the United States. About 120 libraries received the *Proceedings* and Mendel had 40 reprints of his own paper which he seems to have distributed among botanists. Only four of these reprints have been recovered. Hugo Iltis remembered that as an eager young student of natural history at Brünn, he had read through back issues of the *Proceedings*. In 1899 he discovered Mendel's paper and excitedly showed it to his professor. The learned man said, "Oh! I know all about that paper, it is of no importance. It is nothing but numbers and ratios, ratios and numbers. It is pure Pythagorean stuff; don't waste any time on it, forget it." Later Iltis made up for listening to his professor by writing an admirable biography of Mendel.[15]

Perhaps it is more proper to say that Mendel was misunderstood than to claim that he was totally ignored. He seemed to be presenting "mere facts" to his contemporaries, since they had no theoretical structure which would relate his work to what they could call a logical theory. Nor was there any way to relate Mendel's work to the possible physical basis of heredity at the time. Another factor operating against Mendel was that the traits he studied, while admirable was an experimental system, were not the kinds of traits that interested most biologists at the time. Under the spell of evolutionary biology, or for practical reasons, scientists considered traits like size, vigor, strength, milk and beef production in cattle, wool in sheep, speed in race horses, peculiar traits in pigeons, or hereditary genius the only kind worth studying. However, such traits generally do not show simple Mendelian genetics because they are governed by many genes.

Nineteenth century ideas about the mechanism of heredity were not consistent with Mendel's factors. Little had been learned since the time of Hippocrates, for theories like "pangenesis," blending inheritance, and the inheritance of acquired characteristics were still common. Such theories would not suggest the simple, reproducible ratios that Mendel found. Such clear ratios would not be expected in small populations. Thus, it would be unreasonable to expect anyone—other than Mendel—to go to the trouble of looking for such relationships in large populations.

A Footnote to History

Mendel's paper was forgotten, but not unused. It was mentioned 15 times in a paper by Wilhelm Focke, a German botanist. In *Plant Hybrids* (1881),

Focke mentioned Mendel's experiments on peas and the constant ratio found
among his hybrid types. Mendel's paper was also included in the bibliography
of Bailey's *Plant Breeding* (1894), and the Royal Society's *Catalogue of
Scientific Papers*. When George John Romanes, the Oxford biologist, asked
Darwin for help with his article on hybridization for the ninth edition of *Encyclopedia Britannica,* Darwin gave him a reprint of Focke's article, which he
had not read because he found foreign languages too tiring to translate.
Darwin noted that the article had a nice bibliography. Romanes, like Darwin,
wanted to avoid excess work and used Focke's bibliography without relating
it to the text of his article. The bibliography was obviously only a formality,
added to be long and impressive rather than useful.

Mendelian Genetics: Rediscovery

Although Mendel's paper had been cited several times during the nineteenth
century, it made no impact until the beginning of the twentieth century when
it was independently rediscovered by three botanists, Hugo de Vries, Carl
Correns, and Erik von Tschermak. Finally the significance of the work carried
out long ago by the obscure monk was understood and appreciated. During
the intervening years other developments in the study of cell division, fertilization, and primarily August Weismann's theoretical work created a new framework into which Mendel's "ratios and numbers" could fit nicely, be used to
predict patterns of inheritance, and be experimentally verified.

Hugo de Vries (1848–1935)

It seems probable that Hugo de Vries saw Mendel's paper as early as 1899 and
realized its relation to his own work. Apparently de Vries regarded his own
research as more important and hoped to maintain his priority by suppressing
knowledge of Mendel's work.

Although he was born and died in Holland, de Vries was educated and
trained mainly in Germany, studying plant physiology with Ferdinand von
Sachs (1832–1897). After obtaining a position at the University of Amsterdam, de Vries studied the local flora, microorganisms in water supplies, and
turgor in plant cells. He gained an international reputation for his studies on
plant cells by the technique of plasmolysis. Measuring the suction pressure of
cells, he proved that this depended on the concentration of materials in the
cell sap and the elasticity of the cell wall. These experiments led to the work
of J. H. Van't Hoff (1852-1911) on osmosis and the theory of ionic diffusion
of Svante August Arrhenius (1859-1927). In response to a request of the
Prussian Ministry of Agriculture for a series of monographs on cultivated

plants, de Vries contributed work on clover, sugar beets, and potatoes. This work proved helpful in his broader studies of heredity and variation.

In 1889 de Vries published *Intracellular Pangenesis*. Emphasizing the need to study traits as separate units, de Vries proposed a theory of inheritance based on factors in the cell nucleus called "pangenes." Unlike Darwin's gemmules, which circulated throughout the body, de Vries' factors remained intracellular. De Vries was particularly intrigued by large or discontinuous variations rather than the small or continuous variations which interested Darwin.

Many of de Vries' breeding experiments failed for reasons he could not control. In 1876 he began experiments on plant hybridization, crossing varieties of maize with either sugar or starch and black or white grains. The weather conditions were so unfavorable in 1877 that his F_1 hybrids produced no progeny and de Vries was forced to discontinue these experiments. In 1892 de Vries resumed his plant hybridization experiments with hairy and smooth varieties of *Silene alba*. Of 536 F_2 plants, 392 were hairy and 144 were smooth. The next year he crossed poppies with black and white marking. From the black hybrids he observed segregation which gave him 158 black to 43 white. Self-pollination of the F_2 generation showed that the white poppies bred true while the black ones were of two kinds; some bred true but others segregated again to give black and white flowers (1095 black to 358 white).

Clearly de Vries had obtained results similar to Mendel's. He had the 3:1 ratio from his F_2 hybrids, he had observed dominance and segregation, and realized that the black F_2 plants were of two types—one type bred true and the other segregated into black and white offspring. De Vries seems to have proceeded much as Mendel did: firstly, making the hypothesis of unit factors and the ratio of types which would result from the independent assortment as recombination of those factors; and secondly, confirming his hypothesis through experiments in which large numbers of progeny were tested and counted.

Although de Vries had data to support his theory by 1896, he did not publicize his work extensively until the turn of the century. By that time, he had obtained further proof by tests on more than 30 different species and varieties. De Vries had waited too long. Ready now to accept the new laws of segregation and recombination of heredity factors, the scientific world praised Mendel as the founder of genetics, not Hugo de Vries. How unfair it all was, he wrote to a friend, to have prepared his case so carefully and then find Gregor Mendel's papers "coming to so high credit." Reluctantly he acknowledged Mendel's work as "quite good for its time." If Mendel's paper had not become so well known in 1900, the basic laws of genetics would have been known as "de Vries' Laws" instead of Mendel's.[16] Perhaps de Vries felt his own work was much more significant than Mendel's in terms of breadth of experiment, theoretical conceptions (as expressed in his *Intracellular Pangenesis*), and the clear relationship of his work to Weismann's theory of the germ plasm. This must have embittered de Vries and perhaps accounts for his refusal to sign the Mendel Memorial Appeal in 1906. In his book on plant breeding, published in 1907, de Vries did not even mention Mendel's name. Within two months of de Vries' publication, Correns and Tschermak announced further confirmation of Mendelian genetics.

In 1901 de Vries published the first volume of *Mutation Theory*. Here he expounded his theory that evolution occurs not through gradual change, as Darwin believed, but through large discrete steps called "saltations" or "mutations." Experiments with the evening primrose provided the main impetus towards this theory. To determine how mutant plants behave when crossed with the normal variety, de Vries tested many kinds of plants and found that mutations were preserved. He had great hopes for the future of research on mutability. When artificial methods could be used to induce a species to mutate in particularly desirable directions, he predicted there would be "no limit to the power we may finally hope to gain over nature."[17]

Later work with the evening primrose revealed that de Vries had not observed the effect of simple mutations, but was actually seeing the effects of chromosomal rearrangements. The large changes he had seen in primrose had misled him into believing that new varieties or even species could be rapidly produced from existing forms by saltatory transformations.

Codiscovery of Mendel's Laws:
Carl Correns and Erich von Tschermak

Correns and von Tschermak seem to have been more generous towards Mendel than was de Vries. But since they had been anticipated by de Vries anyway they did not have as much at stake. Neither had they been primarily concerned with the laws of heredity in their own researches before 1900.

Carl Correns (1864-1935)

It seems simple justice that Carl Correns should have helped to restore Mendel's place in history, since he had been a student of Nägeli, the botanist who had misled and misunderstood Mendel. Generously, Correns acknowledged that because so much had been discovered since Mendel's time, particularly by Weismann, "the intellectual labour of finding out the laws anew for oneself was so lightened, that it stands far behind the work of Mendel."[18]

Correns was a well-known German botanist who had studied the anatomy and life cycle of mosses before becoming interested in the question of the origin of the endosperm in the seeds of higher plants. The endosperm was supposed to be of purely maternal origin. But in certain cases it seemed as if the pollen influenced the nature of the endosperm. Focke had named the phenomenon "xenia" and had included cases where the pollen seemed to affect the tissue of the ovary, seed, or fruit. This effect was regarded as separate from the expected effect of pollen on the embryo itself. Growing hybrids of maize and peas for several generations led Correns to the Mendelian explanation of inheritance. He later recalled how, after four years of work on pea hybrids on one sleepless night, the explanation of the 3:1 ratio came to him "like lightening." Reading Focke's section on peas led Correns to Mendel's paper, which made him realize that his conclusions were not new after all. After discovering that Mendel had anticipated him by 35 years, Correns was again shaken to find that Hugo de Vries had more recently anticipated his discovery. On the morning of April 21, 1900 he received a reprint of de Vries' work on hybrids. By the evening of April 22, Correns had prepared his paper on "Gregor Mendel's Law." This was immediately sent off to the German Botanical Society and published in May.

Correns and Tschermak suggested the use of "Mendelism" and "Mendelian laws." In his own studies of hybridization, Correns felt that

> finding an explanation for them, I believed myself, as de Vries believes himself, to be an innovator. Subsequently, however, I found that in Brünn during the sixties Abbot Gregor Mendel, devoting many years to the most extensive experiments on peas, had not only obtained the same results as de Vries and myself, but had actually given the very same explanation, so far as this was possible in the year 1866. This paper of Mendel's . . . is among the best works ever written upon the subject of hybrids.[19]

Erich von Tschermak (1871-1962)

It was fitting that von Tschermak too should participate in the rediscovery of Mendel, for his own grandfather (Fenzl) had been Mendel's instructor in systematic botany and microscopy at the University of Vienna. Tschermak was a botanist with a strong interest in practical aspects of plant breeding. His training included work on commercial seed farms as well as at the Universities at Ghent and Vienna. Becoming interested in some aspects of Darwin's work on vegatative vigor, Tschermak began a series of experiments on the growth-promoting effects of foreign pollen and on the xenia question in peas.

After analyzing the results of his experiments, Tschermak discovered the 3:1 ratio for the hybrids between peas with yellow and green cotyledons and for smooth and wrinkled seed. He also noted the 1:1 ratio for backcrosses of peas with green cotyledons and hybrid pollen from the second seed generation. At this point he was led to Mendel's paper by the citation in Focke's book. Reading Mendel's paper, Tschermak was shocked to find that the monk had already carried out such experiments much more extensively and had explained the 3:1 segregation ratio. Tschermak hastened to complete his thesis and handed it to the editors of the institute's journal on January 17, 1900. His work was less complete than Mendel's since he had reared only two generations by the publication of his 1900 paper and could not have proven that the dominants in the F_2 were of two types or that the recessives bred true.

In March of that same year Tschermak was to receive another unwelcome surprise—a reprint from de Vries of his paper "On the Law of Segregation of Hybrids." Since this paper contained the terms dominant and recessive, Tschermak assumed that de Vries was aware of Mendel's work although it was not cited in this particular paper. "Naturally I hurried to the editor's office on the same day," recalled von Tschermak, "in order to obtain the already corrected thesis for immediate publication." Having expedited publication of his paper in the *Journal for Agricultural Research in Austria,* Tschermak also arranged for distribution of reprints of his paper in May before the journal was published.[20]

While his own paper was still in proof, Tschermak received another unpleasant surprise—this time it was Corren's paper "Gregor Mendel's Law." Tschermak quickly prepared an abstract of his paper for publication and sent reprints of his paper ("On Artificial Crossing in *Pisum sativum*") to de Vries and Correns to establish himself as a participant in the rediscovery of Mendel's laws. At a meeting in 1903, Correns and Tschermak amicably agreed to a co-equal status in the discovery. Later, Tschermak found that his role as co-discoverer was often overlooked and protested vigorously.

Although Mendel's work had been sophisticated beyond the capacity of biology in 1865, the simultaneous discovery of Mendel's paper and Mendel's laws in 1900 by de Vries, Correns, and Tschermak shows that rediscovery had become inevitable. Indeed, if the three credited with the rediscovery had not done so at the time, others were very close and certainly would have done so shortly. Clearly, William Bateson was on the edge of such a breakthrough in his own work. Once he became aware of Mendelism, Bateson became the "Paul Revere" of Mendelian genetics.

William Bateson (1861–1926)

Passionate advocacy and popularization of Mendel's work became the special mission of William Bateson. A reprint which de Vries sent to William Bateson led him to seek out Mendel's paper. Reading the old Brünn *Proceedings* in the train on his way to a meeting of the Royal Horticultural Society, Bateson was astonished at the clear-cut nature of Mendel's work and his writing, which Bateson called "a model of lucidity and expository skill."[21] The lecture on "Problems of heredity as a subject of horticultural investigation" was quickly revised to include an account of Mendel's work. Bateson praised Mendel's experiments, his excellent and complete exposition, and the principles deduced from his work. Certainly, Bateson told his audience, Mendel's principles "will play a conspicuous part in all future discussions of evolutionary problems." Thus, after his conversion in a railway car, did Bateson begin his career as "apostle of Mendelism in England."[22]

Bateson was trained in zoology at Cambridge, where his father was Master of St. John's College. Bateson spent the summers of 1883 and 1884 in America studying the development of marine organisms under the direction of William Keith Brooks (1848–1908). Observing marine life along the coast of Virginia and North Carolina, Bateson became the first zoologist to study the complete life history of *Balanoglossus* and identify it as a primitive chordate. Brooks seems to have convinced Bateson that heredity was a subject well worth studying. (Brooks deserves mention as the mentor of both E. B. Wilson and T. H. Morgan.)

When Bateson returned to Cambridge as a Fellow of St. John's College, he took up the study of variation. Like de Vries, Bateson believed that the study of variation and the transmission of variations was the proper approach to the subject of heredity. This was actually the reverse of Mendel's approach to the question. To obtain materials for the study of variation, Bateson travelled as far as Russia and Egypt, where he expected to find parallels between local variations of aquatic species and local conditions. Essentially, he was searching for support of the Lamarckian mechanism of inheritance. In Turkestan, where the lake basins were gradually drying up, he hoped to find a causal relationship between variation and environment. He was quite disappointed in this quest, for the fauna provided no evidence that variations were directly related to conditions of life. Forced to abandon his original theories, Bateson turned to records of variation in man and domesticated animals.

In 1894 Bateson published a remarkable book calling for observations and analyses pertaining to the inheritance of discontinuous variations—*Materials for the Study of Variation.* Although his studies were not the statistical and intensive type that led to Mendel's laws, he did appreciate what was necessary. "If the parents differ in several characters," he wrote, "the offspring must be examined statistically, and marshalled . . . in respect to each of those characters separately." Bateson was struck by the discontinuous nature of many variations. Like de Vries, he found the Darwinian concept of selection acting on small, continuous variations quite unsatisfactory. Bateson became convinced that discontinuity was the more important type of variation among animals and plants which is "in some un-known way a part of their nature, and is not directly dependent upon Natural Selection at all." It seemed to him a "gratuitous assumption" that evolution had proceeded by continuous variations. Although he respected Darwin's achievements, he came to believe that "Darwin's achievement so far exceeded anything that was thought possible before, that what should have been hailed as a long-expected beginning was taken for the completed work." Indeed, his advisors had warned him that it was a waste of time to study variation, since Darwin had already covered the problem. Probably too great a distinction was drawn between "continuous" and "discontinuous" variation. However, Bateson and de Vries did provide examples of variations suitable for genetic analysis.

In his rather controversial book, Bateson called for the organization of "experiments in breeding," although he saw this work as "a class of research which calls perhaps for more patience and more resources than any other form of biological inquiry." Nevertheless, he continued, "such investigations will be undertaken and then we shall begin to know."[23] Unaware that these investigations had already been done successfully by Mendel, Bateson journeyed to Italy to study varieties of butterflies and the Cruciferous spectacle plant. Edith R. Saunders (1865-1945), Bateson's collaborator, raised smooth and hairy varieties of these plants and crossed them in 1896, but did not raise the F_2 generation. In 1897 new experiments were begun with a host of plant species. Unfortunately, bad weather and disease interfered with many of the experiments. While Bateson called for a statistical study of crosses between closely related varieties, the statistics he obtained from his own experiments did not lead him to Mendelian laws directly. Indeed some of his statistics are quite far from theoretical expectations. Although Bateson did not discover Mendel's laws through his own work, he was thinking along statistical lines and was ready to appreciate Mendel's work when de Vries' paper led him to it.

Many of the terms now used in the field were introduced by Bateson—including "F_1" and "F_2 generations," "allelomorph," "zygote," "homozygote," and "heterozygote." Indeed, Bateson coined the word "genetics" from the Greek for "descent" to indicate a new epoch in the understanding of inheritance. In 1909 Wilhelm L. Johannsen (1857-1927) introduced the word "gene" to replace vague terms like "factor," "trait," and "character." To clarify other aspects of the new science, Johannsen coined the terms "phenotype" and "genotype" to denote the appearance of the individual and the actual genetic type, respectively. Further experiments carried out by scientists in Europe and America extended Mendel's work and proved that Mendel's laws were applicable to animals as well as plants. Careful tests showed that Mendel's laws held even where inheritance at first seemed more complicated. "Non-Mendelian" inheritance was revealed as normal Mendelian transmission of factors complicated by interactions between factors. A common example was comb shape in fowls.

After serving as Professor of Biology at Cambridge for one year (1908), Bateson became Director of the John Innes Horticultural Institute at Merton, where he stimulated work of a practical nature. T. H. Morgan of the United States assumed the mantle of leadership in Mendelian genetics from Bateson.

Mendelians vs Biometricians

While Bateson and others were confirming and extending Mendel's work, a new challenge to the simple Mendelian laws of inheritance was launched by scientists called "biometricians." Orthodox Darwinians, such as W. F. R. Weldon, regarded Bateson's *Materials for the Study of Variation* (1894) as a direct challenge. Confusion over the relative importance of continuous and discontinuous variation created much unnecessary controversy between twentieth century geneticists who chose to align themselves as either biometricians or Mendelians, respectively. Much ink was spilled in a wasteful war of words. Although Weldon and Bateson had been friends when at St. John's, Weldon remained convinced that evolution occurred through continuous variations. He and Bateson became bitter enemies over their difference of scientific opinion.

Another aspect of this quarrel involved the work of Francis Galton (1822–1911), who called for a special committee to conduct statistical inquiries into measurable traits of plants and animals. In 1894 the Royal Society organized the committee which later was called the "Evolution Committee." Bateson had declined an invitation to join in 1897, but his opponent Weldon was a member. Eventually Bateson joined the committee and served as secretary. In protest Weldon resigned and founded a new journal, with Professor Karl Pearson of University College London, called *Biometrika*. (Karl Pearson had succeeded Francis Galton as leader of the biometrical school of inheritance.) Weldon used *Biometrika* as a vehicle for his attack on Bateson and Mendelian genetics. Bateson responded with a new book, *Mendel's Principles of Heredity: A Defense with a Translation of Mendel's Original Papers on Hybridization.*

Francis Galton (1822–1911)

Galton can rightly be called the founder of biometry—the application of statistics to biological problems. Unfortunately, he did not set aside the "blending" theory of inheritance that prevailed in the nineteenth century; nor was he able to see the value of analyzing the simple "unit characters" that Mendel had studied. Evolutionists believed the laws of inheritance would emerge from studies of complex traits such as size, vigor, intelligence, and strength.

Galton was Darwin's half-cousin; his grandmother had been the second wife of Erasmus Darwin. Like Charles Darwin, Galton began to study medicine, but found it distasteful. Unlike Darwin, Galton was interested in mathematics.

When his father died, Galton inherited enough money to make working for a living unnecessary. Rather aimlessly, at first, he travelled through the Sudan, Syria, and South West Africa. Discovering a talent for writing, he published some popular works on his travels. Like several other eccentric geniuses, his attention turned to studies of weather and in 1863 he published *Meteorographica*. He was instrumental in establishing the Meteorological Office and the National Physical Laboratory. Like Mendel, his interest in weather was secondary to his interest in heredity, but Galton's primary interest was the inheritance of complex physical and mental characters in man. This led to anthropometric researches, quantitative and statistical studies of human populations, and the Biometric Laboratory at University College, London. The use of fingerprints for identification stems from his work.

Apparently Galton believed he could handle almost any question statistically. Among his studies was an examination of the mortality rates for royalty, which he believed might measure the effectiveness of prayer, since whole kingdoms prayed for their royal families. Similarly, he compared the rates of disasters for ships with and without missionaries. He also tried to design a quantitative scale for the measurement of beauty and love.

In *Hereditary Genius* (1869) Galton demonstrated that the mean of a population reflected "stability of type" from generation to generation. Galton believed the laws of inheritance could be deduced from studies of the distribution of deviations from the average in successive generations. Eventually, Galton formulated a mathematical theory of inheritance. Inheritance was presented in mathematical terms as Galton's "Ancestral Law of Inheritance," in which Heritage = $1/4$ p + $1/8$ pp + $1/16$ ppp, etc. (p = parent, pp = grandparent, etc.). Although blending inheritance is assumed, no ancestral contribution is totally lost, but only diminished with succeeding generations. This law was used to account for reversion to ancestral traits. It was easy for scientists to accept the notion that even a tiny remnant of ancestral material could exert a powerful effect under the proper circumstance. Such effects would be analogous to those of certain poisons and drugs which could produce strong reactions when given in minute amounts.

Unfortunately, Galton further confused the study of inheritance by presenting the old doctrine of blending heredity in a neat, apparently precise, mathematical formula which he promoted to the status of a "law." Ancestral contributions must be influential, he argued, or else "stability of type" would be obtained. If he were wrong then it would be simple to breed giants from giants for "the giants (in any mental or physical particular) would become more gigantic, and the dwarfs more dwarfish, in each successive generation."[24] In "A Theory of Heredity" in the *Journal of the Anthropological Institute* in 1875, Galton seemed close to an appreciation of the continuity of the germplasm. In contrast to Charles Darwin, Galton believed the hereditary material was produced in the reproductive organs, with little if any contribution from the other bodily tissues. Galton subjected Darwin's "gemmules" to an experimental test. If it were true that the gemmules are generated by body tissues and pass through the circulating fluids of the body to the reproductive organs then a blood transfusion should transfer gemmules to a recipient. After transfusing blood between rabbits with different color coats, Galton inbred the "polluted" rabbits. No hereditary effect was found.

It seems most unfortunate that Mendel could not have communicated with the ingenious Galton instead of the conservative Nägeli. Perhaps Galton would have understood. In a discussion of the theoretical ratios for types of offspring of mulattoes—in a letter to Darwin in 1875—Galton worked out essentially Mendelian categories, but Galton never put these speculations to the test. Plant hybridization experiments were contemplated, but Galton felt himself "too ignorant of gardening."

Best known as the founder of eugenics, which he called the science of breeding applied to human populations, Galton urged that defective individuals not reproduce and that well endowed individuals be encouraged to breed. Ironically, his own very eugenic marriage produced no offspring.

Mendelians, Biometricians, and Compromise

As early as 1902, Udny Yule (1873-1949) had suggested that the same mechanism could explain both the simple Mendelian pattern and the complicated cases of apparent blending inheritance, if large numbers of factors were assumed to act together—but the general importance of this work was not appreciated and the battle continued. With the sudden death of Weldon in 1906, the hostilities were somewhat tempered. Bateson was ready to proclaim victory, rather than recognize the necessity for compromise. In *Mendel's Principles of Heredity,* he claimed that the biometricians might hope to contribute to the development of statistical theory, but in studies of heredity their efforts "resulted only in the concealment of that order which it was ostensibly undertaken to reveal." Although the Biometricians were deluded, they had delayed recognition of the value of Mendelism.[25]

Gradually it became apparent that the warring parties could be united by the "multiple gene hypothesis." Ronald A. Fisher (1890-1962), statistician and geneticist, expressed the union of the two schools of thought in 1918 in his paper, "The correlation between relatives on the supposition of Mendelian inheritance." Later Fisher proved that Mendelian genetics could be reconciled with Darwinian natural selection. His book *The Genetical Theory of Natural Selection* (1930) is a classic of population genetics.

Cytology: The Structural Basis of Mendelism

As we have seen in the study of "generation," the microscope, which allowed scientists to see the germ cells and the earliest stages of the developing embryo, seemed at first to encourage more confusion than enlightenment. Similarly in genetics and cytology, microscopic studies did not always clarify long-standing problems. Yet cytological investigations in the late nineteenth and early twentieth century, in conjunction with new theoretical insights and some brilliant guesses, did finally provide a sound structural basis for explaining the workings of Mendel's laws.

Genetics and cytology developed along three lines. First, statistical analysis of patterns of inheritance led to Mendelism and population genetics. Second, microscopic studies revealed significant subcellular structures and their behavior under different conditions—resting, growth, and division. Third, chemical research elucidated the nature of various cellular components. A true understanding of genetics required the confluence and integration of all these studies.

Although many scientists studied the behavior of subcellular entities, such work remained an amorphous collection of confusing facts until August Weismann forged a theoretical framework to give them meaning and direction.

August Weismann (1834–1914) and the Germplasm Theory

A major theoretician of biology, Weismann contributed two important guiding concepts for genetics. The first was his theory of the continuity of the germ plasm and the second was his prediction of and rationalization for the reduction division of the chromosomes. After medical studies at Göttingen, Weismann set up a practice in his home town of Frankfurt-am-Main. Researches in the laboratory of Rudolf Leuckart (1822–1898) convinced him that zoological research was more interesting than medical practice. Leaving his first career behind, he became a Professor of Zoology at the University of Freiburg. A painful eye disorder made microscopic work impossible, and Weismann was forced to take a leave of absence. For a time his sight recovered sufficiently so that he could study the life history of Daphnia and the origin of the germ cells of the hydromedusae. But by the time he was 40 his eyesight was so impaired he had to abandon his experimental work. It was as a theoretician that Weismann made his greatest contributions to science. Examining the major biological problems left in the wake of Darwin, he realized that variation and inheritance were the subjects most in need of reform.

Weismann critically examined all possible theories of heredity in terms of his knowledge of cell structure and reproduction and the facts generally known from breeding experiments. His own work on Daphnia and the hydromedusae suggested that the precursors of the germ cells are distinguishable from the body cells at a very early stage. However, Weismann would certainly have been first to admit that these observations were only suggestive and that they did not constitute the total rationale for his theory of the continuity of the germ plasm. According to this theory, there is a continuous line of descent from the germ cells of one generation to another.

Although he was interested in variation, as was inevitable given the evolution-imbued atmosphere of the time, Weismann recognized that the stability of inheritance between generations was the most salient fact of heredity. Once that was understood, variation could be treated as a special case or a corollary. Although described as more Darwinian than Darwin, Weismann grasped the fact that heredity must first be studied at the level of cell and individual, not in terms of evolving species or populations. This insight led Weismann to reject the inheritance of acquired characteristics. For Weismann, the germ cells were not derived from the body of the parent, but directly from the continuous line of germplasm. Since the rudiments of the reproductive cells were formed at the earliest possible stage in embryology and were subsequently isolated from the body cells (or somatoplasm), it was not possible for events that occurred in adult life to affect inheritance. Every organism is made up of two parts: the "latent" and "patent" materials of germplasm and somatoplasm, respectively. While the germplasm was handed down to successive generations, the somatoplasm died away like the leaves of a tree.

For Weismann, the link between generations was the transmission of heredity through definite chemical entities. His essay, "The Continuity of the Germ-plasm as the foundation of a theory of Heredity," published in 1885, had the following remarkable conclusion:

> . . . 'The Continuity of the Germ-plasm' . . . is founded upon the idea that heredity is brought about by the transference from one generation to another of a substance with a definite chemical, and above all molecular, constitution.[26]

Further, Weismann emphasized that his theory of heredity was not "theoretically superficial and cytologically impractical" like Darwin's, but was a biochemical model. He defined the germplasm as "that part of the germ-cell of which the chemical and physical properties—including the molecular structure— enable the cell to become, under appropriate conditions, a new individual of the same species."

The theory of the continuity of the germplasm provided a framework for many aspects of cell behavior—particularly the division of the nucleus and chromosomes. It also served to stimulate further research on cytology, fertilization, cell division, and reproduction. Whereas division of the somatoplasmic cells (mitosis) required keeping the chromosome number constant, Weismann predicted on theoretical grounds that during the maturation of ovum and sperm there must be a special reduction division to reduce the number of chromosomes by half. Fertilization would restore the normal chromosome complement when the two nuclei fused. This ingenious prediction was confirmed in 1888 by Boveri and Strasburger.

In 1892 Weismann could feel secure that his theories had been vindicated by microscopic observations: "The complex mechanism for cell-division exists practically for the sole purpose of dividing the chromatin, and . . . thus the latter is without doubt the most important part of the nucleus."[27]

In the next chapter we will see how further studies of inheritance and progress in biochemistry brought work on the gene to the molecular level.

Notes

1. W. M. Wheeler (1923). The Dry-Rot of Our Academic Biology. *Science 57:* 61.

2. S. Wright (1959). Genetics and the Hierarchy of Biological Sciences. *Science 130:* 959.

3. S. N. Kramer (1959). *History Begins at Sumer.* New York: Doubleday Anchor, p. 134.

4. Aristotle, *History of Animals,* 606 b 20.

5. Aristotle, *History of Animals,* 607 a 5.

6. For example, see N. N. Riddell 1903. *Heredity and Prenatal Culture Considered in the Light of the New Psychology.* Chicago: Riddell.

7. R. Olby (1966). *Origins of Mendelism.* New York: Schocken, p. 37.

8. Olby, p. 63.

9. Olby, pp. 106–107.

10. E. O. Carlson, ed. (1967). *Modern Biology.* New York: Braziller, p. 41.

11. G. Mendel (1865). *Experiments in Plant Hybridization.* Edited by J. H. Bennett (1965). Edinburgh: Oliver and Boyd, pp. 8–9.

12. R. A. Fisher (1958). *The Genetic Theory of Natural Selection.* 2nd Ed. New York: Dover, p. 9.

13. A. H. Sturtevant (1965). *A History of Genetics.* New York: Harper and Row.

14. C. Zirkle (1968). "Mendel and His Era," in *Mendel's Centenary.* Edited by R. M. Nardone. Washington, D.C.: The Catholic University of America Press, pp. 122–133.

15. Zirkle, p. 124;
H. Iltis (1932). *Life of Mendel.* New York: Norton.

16. Olby, pp. 128–129.
17. H. DeVries (1905). *Species and Varieties: Their Origin by Mutation.* Chicago: Open Court, p. 688.
18. Olby, p. 130.
19. F.A.E. Crew (1966). *Foundations of Genetics.* Oxford: Pergamon, pp. 63–64.
20. Crew, pp. 64–65.
21. W. Bateson (1913). *Mendel's Principles of Heredity.* Cambridge: Cambridge University Press, p. 7.
22. Olby, p. 132.
23. Olby, pp. 133–135.
24. F. Galton (1892). *Hereditary Genius.* Reprinted 1962. London: Fontana Reprint, p. 33.
25. Bateson, pp. 6–7.
26. S. Toulmin and J. Goodfield (1966). *The Architecture of Matter.* New York: Harper Torchbooks, p. 363.
27. Toulmin and Goodfield, pp. 364–365.

References

Babcock, E. B. (1950). *The Development of Fundamental Concepts in the Science of Genetics.* Portugaliae Acta Biologica Series, 1949. Reprint. Washington, D.C.: American Genetics Association.

Baltzer, F. (1964). Theodor Boveri. *Science 144:* 809–815.

Bateson, W. (1913). *Mendel's Principles of Heredity.* Cambridge: At the University Press.

Bateson, B. (1929). *William Bateson, F. R. S., Naturalist, his Essays and Addresses.* Cambridge: At the University Press.

Bodmer, W. F., and Cavalli-Sforza, L. L. (1976). *Genetics, Evolution and Man.* San Francisco: Freeman.

Boyer, S. H. (1963). *Papers on Human Genetics.* Englewood Cliffs, N. J.: Prentice Hall.

Carlson, E. A. (1966). *The Gene: A Critical History.* Philadelphia: Saunders.

Correns, C. (1905). The Birth of Genetics. *Genetics:* supplement to vol. 35.

Crew, F. A. E. (1966). *Foundations of Genetics.* Oxford: Pergamon.

Darlington, C. D. (1969). *Genetics and Man.* New York: Schocken.

Darlington, C. D. (1969). *The Evolution of Man and Society.* New York: Simon and Schuster.

Dobzhansky, T. (1951). *Genetics and the Origin of Species.* 3rd Ed. New York: Columbia University Press.

Dunn, L. C., ed. (1951). *Genetics in the Twentieth Century.* New York: MacMillan.

Dunn, L. C. (1965). *A Short History of Genetics.* New York: McGraw-Hill.

Fisher, R. A. (1929). *The Genetical Theory of Natural Selection.* New York: Dover.

Fruton, J. (1972). *Molecules and Life: Historical Essays on the Interplay of Chemistry and Biology.* New York: Wiley-Interscience.

Glass, B. (1947). Maupertuis and the Beginnings of Genetics. *Quarterly Review of Biology 22:* 196–210.

Garrod, A. E. (1909). *Inborn Errors of Metabolism.* London: Oxford University Press. [Reprinted in H. Harris (1963). *Garrod's Inborn Errors of Metabolism,* London: Oxford University Press, pp. 1–93.]

Iltis, H. (1932). *Life of Mendel.* Translated by E. and C. Paul. New York: Norton.

Iltis, H. (1947). A visit to Mendel's home. *J. Heredity 38:* 163–166.

Jaffe, B. (1958). *Men of Science in America.* New York: Simon and Schuster. (Morgan, in Chapter 16).

Jacobs, F. (1973). *The Logic of Life: A History of Heredity.* Translated by Betty E. Spillman. New York: Pantheon.

Koestler, A. (1971). *The Case of the Midwife Toad.* New York: Random House.

Lerner, I. M., and Libby, W. J. (1976). *Heredity, Evolution and Society.* 2nd Ed. San Francisco: Freeman.

Ludmere, K. (1972). *Genetics and American Society: A Historical Appraisal.* Baltimore: Johns Hopkins University Press.

Luria, S. (1973). *Life: The Unfinished Experiment.* New York: Scribner.

Maupertuis, P. L. (1966). *The Earthly Venus.* Sources of Science, No. 29. New York: Johnson Reprint.

Mendel, G. (1958). *Experiments in Plant Hybridization.* Cambridge, Mass.: Harvard University Press.

Morgan, T. H. (1926). *The Theory of the Gene.* New Haven: Yale University Press.

Nardone, R. M., ed. (1968). *Mendel Centenary: Genetics, Development and Evolution.* Washington, D.C.: The Catholic University of America Press.

Olby, R. (1966). *Origins of Mendelism.* New York: Schocken.

Peters, J. A., ed. (1959). *Classic Papers in Genetics.* Englewood Cliffs, N.J.: Prentice-Hall.

Punnett, R. C. (1950). Early days of genetics. *Heredity 4:* 1–10.

Riddell, N. N. (1903). *Heredity and Prenatal Culture Considered in the Light of the New Psychology.* Chicago: Riddell.

Roberts, H. F. (1929). *Plant Hybridization before Mendel.* Princeton: University Press.

Roper, A. G. (1913). *Ancient Eugenics.* Oxford: Oxford University Press.

Sootin, H. (1959). *Gregar Mendel, Father of the Science of Genetics.* New York: Vanguard.

Srb, A. M., Owen, R. D., and Edgar, R. S. (1965). *General Genetics.* 2nd Ed. San Francisco: Freeman.

Stent, G. S. (1971). *Molecular Genetics.* San Francisco: Freeman.

Stern, C., ed. (1950). The Birth of Genetics. Suppl. *Genetics 35:* (5) pt. 2. (English translations of letters from Mendel to Nägeli and papers of de Vries, Correns, and Tschermak.)

Stern, C. (1953). The Geneticist's Analysis of the Material and the Means of Evolution. *Sci. Monthly 77:* 190–197.

Stern, C. (1962). Wilhelm Weinberg. *Genetics 47:* 1–5.

Stern, C. (1970). The continuity of genetics. *Daedalus 99:* 882–908.

Stern, C. and Sherwood, E. R., eds. (1966). *The Origins of Genetics.* San Francisco: Freeman.

Stern, C. (1973). *Principles of Human Genetics.* 3rd Ed. San Francisco: Freeman.

Stubbe, H. (1972). *A History of Genetics from Prehistoric Times to the Rediscovery of Mendel's Laws.* Cambridge: MIT Press.

Strutevant, A. H. (1965). *A History of Genetics.* New York: Harper and Row.

Sturevant, A. H. (1946). Thomas Hunt Morgan. *Amer. Nat. 80:* 22–23.

Toulmin, S. E., and Goodfield, J. (1966). *The Architecture of Matter.* New York: Harper and Row.

Tschermak-Seysenegg, E. Von. (1951). The rediscovery of Gregor Mendel's Work. *J. Heredity 42:* 163–171.

Udny, Y. (1902). Mendel's Laws and Their Probable Relations to Interracial Heredity. *New Phytologist 1:* 193.

Vries, H. de (1901–3). *The Mutation Theory.* Leipzig: von Veit.

Vries, H. de. (1905). *Species and Varieties. Their Origin by Mutation.* Chicago: Open Court.

Watson, J. D. (1970). *Molecular Biology of the Gene.* 2nd Ed. New York: Benjamin.

Weismann, A. (1889). *Essays upon Heredity and Kindred Biological Problems.* Chicago: Open Court.

Weismann, A. (1893). *The Germ-Plasm; A Theory of Heredity.* Translated by W. N. Parker and H. Ronnefeldt. New York: Scribner.

Wheeler, W. M. (1923). The Dry-Rot of Our Academic Biology. *Science 57:* 61.

Whiting, P. W. (1935). Sex Determination in Bees and Wasps. *J. Heredity 26:* 263–78.

Wright, S. (1959). Genetics and the Hierarchy of Biological Sciences. *Science 130:* 959.

Zirkle, C. (1951). Gregor Mendel and his Precursors. *Isis 35:* 97–104.

Zirkle, C. (1964). Some Oddities in the Delayed Discovery of Mendelism. *J. Heredity 55:* 65–72.

14

GENETICS IN THE TWENTIETH CENTURY: MOLECULAR BIOLOGY, OR ALL THE WORLD'S A PHAGE

The Physical Basis of Heredity

Having assimilated Mendel's genetics, Weismann's theories, and many previously inexplicable microscopic observations, scientists were ready to attack the question of just what the Mendelian factors were and where and when they segregated and recombined. When biologists set aside their preoccupation with evolution and variation, the mystery of inheritance began to yield its secrets. In the twentieth century, through the confluence of microscopic and chemical studies of the cell with statistical and experimental research on stability and variability in heredity, genetics soon exploded into a vital scientific discipline.

A valuable index to prevailing views of the cell is found in Edmund B. Wilson's book *The Cell in Development and Inheritance.* First published in 1896, various editions of the book reflect the state of cytology and genetics before and after the rediscovery of Mendel's laws. At the time scientists knew that sperm and ovum contain equal numbers of chromosomes, the gamates have half the number of chromosomes found in body cells, chromosomes divide longitudinally during mitosis, and a special reduction division halves the number of chromosomes parcelled out to the gamates. However, most scientists believed that all chromosomes were equivalent.

The Sutton-Boveri Hypothesis

Further cytological investigations cleared up some confusing points. Several scientists came close to understanding the relationship between the classical "factors" tabulated in breeding experiments and the behavior of the chromosomes in mitosis and meiosis, but the honor for the clear exposition of the idea that the hereditary factors were actually physically located on the chromosomes belongs to W. S. Sutton (1877-1916). The concept is sometimes called the "Sutton-Boveri hypothesis" to honor Theodor Boveri (1862-1915) for his ingenious proof that chromosomes are not merely equivalent bits of material, but that the chromosomes in a set are qualitatively different from each other. Using quite different methods, Sutton and Boveri demonstrated the individuality of the chromosomes. Boveri's demonstration rested entirely on embryological and cytological studies. Aberrations in the development of doubly fertilized eggs were traced to abnormal combinations of chromosomes in Boveri's classic experiments, described in an article entitled "Multipolar Mitosis as a Means of Analysis of the Cell Nucleus."[1]

In 1903 Sutton concluded a paper, "The chromosomes in heredity," with the prediction that further work would prove "that the association of paternal and maternal chromosomes in pairs and their subsequent separation during the reducing division . . . may constitute the physical basis of the Mendelian law of heredity." Furthermore, Sutton suggested, the random assortment of different pairs of chromosomes could account for the independent segregation of pairs of genes.[2] Such work finally served to unite cytology and genetics in what was to be a very fruitful partnership. Sutton had been a student of Clarence E. McClung who had pioneered the study of the sex chromosomes. In 1902, McClung suggested that a special pair of chromosomes was responsible for determining maleness and femaleness. Sutton's landmark papers were published while he was a graduate student with Wilson at Columbia University. Despite his fine research, Sutton never finished his graduate work. Later he received an M.D. degree and became a surgeon.

Linkage

Although the Sutton-Boveri hypothesis seemed to explain the relationship between segregation of Mendelian factors and chromosome behavior, it also introduced a new difficulty. There must be more factors than there are chromosomes in a cell nucleus. Another apparent difficulty had emerged from classical breeding tests. As more and more traits were studied, it became obvious that independent assortment is not a universal phenomenon. In particular, Bateson and R. C. Punnet in 1905 proved that certain genes in the sweet pea were always transmitted together. Once again, Bateson served as a publicist and stimulated much work on the problem. Although his own approach to this situation, called "reduplication," proved totally incorrect, Bateson found it very difficult to give up his hypothesis and accept linkage.

Both deVries and Sutton had suggested there must be more Mendelian factors than there are chromosomes. Certainly there was a definite relation between chromosomes and unit characters. But did an entire chromosome or only a part serve as the basis of a single gene? The answer must be the latter possibility, wrote Sutton, "for otherwise the number of distinct characters possessed by an individual could not exceed the number of chromosomes in the germ-products. . . ."[3] Therefore, all the traits associated with a given chromosome must be inherited together.

Experimental evidence for the chromosomal basis of heredity and for the arrangement of genes into linkage groups was provided by research in the laboratory of Thomas Hunt Morgan.

Thomas Hunt Morgan (1866–1945) and the Fruit Fly

From the laboratory of Thomas H. Morgan came the experimental system and methods that would solve the vexing problem of the relationship among genes, traits, chromosomes, and the statistics of recombination. Morgan found the fruit fly, *Drosophila melanogaster,* the ideal system. The creature had a short life-cycle, bred prolifically, and was easily maintained in the laboratory. More important, the fly had only four pairs of chromosomes per cell, which varied in size and shape, and had scores of easily recognized inheritable traits.

Probably the first written description of *Drosophila* is Aristotle's reference to a gnat produced from a larva generated in the slime of vinegar. First placed in the genus *Oinopta* (wine drinker), the fly was later given the name "dew lover." The most commonly used species, *Drosophila melanogaster,* which seems to have originated in southeastern Asia, probably arrived in the United States sometime before 1871 as a stowaway in a bunch of bananas. The entomologist, C. W. Woodworth, was probably first to cultivate *Drosophila* in the laboratory. Learning of the advantages of the fly, Professor Castle introduced it to Morgan, who exploited the creature so effectively it was later said that God created it especially for Morgan. The fruit fly breeds rapidly, having a life cycle of 2 weeks, with 10 days from egg to adult. It is only about 1/8 of an inch in size—small enough so that thousands can be kept in a few milk bottles on banana mash, but large enough to be easily studied with a low-powered microscope. A single pair of parents will quickly produce hundreds of offspring. By 1925 Morgan had identified about 100 different genes in this Lilliputian creature.

Born in Lexington, Virginia, Morgan attended the State College of Kentucky and did his graduate work at Johns Hopkins University. Morgan held professorships at Bryn Mawr, Columbia University, and California Institute of Technology. While his first interest was experimental embryology, a visit to Hugo de Vries' experimental garden aroused an interest in variation, particularly the large, discontinuous type called mutation. Like Bateson and de Vries, he was skeptical of the orthodox Darwinian theory of the primacy of continuous variation as the raw material of evolution. This interest in mutations led to breeding experiments on various animals—mice, rats, pigeons, lice, and finally *Drosophila*. Initially, attempts to induce mutations were quite fruitless.

Although Morgan was awarded the Nobel Prize in 1933 as the founder of the gene theory, he was originally quite skeptical about the universality of Mendel's work, especially after reading Lucien Claude Cuénot's (1866-1951) work on mice. Interpretation of Cuénot's breeding experiments was complicated because of multiple alleles.

In 1910 Morgan sent a paper to *American Naturalist* in which he argued that Mendelian factors could not possibly be carried by the chromosomes because if they did, characters in the same chromosome would have to "Mendelize" together. Even before this paper appeared in print, Morgan's experiments on flies convinced him of the validity of Mendel's laws. Furthermore, he found clear evidence that apparent deviations from the law of independent assortment were due to linkage—that is, two characters were found which were carried on the same chromosome. Linkage was first determined for the sex-linked traits "white eye" and "rudimentary wings." Studies of double mutants revealed another phenomenon—exchange of genes between homologous chromosomes.

The "white eye" and "rudimentary" mutations appeared almost exclusively in male flies, because the trait is carried by the X chromosome. (males = XY, females = XX) Therefore, reciprocal crosses involving such mutants did not produce identical results. The two mutants provided a perfect test for the question of whether genes on the same pair of chromosomes could recombine. Crosses between the two mutants clearly showed that recombination did occur. Because the genes for these traits were located quite far apart on the X chromosome, "crossing-over" was common. These studies supported F. A. Janssens' (1863-1924) "chiasmatype hypothesis." In 1909 Janssens had noted that members of a pair of chromosomes sometimes seemed to stick together and suggested that the physical exchange of segments might accompany the formation of chiasmata. Because the genes for white eye and rudimentary were far apart on the X chromosome, the deviation from expected Mendelian ratios was not enough to provide unequivocal proof for both linkage and crossing over. However, in 1911 studies of the very low recombination between "white eye" and "yellow body" provided a stronger case. ("Crossing over" describes the occasional interchange of genes between homologous chromosomes.)

Morgan realized that the degree of linkage established by mating tests could serve as a measure of the distance between genes on a chromosome. In 1911 he proposed the "chromosome theory of inheritance" which states that the extent of recombination found between genes on the same chromosome is a measure of their spacial separation. The idea of a "chromosome map" led to a burst of activity and experimentation. A. H. Sturtevant (1891-1971), who was still an undergraduate in 1911, realized after a conversation with Morgan that variations in the strength of linkage between genes was an index of their linear sequence on the chromosome. That very night Sturtevant drew up the first chromosome map. In 1913 he published his results in the *Journal of Experimental Zoology*: "The linear arrangement of six sex-linked factors in Drosophila, as shown by their mode of association."

The chromosome theory or, as Morgan called it, "the theory of the gene," was not at first accepted by all scientists, but like any real scientific break-through, it produced a flood of experimentation. After visiting Morgan's laboratory in 1922, Bateson abandoned his doubts about the chromosome theory of inheritance and addressed his "respectful homage before the stars that have arisen in the West." Although a popular concept of the chromo-some theory portrayed the genes very simplistically as beads on a string, Morgan saw the theory of the gene as a powerful and sophisticated generaliza-tion. In 1928 he outlined it as follows:

> The theory states that the characters of the individual are referable to paired elements (genes) in the germinal material that are held together in a definite number of linkage groups; it states that the members of each pair of genes separate when the germ-cells mature in accordance with Mendel's first law, and in consequence each germ-cell comes to contain one set only; it states that the members belonging to different linkage groups assort independently in accordance with Mendel's second law; it states that an orderly interchange—crossing over—also takes place, at times, between elements in corresponding linkage groups; and it states that the frequency of crossing-over furnishes evidence of the linear order of the elements in each linkage group and of the relative position of the elements with respect to each other.[4]

Through his extensive writings and his students, Morgan exerted a great in-fluence on the development of genetics and cytology. His students, Calvin B. Bridges (1889-1938), A. H. Sturtevant, Curt Stern, and H. J. Muller established themselves as pioneers in the field. Muller's work on induced mutation was particularly significant.

Hermann Joseph Muller (1890-1967) and Mutation Theories

It was Hugo de Vries who first turned the attention of geneticists to the ex-perimental study of mutation. Although his idea of evolution through macro-mutations—variations of such magnitude that new species were formed directly—had to be abandoned, his mutation theory did point out the importance of "discontinuous variation." Although work in Morgan's group showed the value of mutants in genetic analysis, the natural rate of mutation was too low for practical quantitative studies of mutation as a process. The first successful systematic studies of mutation in *Drosophila* were those of Hermann Joseph Muller. Muller was born in New York and studied biology at Columbia. Very early in his education he decided that genetics was the field for a lifetime commitment. After a short time at Cornell Medical College, he returned to Columbia (1912–15) where he was appointed as lecturer. He then spent some time at the Rice Institute with Sir Julian Huxley. In 1932 Muller went to Berlin and then on to Russia for genetic research at the Academy of Sciences. Disillusioned by the Lysenko affair, which led to the downfall of Nikolai Ivanovitch Vavilov (1887–1943), Muller left Russia. After three years at Edinburgh he returned to the United States and finally became Professor of Zoology at the University of Indiana.

First, Muller realized that the term "mutation" had been used for several distinct phenomena which were totally unrelated from the genetic point of view. Although "mutation" referred to the sudden appearance of a new genetic type, Muller found that some so-called mutations were special cases of Mendelian recombination, some were due to abnormalities in chromosome distribution, and others were caused by changes in individual genes or hereditary units. Muller argued that "in the interests of scientific clarity" the term should be limited in usage and redefined as an "alteration of the gene." Studies of the evening primrose, which once formed the backbone of mutation theory, had to be set aside since these did not involve true mutations, but were caused by abnormalities in the distribution of chromosomes.[5] Beginning a full-scale attack on the problem of mutation—defined as a heritable change in a gene—Muller tried to increase the frequency of mutations in *Drosophila* with various agents. When he began his work, colleagues in his laboratory had painstakingly identified at least one hundred mutant loci.

In 1927 Muller announced that X-rays were potent mutagenic agents. Where other geneticists had been limited to the few spontaneous mutants appearing in laboratory populations, Muller produced several hundred mutants in a short time. He found that most of these induced mutations were "stable in their inheritance, and most of them behave in the manner typical of the Mendelian chromosomal mutant genes found in organisms generally." Muller's report was quickly confirmed by Lewis J. Stadler, who found that irradiation of barley seeds with X-rays (or radium) induced mutations.

For his research on the effect of X-rays on mutation rates, Muller won the Nobel Prize for medicine in 1945. However, his interest in genetics went far beyond laboratory studies of the fruit fly. Ever since his undergraduate days Muller had a deep interest in evolution and human genetics, including a concern for the preservation and improvement of the human gene pool. Most mutations, he noted, were stable, deleterious, and generally recessive. Knowledge of mutation, therefore, had profound implications for eugenics and human reproduction. According to Muller,

without selection . . . undesirable genes will inevitably accumulate, until
the germ plasm becomes . . . riddled through with defect . . . and
progress through selection of desirable recessive traits can never more be
effected, since each of them will have become tied up with a lethal.[6]

Although he attempted to determine the relative importance of environment
and heredity in human development (mainly by studies of twins) he predicted
"a complete and permanent collapse of the evolutionary process" unless man
or nature resorted to periodic inbreeding and selection.

With the success of mutation studies, Muller saw the question of the basic
mechanism of evolution translated into the problem of the "character, frequency, and mode of occurrence of mutation." Because he saw eugenics as a
special branch of evolutionary science, mutations must be of fundamental concern to eugenics. Certainly Muller raised many serious questions about human
evolution and the human "load of mutations," but many scientists questioned
the wisdom of attempting to apply selective or eugenic measures to human
beings.

What Is the Gene?

The work of Morgan and others on the genetics of *Drosophila* had far-reaching
consequences. More and more detailed studies of the mechanism of inheritance,
especially the production and analysis of mutations, led to a quest for understanding the actual physical nature of the gene. Even the interest in mutations
did not obscure a fundamental feature of the gene—its stability. Genes normally produce tens of thousands of accurate copies. Errors or mutations are
extremely rare, but once they occur these new forms of the gene are stable in
their turn. After a period in which concern for genetics overshadowed interest
in evolution, the study of mutations was recognized as an integral part of the
Darwinian puzzle, since the mutational event was the source of variation.
Natural selection acted ultimately to choose among the infrequent, but critical
mutations which affect the fitness of the individuals carrying them.

Several different lines of investigation led to an understanding of the gene.
Physicists interested in the mechanism of heredity surveyed the existing
evidence bearing on the problem and speculated about the nature of the gene.
After Muller's discovery of the mutagenic effect of X-rays, other kinds of
radiation (such as ultraviolet light) were found to have a similar effect. During the 1930s, geneticists and physicists began to collaborate in an attack on
the gene. For example, the geneticist, Nikolai Timofeev-Ressovsky, the
physicist, Karl Zimmer, and the theoretical physicist, Max Delbrück, drew
attention to the applicability of the "hit" and "target" theories of radiobiol-
ogy to the mutational process. In 1935, in a paper entitled "On the Nature

of Gene Mutation and Gene Structure," they treated the "gene molecule" from a quantum-mechanical point of view. Delbrück had been influenced by Niels Bohr's (1885-1962) 1933 essay "On Light and Life." Erwin Schrödinger (1887-1961) later made Delbrück's speculations quite well known through his book *What is Life?* (1944), a work which has been called the "Uncle Tom's Cabin of Molecular Biology." Schrödinger suggested that the understanding of the hereditary substance would not come directly through physics, but from advances in biochemistry "under the guidance of physiology and genetics."[7]

In the 1930s other workers returned to cytology in the hope of elucidating the nature of the hereditary material. John Belling's studies of Jimson weed and his discussion of chromosome rings (1927) encouraged the resurgence of cytological studies. However, just as *Drosophila* provided advantages for breeding experiments, its giant salivary gland chromosomes seemed specially designed for cytologists. Calvin Bridges and others demonstrated a relationship between bands on the chromosomes and the linear sequence of gene loci on linkage maps. Bridges also found evidence that certain peculiar traits in *Drosophila,* such as the mutation which affects eye size, were due to repeats of a chromosomal section. Therefore, certain traits were not due to true mutations, but to duplication or triplication of the same chromosomal material and the effect of certain genes was related to their position on the chromosome. Linkage maps of maize and Barbara McClintock's cytological studies provided further support for the chromosome theory. Thus, the correlation between chromosomal structure and measurements of linkage was well documented for representatives of the plant and animal kingdoms. During the same decade, development of ultraviolet microspectrophotometry and special staining techniques extended the value of cytological studies. Ingenious use of such techniques brought cytochemists closer to understanding the chemical nature of the chromosomes. However, the complexity of the chromosomes made the interpretation of such studies very difficult.

Slow to develop, but eventually rich in contributions to genetics, were direct biochemical studies of the genetic material and the chemical basis of mutations.

The Chemical Nature of the Gene

Elucidation of the chemical nature of the gene was retarded by the failure of chemists and geneticists to share their work and appreciate each others' progress. While chemists tended to regard the role of DNA with indifference, many classical geneticists ignored the chemistry of the gene—assuming that its secret would be revealed through ever more sophisticated genetic analyses. Furthermore, some geneticists could not understand how any particular

particle or molecule could possibly be the genetic material. In a review of
The Mechanism of Mendelian Heredity (1915) by Morgan and others, Bateson
(1916) expressed total incredulity:

> . . . it is inconceivable that particles of chromatin or of any other sub-
> stance, however complex, can possess those powers which must be as-
> signed to our factors [i.e., genes]. . . . The supposition that particles of
> chromatin, indistinguishable from each other and indeed almost homo-
> geneous under any known test, can by their material nature confer all
> the properties of life surpasses the range of even the most convinced
> materialism.[8]

Yet other geneticists were sure that the gene must be some particular molecular arrangement. Both Weismann and de Vries conceived of the gene as a
special chemical entity. In 1910 de Vries urged scientists to focus their attention on the hereditary factors. "Just as physics and chemistry go back to
molecules and atoms," he wrote, "the biological sciences have to penetrate to
these units in order to explain, by means of their combinations, the phenomena of the living world."[9] While some scientists found studies of the position effect so compelling they could only conclude that "'genes' had no real
existence other than as points arranged in a particular pattern on a chromosome," (Goldschmidt, 1938), others saw the gene as "a minute particle, probably a single large molecule, possessing the power of reproduction."[10] Some
biologists proposed an analogy between the action of hereditary factors and
the ferments, or "living proteins." In the decade before 1920, scientists argued over the possibilities that genes might be enzymes or that genes might
make enzymes. However, before biochemists had clarified the structures and
properties of proteins, enzymes, and nucleic acids, arguments over their functions in heredity inevitably generated more heat than light.

Progress in elucidating the chemical nature of the gene seems in retrospect
excruciatingly slow. Although the nucleic acids are now known to be the
material basis of heredity, for many years they were the stepchildren of biochemistry. Appreciation of their role as the physical basis of inheritance has
been called a Cinderella story.

Johann Friedrich Miescher (1844-1895)

In many respects, the story of Johann Friedrich Miescher, the discoverer of
the nucleic acids, is similar to that of Mendel. Mendel was rediscovered, his
work vindicated, his genius eulogized, and his name attached to the fundamental laws of genetics, whereas Miescher remains a forgotten man. Indeed,
the "discovery" of DNA is often treated as the great scientific event of the

twentieth century. The man who prepared "nuclein" more than 100 years ago has been generally ignored.[11]

After receiving his M.D. degree in 1868, Miescher went to Tubingen to study physiological chemistry with Adolf Strecker (1822-1871) and Felix Hoppe-Seyler (1825-1895). Miescher was assigned to studies of the chemistry of pus cells, which he obtained by washing out bandages used in the surgical clinic. Attempting to prepare pure nuclei, Miescher extracted fatty materials from the pus cells with alcohol and then treated them with an acid extract of pig gastric mucosa (a crude pepsin preparation which removed proteins). After this treatment, Miescher found in the remaining nuclei a strong organic acid with an unusually high amount of phosphorous and very low content of sulfur (14% N, 2% S, 3% P). The solubility properties of this organic acid and its resistance to pepsin suggested that it was a new cell constituent. Hoppe-Seyler was at first dubious about some of Miescher's observations, but soon he too had found similar material in yeast and other cells. Hoppe-Seyler suggested that this new material "perhaps may play a highly important role in all cell development." The new material was called "nuclein." From 1871 to 1873 Miescher continued his studies of nuclein from sperm of the Rhine salmon. (Basle at the time was a center of the salmon fishing industry.) This system was not only more pleasant to work with than pus cells, but also very advantageous since sperm heads are essentially nothing more than cell nuclei and a very good source of DNA. According to Miescher's analyses, he had identified a new material with the formula $C_9H_{21}N_5O_3$ which he called protamine. Later analyses showed the material to be an organic base with the composition $C_{16}H_{28}N_9O_2$. An acid-insoluble component was found to be 13% N, 9% P with the formula $C_{29}H_{49}N_9P_3O_{22}$.

Nuclein was regarded as a colloidal substance because it did not pass through a parchment filter. Since it was unstable, great care was needed to isolate it. The preparation had to be done quickly and kept in the cold. To prepare nuclein, Miescher began work at 5 A.M. and worked rapidly in an unheated room. The final preparation could be preserved under absolute alcohol. Although Miescher recognized the problems inherent in his studies, especially the problem of the "purity" of his nuclein preparations, he was discouraged by the cruel abuse heaped upon his work. Miescher felt he had struggled with biological substances so complex, creating such difficulties, that "the real chemists shun them."

While Mendel was ignored during his lifetime, Miescher did not go unnoticed by his contemporaries. Much of the criticism of his work was justified, but not very constructive. Some English chemists claimed that nuclein was "nothing but an impure albuminous substance." The French chemist Adolphe Wurtz argued that Miescher's results were "somewhat vague from a chemical point of view." If further research had lead to purer and better characterized nucleic acid, this would have been a valuable outcome of such criticism. Even the eminent Hoppe-Seyler was attacked for his work on nuclein. Carl von Nägeli and Oscar Loew (1844–1941) attacked his yeast nuclein work as a misinterpretation of what was merely albumin contaminated with K_2PO_4 and $MgPO_4$.

Stung by harsh criticism, cautious, and reluctant to publish preliminary data, Miescher wrote little on the chemistry of nucleins after 1874. The physiology of the Rhine salmon and of breathing seemed preferable topics. About five years before his death, Miescher resumed his chemical work on nuclein. These studies were published posthumously. Just what physiological function Miescher would ascribe to nuclein is not entirely clear. Before cytologists clarified the role of the sperm nucleus in fertilization, Miescher denied the idea that sperm "might be carriers of specific substances that act as fertilizing agents by virtue of their chemical properties." But, in the same paper he argued that if "we wish to assume at all that a single substance, acting as a ferment or in another manner as a chemical sensitizer, is the specific cause of fertilization, one must unquestionably think of nuclein." Yet Miescher denied the existence of a specific fertilization substance. However, after Miescher had denigrated the role of chemical phenomena in fertilization, progress in microscopy unequivocally established the role of the nucleus in fertilization and heredity. Work with dye stuffs revealed the specific localization of cellular components. Walter Flemming (1843–1905) suspected that nuclein was an important component of the cell nucleus, but used the term "chromatin" because further chemical tests of the relationship between nuclein and chromatin were needed (1882). Privately Miescher speculated on the possible role of nuclein in the transmission of heritable traits. He thought its atoms could form "isomers" or "alternative spatial arrangements" which could account for variations, but his contemporaries could not see how enough variation could exist in such molecules to explain heredity.[12]

With other scientists taking up the study of nuclein its unique chemical properties became more apparent. However, proteins were better known and protein seemed the more logical chemical to serve as the physical basis of heredity. Although Emil Fischer had carried out elegant work on purines (a component of the nucleic acids), his synthesis of a polypeptide chain with 30 amino acids was more dramatic. With 30 different residues one could generate 2.635×10^{32} different combinations. It seemed obvious that even such moderately sized polypeptides represented a fantastic amount of variety.

Chromatin and Nuclein

By the end of the nineteenth century there were speculations that chromatin was identical to nuclein and thus identical to the physical basis of heredity.

In 1895 Edmund B. Wilson pointed out that the chromosome complements contributed by the two sexes were precise equivalents and the two sexes play an equal role in heredity—even though different species vary in other aspects of reproduction and development. Because the physical basis of heredity must reside in something that the sexes contribute equally in all types of reproduction it must reside in the chromatin. Since chromatin seemed to be essentially the same as Miescher's nuclein, this substance must be the genetic material. Wilson concluded that "inheritance may, perhaps, be effected by the physical transmission of a particular chemical compound from parent to offspring." [13]

In 1885 Oscar Hertwig (1849–1922) suggested that nuclein was probably responsible for fertilization and the transmission of the hereditary characteristics. He regarded fertilization as a physicochemical and morphological process, having nothing to do with ancient vague ideas of "essences," "vapors," or "ferments." Hertwig opposed Liebig and others who claimed that fertilization was a kind of fermentation process where the sperm merely served as a kind of catalyst. Studying nuclein from thymus and yeast, Albrect Kossel (1853–1927) proved that there were two kinds of nucleic acid. The two types, which for a time were known as "thymus nucleic acid" and "yeast nucleic acid," are now designated deoxyribose nucleic acid (DNA) and ribose nucleic acid (RNA). Both nucleic acids contain the bases adenine, cytosine, and guanine, but in DNA the fourth base is thymine and in RNA it is uracil. The two forms of nucleic acid also differ in their sugar component (either deoxyribose or ribose).

After this very promising beginning, the path to the understanding of the chemical nature of the gene was diverted from its proper course. Ironically this detour was the result of improved chemical analyses of the nucleic acids. Between 1910 and 1930 chromatin seemed to have lost its claim to be taken seriously as the heredity material. Most of the papers on the chemistry of chromatin written during this period did not discuss its possible biological role at all. Biologists too were becoming disillusioned about chromatin. It seemed to behave very peculiarly during the cell cycle. At times it even seemed to disappear. This did not seem consistent with the characteristics proper to the genetic material—particularly the need for stability in transmission from generation to generation. Finally, work carried out by P. A. Levene seemed to rob nuclein of another trait needed for the genetic material—complexity.

Phoebus Aaron Levene (1869–1940)
and the Tetranucleotide Hypothesis

In 1891 P. A. Levene received his M.D. degree from St. Petersburg Imperial Medical Academy, left Russia, and emigrated to New York. While practicing medicine on the East Side, Levene studied chemistry at Columbia University. In 1896 overwork and tuberculosis necessitated a long period of recuperation. During this convalescence Levene decided to abandon his medical practice for chemistry. After further study in Germany and work in a New York hospital, he joined the newly formed Rockefeller Institute for Medical Research (1905). He continued his association with that institution until shortly before his death. His researches included almost every important class of biological compounds: cerebrosides, chrondroitin-sulfates, hexosamines, sugars, phosphate esters of sugars, and the nucleic acids. Following up on the work of Emil Fisher, Levene elucidated the fundamentals of nucleic acid chemistry. Levene published his research results rapidly, but in a quite fragmentary fashion which resulted in more than seven hundred papers in the *Journal of Biological Chemistry*.

Nucleic acid studies were in a very sorry state when Levene began to investigate their chemistry in 1900. His chemical work helped to form a clearer picture of the nucleic acids and distinguish between them and the proteins. Analyses of the nucleic acids suggested that the four bases were present in equimolar amounts in the nucleic acids from all sources. These data led to the "tetranucleotide interpretation"—a theory which began as a working hypothesis but soon turned into a paradigm of biochemistry. A minor revolution in biochemical concepts was needed to overturn the tetranucleotide interpretation—and so was a lot of hard chemical work.

To appreciate the difficulty of the situation, we must recall that the tetranucleotide theory operated at several levels of explanation. Essentially it meant that there were equal amounts of all four bases in DNAs from all sources. On a more sophisticated level of organization, it meant that the polynucleotides were combinations of a fixed and unchangeable sequence of units which were themselves a combination of four nucleotides. Finally, this interpretation of the data meant that DNA was a repetitious polymer, analogous to glycogen, and therefore incapable of generating the diversity that was essential in the genetic material.

Erwin Chargaff and the Demise
of the Tetranucleotide Hypothesis

Although several studies conducted in the 1940s suggested that DNA might be the hereditary material, such interpretations were dismissed because nucleic acids were erroneously thought to be simple linear tetranucleotide polymers. Thus, even though the Feulgen stain and UV microscopy had localized DNA in the chromosomes, most scientists preferred to think of proteins as the genetic material. Between 1946 and 1950 Erwin Chargaff produced the chemical studies that revolutionized attitudes towards DNA.

Dr. Erwin Chargaff. (From *Annual Review of Biochemistry* **44, 1975.**)

Chargaff was born August 11, 1905 in Czernowiz, which was at the time a provincial capital of the Austrian monarchy. Educated in one of the "excellent gymnasiums" of Vienna, he found the instruction "limited in scope, but on a very high level." His parents had been moderately well off, but all they had was lost in the Great Inflation. (An insurance policy his father bought in 1902 was redeemed 20 years later. It was then worth the price of one trolley ticket.) Entering the university to postpone the "unpleasant decision" about occupation by four years and to acquire the "docteur" degree ("without which a middle class Austrian of my generation would have felt naked"), Chargaff chose the chemistry faculty for what he recalls were essentially trivial reasons. He knew nothing of chemistry because he had never studied it before, but it seemed the only natural science likely to bring some hope of employment. Also he thought it would help him with a rich uncle who owned alcohol refineries. Because students had to pay for their own chemicals and apparatus, Chargaff chose a professor "whose problems were known to require neither much time nor money." When he graduated in 1928, he could expect no position in Austria, a nation which exported many of its academically trained people. After many vicissitudes Chargaff found himself at Columbia University, with a yearly grant of $6000, beginning his studies of the nucleic acids.[14]

From the outset, Chargaff assumed that the nucleic acids might be as complicated and highly polymerized as the proteins. He made precise analyses of nucleic acids from many sources, such as beef thymus, spleen, liver, human sperm, yeast, and tubercle bacilli. Using the methods which first became available after World War II—paper chromatography, UV spectrophotometry, and ion-exchange chromatography—he proved unequivocally that the four bases were not present in equal amounts. He also demolished the notion that all DNAs were identical to calf thymus DNA. However, the composition of DNA from different organs of the same species was constant and characteristic of the species. Although at the time no theory could account for his observation, Chargaff emphasized the finding that the molar ratios of total purines to total pyrimidines, and the ratio of adenine to thymine, and guanine to cytosine were always about one. From Chargaff's work it was clear that, like the proteins, nucleic acids are complex, interesting, and unique compounds.

Further research proved that the nucleic acids are macromolecules—long-chain structures of high molecular weight—as revealed by measurements of viscosity and the rates of diffusion and sedimentation of carefully prepared solutions of DNA. In the 1930s, Einar Hammarsten (1889–1968) and others proved that DNA behaved like a long thin rod (length about 300 times the width) with an apparent molecular weight between 500,000 and 1,000,000. These figures were subsequently refined and DNAs with molecular weights of about 3,000,000 and a length-to-width ratio of about 700 were prepared. As Miescher had recognized, the physical properties of "nuclein" depend very much on the methods used in preparation.

Other chemical studies of the properties of nucleic acids resulted from research on the "filterable viruses" which attacked plants and bacteria. By the 1930s it was known that these strange "contagious living fluids" were submicroscopic, filterable infectious agents which multiplied in living cells. In 1936 Wendell Meredith Stanley (1904–1971) applied the methods recently used to crystallize proteins to prepare tobacco mosaic virus. He found it could be recrystallized repeatedly "with retention of constant physical, chemical, and biological properties."[15] Later workers found that the infective material was a nucleoprotein, but until 1957 it was unclear which component was infectious.

While chemists, geneticists, and physicists were issuing learned pronouncements during the 1940s as to why the gene must be a protein, or perhaps a nucleoprotein, another significant line of research on the transforming ability of nucleic acids was generally being ignored.

DNA as the Genetic Material

The first well-known series of experiments to challenge the assumption that
genes must be proteins or nucleoproteins was carried out in the laboratory of
Oswald T. Avery (1877–1955). In some respects Avery's work was a refinement of research previously conducted by Fred Griffith.

Fred Griffith (1877–1941)

Various strains of *Diplococcus pneumoniae* were known to cause pneumonia
in man. These bacterial strains could be differentiated by measuring the ability of bacteria-free filtrates of young cultures to precipitate antibodies produced by rabbits which had been injected with the individual strains of the
bacteria. By 1920 specific serological types of pneumococcus had been quite
well established. In the 1920s Michael Heidelberger and Avery proved that
various strains differed in the polysaccharide capsule that envelopes the bacterium. The encapsulated forms produced smooth (S) and shiny rounded
colonies when grown in vitro. Variants could be isolated which were attenuated in virulence and produced rough (R) colonies which lacked the slimy
polysaccharide capsule. In the 1930s Avery's group chemically characterized
the polysaccharide capsular materials.

A paper—valuable in retrospect, but for many years more puzzling than
enlightening—was published in 1928 by Dr. Fred Griffith. Griffith described
some rather peculiar observations on the behavior of pneumococci, which suggested they could undergo some kind of transmutation of type. From the
blood of mice injected with living Type II (R) pneumococci together with
heat-killed Type III (S) cells, Griffith isolated living Type III bacteria. This
indicated that the living nonvirulent (R) bacteria had somehow acquired something from the dead Type III (S) which transformed the Type II (R) into
deadly organisms with the Type III (S) polysaccharide capsule. Although it is
now obvious that Griffith had observed genetic transformation, he does not
seem to have realized that the transfer of hereditary material was involved in
his observations. Griffith was a physician researcher of the Ministry of Health
in London. He was a well-known bacteriologist and epidemiologist, and had
studied the various forms of bacteria for many years. He did not follow up
his work on the transformation of bacterial strains, nor did he live to see the
meaning of his work finally revealed. In 1941 a bomb blew up the dilapidated
laboratory where he and his colleague William Scott were working, killing
them both.

Several scientists soon confirmed Griffith's observations and proved that
pneumococcal transformations could be effected in the test tube as well as in
the animal body. The agent responsible was found to pass through a filter

known to hold back bacteria. By 1933 James Alloway, an associate of Avery, had prepared crude acqueous solutions of the transforming principle. Naturally, attempts were made to purify this material. While these experiments intrigued bacteriologists attempting to modify the virulence of pathogens, Sir Peter Medawar[16] believes most biologists and geneticists were quite unaware of this work. However, it proved impossible to ignore a paper published by Oswald Avery, Colin Macleod, and Maclyn McCarty in 1944.

Oswald T. Avery (1877–1941)

Having isolated the active transforming principle from Type III (S) pneumococci, Avery identified it as "a highly polymerized and viscous form of sodium desoxyribonucleate." Tests for the presence of protein proved negative. Furthermore, the transforming principle was inactivated by crude preparations of "desoxyribonucleodepolymerase" and resistant to enzymes which attack ribonuclease or proteins. Although he noted that the transforming principle had been compared to genes and viruses, Avery was cautious in drawing conclusions about the genetic role of DNA. The possibility that the biological activity of the preparation was due to minute amounts of some other substance associated with DNA could not be rigorously excluded. However, he continued, if DNA "actually proved to be the transforming principle, as the available evidence strongly suggests, then nucleic acids of this type must be regarded not merely as structurally important but as functionally active in determining the biochemical activities and specific characteristics of pneumococcal cells."

"If the results of the present study are confirmed," Avery continued, "then nucleic acids must be regarded as possessing biological specificity the chemical basis of which is as yet undetermined."[17] Avery's work was, indeed, quickly confirmed and soon DNAs with transforming properties were isolated from other bacteria. Although Sir Peter Medawar called Avery's article the end of the "dark ages" of DNA, the darkness was not wholly dispelled until 1952.

Because of misconceptions about the chemical nature of DNA, many scientists were unwilling to accept evidence that DNA, rather than protein, could serve as the physical basis of inheritance. Until the triumph of the Watson-Crick model of DNA, Avery and others could not reconcile the biological activity of their DNA preparations with the "facts" of DNA chemistry. During the intervening years, other discoveries prepared the way for a new understanding of the mechanism of heredity at the molecular level. Bacteriophages were found to be capable of transferring heritable bacterial traits; this phenomenon, studied by Norton Zinder and Joshua Lederberg, was called transduction. By 1947 DNA structures analogous to the nuclei of higher cells had been discovered in bacteria and Lederberg and Tatum proved that genetic recombination occurs as a result of bacterial "matings." In many ways bacteria and bacteriophage (commonly called "phage") were more favorable systems for an attack on the chemical nature of the gene than higher organisms. Indeed, viruses consist of little more than genes dressed in protein coats.

Because many scientists continued to cling to the hope that minute amounts of contaminants in the nucleic acids might be the actual genetic material, much work was devoted to purification of transforming factors. Up until 1952 a cautious attitude as to the role of DNA persisted. Even when highly purified DNA preparations (less than 0.02% protein) proved effective transforming agents, scientists did not accept this as a conclusive demonstration that DNA itself was the transforming agent, since minute amounts of some other substance might remain attached to the nucleic acid. James Watson seems to have been unique in that as early as 1951 he wanted to show that DNA was the heredity material. Not until the Hershey-Chase experiments of 1952 did the general scientific community take DNA seriously.

The Hershey and Chase Experiment

Between 1951 and 1952 Alfred Hershey and Martha Chase carried out the experiments which supplied convincing evidence that DNA was the hereditary material. The famous blender experiment was stimulated by reports that DNA phosphorus is transferred from parental phages to their offspring. Radioactive sulfur was used to label phage proteins and then infected bacterial cells were spun in a blender and centrifuged to separate infected bacteria from smaller particles.

Perhaps Hershey began these experiments rather skeptical about the idea that the nucleic acid would enter the bacterial cell. Thomas F. Anderson remembered a day at Cold Spring Harbor Laboratory when he and Hershey discussed the "wildly comical possibility that only the viral DNA finds its way into the host cell, acting there like a transforming principle in altering the synthetic processes of the cell." Obviously it did not seem a very serious possibility until 1952 when Hershey and Chase proved the joke was "not only ridiculous but true."[18]

Finding most of the phage DNA remaining with the bacterial cells and the phage protein released in the supernatant fluid, Hershey and Chase concluded that the protein served as a protective coat needed for adsorption to bacteria and as an instrument for the injection of the phage DNA into the bacteria, but had no function in the growth of intracellular phage. Because their technique for separating labelled phage protein coats from injected bacterial cells introduced a certain degree of uncertainty, Hershey and Chase did not claim their data provided unequivocal proof that the DNA was the genetic material. The conclusions were stated tentatively:

We infer that the sulfur-containing protein has no function in phage multiplication, and that DNA has some function. Our experiments show clearly that a physical separation of the phage T2 into genetic and non-genetic parts is possible. . . . The chemical identification of the genetic part must wait, however, until some of the questions asked above have been answered.[19]

Obviously the chemical evidence in the Hershey-Chase experiment did not provide compelling evidence for the genetic role of DNA. Studies of the transforming principles which had been compared to genes and viruses had pointed in the same direction as the bacteriophage experiment. According to Medawar, the idea that the transforming principle was nucleic acid "aroused much resentment, for many scientists unconsciously deplore the resolution of mysteries they have grown up with and have therefore come to love."[20] Thus, it was fortunate that Hershey was a member-in-good-standing of a closely knit group of innovative researchers, led by Max Delbrück, called the "phage group."

Although some scientists felt the Hershey-Chase data should be cautiously interpreted, among the members of the phage group, this work was rather quickly accepted as good evidence for the genetic role of DNA. A junior member of the phage group, named James Dewey Watson, accepted this work as "a powerful new proof that DNA is the primary genetic material."[21] Because the U. S. State Department refused Salvador Luria permission to attend the meeting of the Society for General Microbiology held at Oxford in April, 1952, Watson presented the Hershey-Chase experiments and defended the interpretation that the data proved DNA was the genetic material—probably much more vigorously than Hershey himself would have.

Thus by 1953, phage workers had for the first time what Delbrück and Hershey called a "party line."[22] Although DNA was gaining acceptance as the genetic material, because its structure remained a mystery, it was impossible to establish the relationship between what the gene was chemically and how the gene acted as the vehicle of inheritance. In 1953, James Watson and Francis Crick described a model of the structure of DNA which immediately suggested explanations for the biological activity of the nucleic acids.

The Adventures of Watson and Crick

The elucidation of the three dimensional structure of DNA by James Watson and Francis Crick has been hailed as one of the greatest achievements of twentieth century biology, comparable to the achievements of Darwin and Mendel in the previous century. Recognition of the architecture of DNA led to an avalanche of new discoveries concerning the mechanism of inheritance. In the Watson-Crick structure, scientists could anticipate how DNA "copies itself with each cell generation, how it is used in development and function, and how it undergoes the mutational changes that are the basis of organic evolution."[23]

Since Watson and Crick based their model on a combination of guesswork and inspection of Rosalind Franklin's X-ray data, a brief description of X-ray studies is appropriate. Several X-ray crystallographers attempted to determine the molecular dimensions (bond lengths, bond angles) of DNA and its constituent moieties. Outstanding for his pioneering work (and his own estimate of his worth) was William T. Astbury (1898-1961). Assessing himself as "the alpha and omega, the beginning and end of the whole thing," Astbury's work was indeed the beginning of one very fruitful approach to the structure of DNA. The X-ray diffraction patterns of DNA suggested an arrangement of long polynucleotide chains with "flattish nucleotides standing out perpendicularly to the long axis of the molecule to form a relatively rigid structure." In 1947, about the time Chargaff was demolishing the tetranucleotide hypothesis, Astbury concluded from X-ray studies that the "degree of perfection of the X-ray fibre diagram" supported Levene's theory of the regularity of DNA composition.[24] Although wrong in his speculations about the regularity of the sequence of tetranucleotides, Astbury made it clear that DNA had a crystalline structure. But just what it was remained a mystery until James Watson teamed up with Francis Crick.

James D. Watson

In 1951 James Watson, who seemed a typical enough American post-doctoral researcher, met Francis Crick, 12 years older but still a graduate student. Two years later both Watson and Crick were world famous. Both attributed their success to their special relationship—their ability to complement, criticize, and stimulate each other's ideas. The contrast between these two scientists could hardly have been more striking.

A former Quiz Kid, Watson entered Chicago University when only 15 years old. This experimental program of early intake permitted Watson to take integrated courses in the biological sciences. Although enthusiastic about ornithology, he did not seem especially interested in anything else. Paul Weiss, who taught embryology and invertebrate zoology, recalled Watson as apparently "completely indifferent to anything that went on in class; he never took any notes and yet at the end of the course he came top of the class."[25] Although his formal training in genetics was negligible at the undergraduate level, Watson recalls having made up his mind to go into genetics. Reading Schrödinger's *What is Life?* "polarized" him "toward finding out the secret of the gene."[26] Since Cal Tech and Harvard did not want him, he did his graduate work at Indiana University where H. J. Muller, Tracy Sonneborn, and Salvador Luria were on the faculty. However, in 1947 he enrolled in a summer course

for advanced ornithology and systematic botany at the University of Michigan. Fernandus Payne, Dean of Sciences and chairman of the Zoology Department at Indiana University recommended his department's graduate work in genetics and experimental embryology, but suggested Watson go elsewhere if his major interest was still the study of birds.

Although Muller seemed the obvious choice as a thesis advisor, Watson made up his mind that "Drosophila's better days were over and that many of the best younger geneticists, among them Sonneborn and Luria worked, with microorganisms." Watson decided that phage were preferable to paramecia, and decided to work under Luria's direction. Luria's friendship with Max Delbrück, a hero in *What is Life?*, convinced Watson he had made the right choice. Having met Delbrück, Watson sometimes worried that his "inability to think mathematically might mean I could never do anything important,"[27] but, lucky in finding other minds to complement his own and clever at picking other people's brains, Watson was able to find the facts he needed without being forced to learn a whole new field when there was an easier way. As a graduate student Watson became a junior member of the phage group—a special inner circle of scientists among whom information passed "by a sort of beating of tom-toms, while others await the publication of a formal paper in a learned journal."[28]

While the *Double Helix* suggests that Watson left Indiana University totally innocent of chemistry, physics, and the scientific literature, his doctoral dissertation on the effects of X-rays on phage replication suggests otherwise. Receiving his doctorate at 22, Watson was sent to Europe for further training. From meetings with luminaries of the Luria-Delbrück circle he was convinced that "Europe's slower paced traditions were more conducive to the production of first-rate ideas." After a year in Copenhagen with Herman Kalckar and Ole Maaløe, Watson transferred to Perutz's group at the Cavendish Laboratory to work on the molecular structure of nucleic acids extracted from plant viruses.

At Cambridge, Watson found a kindred soul in Francis Crick, who shared his enthusiasm for genetics and its most essential part—"what the genes were and what they did."[29]

Francis Crick

Although the Crick family was considered well-to-do middle class, the boot and shoe factory run by Francis Crick's father and uncle failed during the economic slump before World War II. Even as a school boy at Mill Hill, Francis was remarkable for his shrill voice and loud laughter. Leaving Mill Hill in 1934, Crick had distinguished himself in physics and mathematics, and had a pass in chemistry. No one in Crick's family shared his enthusiasm for science. Indeed the family called him "crackers" for what they saw as an excessively narrow range of interests. Earning his bachelor's degree at

University College, London, in 1937, Crick disappointed friends and family when he failed to take a first in physics. His graduate work was interrupted by World War II when Crick joined the Admiralty Research Laboratory at Teddington. The apparatus on which he had worked for two years as a graduate student was destoyed by a German bomb (actually, a naval mine used as a bomb).

Although Crick's genius was appreciated in the Mine Design Department, his tendency to tell others they were talking nonsense caused some tensions. After the war, Crick remained with the Naval Intelligence Division in London, although he planned eventually to do basic research in particle physics or physics applied to biology. Reading Schrödinger's *What is Life?* convinced Crick that "fundamental biological problems could be thought about in precise terms, using the concepts of physics and chemistry."[30] Ironically, by the time Watson and Crick described the structure of DNA and its genetic implications, Schrödinger seems to have lost interest in the problem. (Crick sent Schrödinger a reprint of the 1953 *Nature* paper, but received no reply to his letter.)

Rejecting his family's strict Congregationalist pattern, Crick happily adopted an anti-vitalist stance and endeavored "to show that areas apparently too mysterious to be explained by physics and chemistry could in fact be so explained." Therefore, he intended to study either the brain or molecular biology, "the borderline between the living and the non living." In September, 1947, Crick resumed graduate studies under the supervision of Dr. Honor Fell at the Strangeways Research Laboratory. His attempt to work under Prof. J. D. Bernal on X-ray crystallography of proteins and nucleic acid had failed. Unfortunately, research at Strangeways had at the time gone from arthritis to determinations of the viscosity of protoplasm. From 1947 to 1949 Crick embarked on an intensive reading program to teach himself biology and physics. In 1949 Crick escaped from "inconsequential" work on cytology and joined Max Perutz at the Cavendish Laboratory to study the structure of proteins and related molecules by X-ray techniques. One year later Crick gave a seminar titled "What Mad Pursuit" (suggested by John Kendrew), which exposed errors in the work of Perutz, Kendrew, and Bragg on the hemoglobin molecule. Relations with Sir Lawrence Bragg became quite tense after this. Although in his mid-thirties, Crick could not settle down to one topic long enough to construct a doctoral thesis. The world famous DNA paper was published while Crick was still struggling to make his "ragbag" of topics into a presentable thesis.[31]

Crick saw Watson as "the first person I had met who thought the same way about biology as I did." In Crick, Watson saw the "brightest person I have ever worked with." In personality the two were poles apart. Watson was described as a "loner," quiet, and introverted. In contrast, Crick was famous for his loud voice and laugh, an extrovert whom Watson never saw in a "modest moment." Their collaboration illustrates Watson's talent for establishing useful scientific relationships. Watson's rather strange relationship with Maurice Wilkins, cold and devious as it was, is another example of an association that was ultimately productive. The Watson-Crick collaboration was a fortunate one, for Crick doubts that either he or Watson could have discovered the structure of DNA alone. Speculating about the probable course of history if Watson had not come to Cambridge, Crick wrote:

> the structure was there waiting to be discovered—Watson and I did not invent it. It seems to me unlikely that either of us would have done it separately, but Rosalind Franklin was pretty close. . . . Wilkins, after Franklin left, could well have got there in his own good time. Whether Pauling would have made a second attempt . . . I am less certain . . . Would the biochemists have hit on it in the end? And, if so, what difference would it have made?[32]

Watson and Crick arrived at the structure of the double helix by a combination of guessing, model building, and the unacknowledged exploitation of Rosalind Franklin's X-ray crystallographic data.[33] In their first *Nature* paper of 1953, Watson and Crick described their "radically different structure for the salt of deoxyribose nucleic acid." Incorporating the usual chemical assumptions, the "novel feature" of their structure was "the manner in which the two chains are held together by the purine and pyrimidine bases."[34] In the proposed double helix, the planes of the bases were perpendicular to the fibre axis. The bases were joined by hydrogen bonding automatically determined always paired in this manner to a pyrimidine on the other chain. According to Watson and Crick, adenine always paired with thymine and guanine with cytosine. These bonding rules explained Chargaff's data concerning the molar ratios of purines to pyrimidines. Although any sequence of bases was possible on one chain, the rules of hydrogen bonding automatically determined the sequence on the other chain. The Watson-Crick double helix immediately suggested how DNA could exist as a stable crystalline molecule while still providing for variety and mutability.

Despite the irresistible elegance of the Watson-Crick model, many scientists felt the paper had a rather hollow tone. The X-ray data available in the literature at the time were, as the authors admitted, "insufficient for a rigorous test of our structure." More exact data appeared in the two communications which followed. One paper by Wilkins, Stokes, and Wilson, discussed the available X-ray data for calf thymus DNA, and nucleoprotein preparations from sperm heads and T_2 bacteriophage. The second paper by Rosalind Franklin and R. G. Gosling presented the most refined X-ray diffraction patterns of DNA then available. Further research provided additional evidence for the Watson-Crick model. Ten years after the publication of their papers, Watson, Crick, and Wilkins were awarded the Nobel Prize for Medicine and Physiology.

Although Watson and Crick only alluded to it in their first paper, the double helix could theoretically explain, at the molecular level, how DNA reproduces and makes more copies of DNA exactly like the original. In a sentence Gunther Stent described as "one of the most coy statements in the literature of science,"[35] Watson and Crick noted: "It has not escaped our notice that the specific pairing we have postulated immediately suggests a possible copying mechanism for the genetic material." Crick called this sentence a "compromise." Although Crick wanted to discuss the genetic implications of the model, his co-author "suffered from periodic fears that the structure might be wrong and that he had made an ass of himself." Crick insisted on including the suggestion because otherwise someone else would do so, assuming they had been "too blind to see it."[36]

Shortly after their first letter to *Nature,* Watson and Crick elaborated on the implications of their model in a second communication. Although the phosphate-sugar backbone of the DNA model was completely regular, any sequence of base pairs could fit into the structure. "It follows that in a long

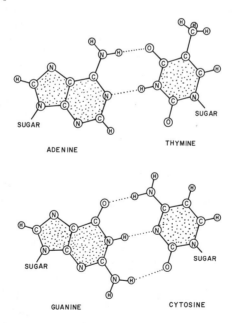

Diagram of the complementary base pairs found in the DNA double helix. Adenine pairs with thymine and guanine pairs with cytosine.

molecule many different permutations are possible," they wrote, "and it therefore seems likely that the precise sequence of bases is the code which carries
the genetical information."[37] Since the sequence of one chain was the complement of the other, when the two strands separated, each could serve as the
template for construction of a daughter DNA chain.

Why were Watson and Crick so successful when they began the race for the
double helix far behind their competitors Franklin and Wilkins, at King's
College in London, and the great chemist, Linus Pauling in California? In
retrospect it is easy to see that Pauling's approach to the DNA problem was
simply wrong and the group at King's too split by personal incompatabilities
to work effectively. Watson and Crick recognized the biological significance
of DNA and the potentialities of X-ray data and model building. They also
had the great advantage of being at the heart of an active grape-vine from
which they picked ripe data and the progress reports of their competitors in
London and even Pasadena (through Peter Pauling). Together and separately,
they picked the brains of colleagues and visitors such as Wilkins and Chargaff
for critical insights. Despite their obvious advantages, for two years they
stumbled along the wrong path, making all possible incorrect choices for the
structure of DNA. When they finally put together Chargaff's data on the base
ratios with an elementary knowledge of the chemistry of the bases, they were
able to build a reasonable, "pretty," and biologically interesting model of DNA
in about three weeks.

It took about five years for experimental tests to prove the essential validity of the DNA replication scheme proposed in 1953. In 1958, Mathew
Meselson, a graduate student of Pauling, and Frank W. Stahl, a postdoctoral
associate of Delbrück, traced the fate of parental strands of DNA labelled
with N^{15} by equilibrium density gradient sendimentation. They proved that
DNA replication was semiconservative, that is, each old strand became associated with a newly synthesized strand of DNA. In 1956, Arthur Kornberg
isolated an enzyme from *Escherichia coli* which catalyzed the synthesis of
DNA. In 1959, Kornberg shared the Nobel Prize with his former mentor,
Severo Ochoa, for their work on the biological synthesis of the nucleic acids.

Watson's book, *The Double Helix* remains the only account of the story
written by one of the participants. Many scientists were dismayed by the un-
flattering portrait of science that emerged from Watson's book. John Maddox,
then editor of *Nature,* said that scientists who felt insulted had a duty to do
more than take James Watson off their Christmas card lists. Scientists have
not rushed forward to rewrite Watson's history. However, Anne Sayres
(1975) has attempted to rescue Rosalind Franklin (1920-1958) from the
harsh and apparently totally fallacious picture painted by Watson. Franklin
died of cancer in 1958, only 38 years of age. She did not know the Nobel
Prize would be awarded for the discovery of the DNA double helix nor did

she realize how her data had been used by Watson and Crick, since Watson's book was not published until 10 years after her death.

How Genes Work

The Central Dogma

Satisfied that the chemical nature of the gene had been discovered, scientists were able to link studies of how the gene worked to knowledge of what the gene was. Watson and Crick played an important role in formulating the general principles explaining how information stored in DNA is replicated and passed on to daughter molecules and how information was transferred from DNA into the workings of the cell.

Even before the structure of DNA had been worked out, Watson wrote himself a message: "DNA → RNA → protein." Certain that DNA was the template for RNA chains, he saw RNA chains as "likely candidates for the templates for protein synthesis." The arrows in his slogan "did not signify chemical transformations, but instead expressed the transfer of genetic information from the sequences of nucleotides in DNA molecules to the sequences of amino acids in proteins." Although he knew that "a slogan was no substitute for the DNA structure," Watson taped the slogan on the wall above his desk.[38] The flow of information from DNA to RNA to protein has been called the "central dogma" of molecular biology. According to this scheme, the flow of information is a one-way path which is never reversed. Before the necessary evidence had been obtained, the central dogma served as a stimulus for unravelling genetic mechanisms at the molecular level. However, long before Watson taped the scheme to his wall other scientists had a glimpse of the way the gene works through studies of altered gene products—i.e., mutations manifested as metabolic disorders due to defective enzymes. Although Beadle and Tatum are well known for their work on genetic control of biochemical reaction in the red bread mold, Sir Archibald Garrod had come close to the "one gene-one enzyme" theory through work in that most refractory but most interesting of all systems—*Homo sapiens.*

Archibald Garrod (1857–1936) and Inborn Errors of Metabolism

Garrod's report of his studies of alcaptonuria in 1902 may be regarded as the beginning of biochemical genetics. Persons with alcaptonuria cannot metabolize tyrosine completely and excrete homogentisic acid in their urine. The condition was easily discovered since the urine in such individuals darkens when exposed to air and causes deep stains in fabrics. From tests of alcaptonuric individuals Garrod concluded that "alkaptonuria is not the manifestation of a disease but rather of the nature of an alternative course of metabolism, harmless and usually congenital and lifelong." Other human "chemical abnormalities"—such as albinism, cystinuria, and pentosuria—seemed to be comparable conditions. Garrod concluded that these disorders were due to alterations in metabolism and called them "inborn errors of metabolism." Furthermore, these disorders seemed to be inherited as Mendelian recessives. Data collected on the incidence of such disorders in certain families revealed that a large proportion of the affected individuals were the children of first cousins. Matings of first cousins were likely to reveal recessive traits. Thus, Garrod's work showed that Mendelian genetics applied to human beings. Somewhat later Garrod traced the defect in alcaptonuria to "the penultimate stage of the catabolism of the aromatic protein fractions." Furthermore, he concluded that "the splitting of the benzene ring in normal metabolism is the work of a special enzyme" which was missing in congenital alcaptonuria.[39]

Not only did this work indicate that Mendelian genetics applied to man, but it also suggested a relationship between genes and enzymes. Individuals with the disease had inherited two recessive genes which meant that they lacked the "ferment" to complete the metabolism of particular compounds. In the case of alcaptonuria, Garrod proved that his patients could metabolize precursors of homogentisic acid, but that the pathway was interrupted at that point. Therefore homogentisic acid accumulated and was excreted in the urine.

Since Garrod's reports, many other human diseases have proved to be "inborn errors of metabolism." In many cases the inherited disease has been unequivocally linked to a missing or defective enzyme. Garrod's studies provided an early form of the "one gene-one enzyme" hypothesis. Beadle and Tatum later admitted that their work would have been less circuitous if they had known of Garrod's studies before they began their own.

Beadle and Tatum: One Gene—One Enzyme

Although Garrod's work and his predictions have been amply supported, few geneticists would have considered following in the footsteps of this brilliant clinician. The trend in modern genetics has been to use smaller and smaller and more and more simple creatures for research. Human beings hardly seemed a likely "experimental animal" to sophisticated geneticists. Garrod's work, unfortunately, was largely ignored. Attempting to elucidate the mechanism by which genes controlled heredity, George Beadle and Edward Tatum turned to the red bread mold, Neurospora. Trained as a geneticist, Beadle attempted to analyze the biochemical basis of heredity through studies of the genetic control of eye-color pigment in fruit flies. In 1936 Beadle and Boris Ephrussi showed that a larval eye taken from the mutant "vermilion" implanted in the abdomen of a wild-type larva developed into an adult structure with the wild-type eye color. Tests with various mutants proved that the synthesis of eye-pigment was controlled by several genes. Some of the biochemical intermediates in the biosynthetic pathway to pigment were identified.

Finding Drosophila unsatisfactory for unravelling the biochemistry of the gene-gene product relationship, Beadle and Tatum considered the possibility of reversing the procedures generally used to identify specific genes with particular chemical reactions. Instead of taking a mutant as their starting point and then searching for the chemical reaction it controlled, they decided to begin with known chemical reactions and look for the genes controlling them. Therefore, they needed an organism with well-known biochemical pathways. By using X-rays they could produce many mutations, but would deliberately select for mutations which affected the synthesis of specific vitamins. The red bread mold, Neurospora, was selected because it had a fairly short life cycle and techniques for genetic analysis had already been worked out. Furthermore, the wild-type mold could be grown on a simple synthetic medium containing only a carbon source, certain inorganic salts, and biotin. After X-raying the mold, Beadle and Tatum were rewarded with numerous mutants which had lost the ability to synthesize certain organic substances. Their preliminary results indicated that they had indeed discovered a valuable tool for learning how genes regulate development and function. The salient feature of their analysis was proof that mutant strains differed from the parental type by an alteration in a single gene. Therefore, the gene acted by "directing the final configuration of a protein molecule and thus determining its specificity." In 1945, when Beadle enunciated the one gene-one enzyme hypothesis, the chemical nature of the gene was still unknown. After 1952, the problem became one of explaining how information encoded in DNA could be used to direct the synthesis of proteins. The central dogma, after considerable elaboration from Watson's cryptic note (DNA → RNA → protein), guided research in this area.

Sickle Cell Anemia—A Molecular Disease

Despite the prominence of bacteriophage studies, it was the analysis of another human disease that led to the first significant breakthrough in proving the relationship between a heritable gene mutation and a molecular alteration in the gene product. This was the study of sickle cell anemia carried out by Linus Pauling and Vernon M. Ingram.

Linus Pauling, who won two Nobel Prizes (one for Chemistry and one for Peace), also made two major contributions to biology. The first was his elucidation of the alpha helix structure in proteins; the second was proof that genetic diseases in man can be traced to abnormal proteins formed by mutant genes. The red blood cells of about eight percent of American blacks possess the capacity to undergo a reversible change in shape called "sickling" in response to alterations in the partial pressure of oxygen. Although this tendency towards sickling generally has no pathological consequences, in about one in forty individuals a severe chronic anemia occurs. Research in Pauling's laboratory in 1949 proved that sickling was caused by an alteration in the hemoglobin molecule of individuals with sickle cell anemia. Pauling suggested that the alteration in the chemical and physical properties of sickle cell hemoglobin was caused by "a difference in the number or kind of ionizable groups in the two hemoglobins."[40] From Pauling's work it was apparent that sickle cell anemia is a "molecular disease." This seems to represent the ultimate level of resolution of pathology from the vague humors of Hippocrates and Galen, to organ, tissue, cell, protein, altered amino acid sequence, and hence altered nucleotides in the genome.

Although Pauling's work proved that sickle cell anemia was a molecular disease, the precise nature of the alteration in the molecule was still unknown. In 1957, Vernon Ingram published an account of the chemical difference between normal and sickle cell hemoglobin. Ingram believed he could determine the amino acid sequence of hemoglobin, as Frederick Sanger had done for insulin. Of course dealing with a much larger protein presented many difficulties. However, Ingram was able to prove that sickle cell hemoglobin differed from the normal type by only one amino acid among about 300 amino acids in the hemoglobin molecule: a valine residue had been substituted for the normal residue, glutamic acid.

Finding a change in one amino acid in the hemoglobin molecule, Ingram postulated that a corresponding change—a true point mutation—had occurred in the gene coding for hemoglobin. Since geneticists had already shown that sickle cell anemia was transmitted as a simple Mendelian recessive, for the first time Mendelian inheritance, mutation, and the precise alteration in a gene product had been linked together—and for a disease of no small importance. Obviously, not all significant studies of molecular genetics need take place in bacteriophage.

The Post Watson-Crick Period

Since the discovery of the DNA double helix, biological research has proceeded at an unprecedented pace. Even a sketch of the story of the life sciences since 1953 would fill a volume many times larger than an account of events up to 1953. Properly speaking, such recent events do not belong in a general history of biology, because they constitute the content of present-day science courses and are already well covered in modern textbooks of genetics, molecular biology, and molecular genetics. Furthermore, the frontiers of research advance so rapidly it is difficult to assess the real value and relative importance of recent discoveries.

An understanding of how rapidly ingenious predictions and seemingly universal "laws" can turn into another chapter in the history of half-baked ideas can be gained from an outline of the recent history of the "central dogma." We have already seen how Watson drew up a note indicating the direction of information flow within the cell, which read "DNA → RNA → protein." This was done well before there was any hard evidence to support the whole sequence. As they eventually went their separate ways, Watson, Crick, and many others probed the mechanism by which information encoded in DNA is transcribed into RNA and finally translated into proteins. In 1958, Crick published an elegant and influential piece of theoretical biology entitled "On Protein Synthesis."[41] Crick noted that proteins are the macromolecules that do most of the jobs needed for life. Proteins, he said "can do almost anything." Mainly they are enzymes, but also they play a structural role. Yet it is DNA that fascinates Crick. It is DNA as the genetic code, ultimately responsible for the synthesis of proteins, that is so important. In looking at the genetic map of DNA and the amino acid sequence of proteins, Crick made many predictions about how DNA could direct the synthesis of proteins. In this essay Crick presented the famous "adaptor hypothesis," discussed the logic of colinearity between the sequence of bases in nucleic acids and amino acids in proteins, and elaborated on the "central dogma."

According to Crick, the central dogma states that information is transferred from nucleic acid to nucleic acid or nucleic acid to protein, but not from protein to nucleic acid. ("Information" in this context means the precise determination of sequence.) When information passes into protein it cannot get out again. It seems strange that a materialist like Crick would use the term "dogma"—a term heavily laden with religious connotations—for a scientific hypothesis. In this case, as in so many others the evolution of a scientific idea began as a heresy and ended up as a dogma.

Reverse Transcriptase

In 1975 the Nobel Prize committee in all its wisdom officially recognized the end of an era. The central dogma no longer reigns supreme in splendid infallibility—at least not in the form originally suggested by Watson. The prize for Physiology or Medicine was awarded to Howard Temin, Renato Dulbecco, and David Baltimore, for discoveries concerning the interaction between tumor viruses and the genetic material of the cell. Primarily, the new laureates were cited for providing the conceptual foundation and technology for examining a possible relationship between viruses and human cancer.[42]

Through his studies of animal viruses, Howard Temin had proved that information did not invariably flow from DNA to RNA, but that an enzyme called "reverse transcriptase" exists which can direct the transcription of RNA into DNA. Howard Temin studied Rous sarcoma virus (RSV) as a graduate student at Caltech in Dulbecco's laboratory. In 1958 Temin and Rubin developed the first reproducible *in vitro* assay for a tumor virus. At the University of Wisconsin (1960) Temin continued work with RSV. He noted that inhibition of DNA synthesis, and inhibition of DNA-dependent RNA synthesis, blocked RSV infection. This is surprising, because the RSV genome is single-stranded RNA. In 1964 Temin proposed that there is a DNA intermediate in RSV infection. This is called the "provirus hypothesis": after infection a DNA provirus is synthesized which contains all of the genetic information of the RNA viral genome. Progeny viral RNA would be synthesized from the provirus. This hypothesis could also account for integration of RSV into the host chromosome.

Temin's hypothesis was not readily accepted. The use of various inhibitors suggested possible artifacts, but the major impediment seems to have been conflict with the central dogma of molecular genetics. Temin continued this work in the 1960s, but made little impact on the prevailing skepticism. In 1970, Temin and Satoshi Mizutani did what could be called the "critical experiment." They proved that virions of RSV contain an enzyme that can transcribe single-stranded viral RNA into DNA. David Baltimore discovered the same enzyme in Rauscher murine leukemia virus and RSV. With the new enzyme, an obvious mechanism for the formation of a DNA intermediate was revealed.

Crick has argued that reverse transcriptase does not really contradict central dogma, in as much as the transcription of information from one form of nucleic acid to another proceeds through essentially the same base pairing process.

The RNA tumor viruses and the enzyme, reverse transcriptase, have now become basic tools for studying the molecular biology of mammalian cells. If these discoveries do not overturn sophisticated versions of the central dogma, perhaps they will be valuable in exposing the deficiencies in an informal version of the dogma which holds that "what is true for a phage is true for an elephant." While the molecular genetics of many microbial organisms is known in great detail, many aspects of cell growth, differentiation, and regulation in higher organisms are still obscure and the techniques of investigating these problems are quite primitive.

More immediately, scientists anticipate using reverse transcriptase and radioactive DNA probes to detect viral genetic information in normal and malignant cells. With these new tools, DNA copies of messenger RNA can be prepared, isolated, and purified. This last possibility, in particular, gives rise to the awesome possibilities of genetic manipulation in plants, animals, and human beings. Messenger for particular enzymes could serve as the template for DNA to be used in "genetic surgery" which might cure genetic diseases. In agriculture and industry there are other exciting possibilities: Bacteria could be engineered to produce hormones, antibiotics, silk, etc. Nitrogen-fixing genes might be added to food crop species, obviating the need for nitrogen fertilizers. On the other hand, the technique could be used to add genes for toxins to common bacteria as weapons of germ warfare. Even without evil intentions, such techniques could cause many problems. Imagine the consequence if bacteria capable of pumping out large amounts of silk or insulin escaped from the laboratory and became established in the human gut.

Synthesis of Artificial Genes

It would be imprudent to try to predict the future course of the life sciences. Discoveries in the field have a way of happening and becoming exploited or exploded more rapidly than one would ever expect. In a fairly short time, biology has experienced great triumphs and come into possession of such awesome possibilities that biologists are now facing the loss of innocence and sense of danger that overwhelmed physicists when the atomic bombs were dropped on Hiroshima and Nagasaki. It is more satisfying to record the triumphs than to repeat the warnings. There is much to admire in a science that in the twentieth century has progressed from the first glimmerings of appreciation of the Mendelian "factors" of inheritance to the *in vitro* synthesis of a functional gene. After nine years of work by 24 postdoctoral fellows under the direction of Nobel laureate Har Gobind Khorana, the gene for tyrosine transfer-RNA, which contains 86 nucleotides, was synthesized and found to be biologically active. Shortly after Khorana's success, another team of researchers announced the synthesis of a mammalian gene, the gene for rabbit hemoglobin, which has about 650 nucleotide units.[43]

As biologists look into the future, they see both great promise for promoting human welfare and a terrible potential for destruction. It would be appropriate here to recall the words of Louis Pasteur concerning the battle between the "law of blood and death, ever imagining new means of destruction" and the "law of peace, work and health, ever evolving new means for delivering man from the scourges which beset him." Like Pasteur, we must dedicate our science to the workings of the second law and thereby justify his invincible faith that "Science and Peace will triumph over Ignorance and War."[44]

Notes

1. B. H. Willier and J. M. Oppenheimer (1974). *Foundations of Experimental Embryology.* New York: Hafner, pp. 74-97.

2. W. S. Sutton (1903). The Chromosomes in heredity. *Biological Bulletin 4:* 231-251.

3. Sutton, p. 240.

4. T. H. Morgan (1928). *The Theory of the Gene.* New Haven: Yale University Press, p. 25.

5. H. J. Muller (1923). Mutation. *Eugenics, Genetics and the Family 1:* 106-112.

6. H. J. Muller (1927). Artificial transmutation of the gene. *Science 66:* 84-87.

7. E. Schrödinger (1944). *What is Life?* Cambridge: Cambridge University Press, p. 68.

8. W. Bateson (1916). Review of *The Mechanism of Mendelian Heredity* (by T. H. Morgan et al.). *Science 44:* 536-543.

9. H. DeVries (1910). *Intracellular Pangenesis.* Translated by C. S. Gager. Chicago: Open Court, p. 13.

10. J. S. Fruton (1972). *Molecules and Life.* New York: Wiley-Interscience, especially, "The Chemical Nature of the Gene," pp. 225-261.

11. F. H. Portugal and J. S. Cohen (1977). *A Century of DNA.* Cambridge, Mass.: MIT Press.

12. Fruton, "Nuclein and Chromatin," pp. 183-193; "Nuclein and Heredity," pp. 193-204.

13. E. B. Wilson (1895). *The Cell in Development and Inheritance.* New York: MacMillan, p. 4.

14. E. Chargaff (1975). A Fever of Reason, the Early Way. *Annual Review of Biochemistry 44:* 1-18. [Chargaff has also written an autobiography: E. Chargaff (1978). *Heraclitean Fire.* New York: Rockefeller University Press.]

15. W. M. Stanley (1935). Isolation of a Crystalline Protein Possessing the Properties of Tobacco-Mosaic Virus. *Science 81:* 644-645.

16. P. Medawar (1973). *The Hope of Progress.* New York: Anchor, pp. 98-100.

17. O. T. Avery, C. M. MacLeod, M. McCarty (1944). Studies on the Chemical Nature of the Substance Inducing Transformation of Pneumococcal Types. *J. Exper. Med. 79:* 137-158.

18. J. Cairns, G. S. Stent, and J. D. Watson (1966). *Phage and the Origins of Molecular Biology.* New York: Cold Springs Harbor Laboratory of Quantitative Biology, p. 76.

19. A. D. Hershey and M. Chase (1952). Independent Functions of Viral

Protein and Nucleic Acid in Growth of Bacteriophage. *J. Gen. Physiol. 36:* 39–56, p. 54.

20. Medawar, p. 99.

21. J. D. Watson (1968). *The Double Helix.* New York: Atheneum, p. 119.

22. Cairns et al., p. 100.

23. G. W. Beadle (1969). "Genes, chemistry, and the nature of man," pp. 1–13, in *Biology and the Physical Sciences.* Edited by S. Devons. New York: Columbia University Press, p. 2.

24. Fruton, p. 220.

25. R. Olby (1974). *The Path to the Double Helix.* Seattle: University of Washington Press, p. 297.

26. Cairns et al., p. 239.

27. Cairns et al., p. 240.

28. Medawar, p. 103.

29. Olby, p. 310.

30. R. Olby (1970). Francis Crick, DNA, and the Central Dogma. *Daedalus 99:* 938–987, p. 943.

31. Olby (1970 and 1974).

32. Olby (1974), p. vi.

33. A. Sayres (1975). *Rosalind Franklin and DNA.* New York: Norton.

34. J. D. Watson and F. H. C. Crick (1953). A. Structure for Deoxyribose Nucleic Acid. *Nature 171:* 737–738.

35. G. S. Stent (1970). DNA. *Daedalus 99:* 909–937, p. 920.

36. F. H. C. Crick (1974). *Nature: 248,* p. 765.

37. J. D. Watson and F. H. C. Crick (1953). Genetical Implications for the Structure of Deoxyribonucleic Acid. *Nature 171:* 964–967.

38. Watson, 1968, p. 153.

39. A. E. Garrod (1909). *Inborn Errors of Metabolism.* Reprinted 1963. London: Oxford University Press, p. 80.

40. L. Pauling, H. A. Itano, S. J. Singer, and I. C. Wells (1949). Sickle Cell Anemia, a Molecular Disease. *Science 110:* 543–548.

41. F. H. C. Crick (1958). On Protein Synthesis. *Symp. Soc. Exp. Biol. 12:* 138–163.

42. W. Eckhart (1975). The Nobel Prize for Physiology or Medicine. *Science 190:* 650

43. T. H. Maugh (1976). The Artificial Gene: It's Synthesized and It Works in Cells. *Science 194:* 44.

44. R. Dubos (1960). *Pasteur and Modern Science.* New York: Doubleday Anchor, pp. 147–148.

References

Astbury, W. T. (1947). X-Ray Studies of Nucleic Acids. *Symp. Soc. Exp. Biol. 1:* 66–76.

Avery, O. T., MacLeod, C. M., and McCarty, M. (1944). Studies on the Chemical Nature of the Substance Inducing Transformation of Pneumococcal Types. *J. Exper. Med. 79:* 137–158.

Bateson, W. (1922). Evolutionary Faith and Modern Doubts. *Science 55:* 55–61.

Bateson, W. (1916). Review of *The Mechanism of Mendelian Heredity* (by T. H. Morgan et al.). *Science 44:* 536–543.

Bateson, W. (1909). *Mendel's Principles of Heredity.* Cambridge: University Press.

Bateson, W. (1913). *Problems of Genetics.* New Haven: Yale University Press.

Bateson, W., and Saunders, E. R. (1902). The Facts of Heredity in the Light of Mendel's Discovery. *Reports to the Evolution Committee of the Royal Society of London 1:* 125–160.

Beadle, G. W. (1945). Biochemical Genetics. *Chem. Revs. 37:* 15–96.

Beadle, G. W. (1945). The Genetic Control of Biochemical Reactions. *Harvey Lectures 40:* 179–194.

Beadle, G. W. (1963). *Genetics and Modern Biology.* Philadelphia: American Philosophical Society.

Beadle, G. W. (1969). "Genes, Chemistry, and the Nature of Man," in *Biology and the Physical Sciences.* Edited by S. Devons. New York: Columbia University Press, pp. 1–13.

Beadle, G. W., and Ephrussi, B. (1936). The Differentiation of Eye Pigments in Drosophila as Studied by Transplantation. *Genetics 21:* 225–247.

Beadle, G. W., and Tatum, E. L. (1941). Genetic Control of Biochemical Reactions in Neurospora. *Proc. Natl. Acad. Sci. 27:* 499–506.

Bohr, N. (1933). On Light and Life. *Nature 131:* 421–423, 457–459.

Benzer, S. (1955). Genetic Fine Structure and Its Relation to the DNA Molecule. *Brookhaven Symposia in Biology 8:* 3–5.

Cairns, J., Stent, G. S., and Watson, J. D. (1966). *Phage and the Origins of Molecular Biology.* New York: Cold Spring Harbor Laboratory of Quantitative Biology.

Chargaff, E. (1950). Chemical Specificity of Nucleic Acids and Mechanism of their Enzymatic Degradation. *Experientia 6:* 201–209.

Chargaff, E. (1968). What Really is DNA? Remarks on the Changing Aspects of a Scientific Concept. *Progress in Nucleic Acid Research and Molecular Biology 8:* 297–333.

Chargaff, E. (1975). A Fever of Reason, the Early Way. *Annual Review of Biochemistry 44:* 1–18.

Chargaff, E. (1978). *Heraclitean Fire.* An Autobiography. New York: Rocke-feller University Press.

Crick, F. H. C. (1958). On Protein Synthesis. *Symp. Soc. Exp. Biol. 12:* 138–163.

Crick, F. H. C., and Watson, J. D. (1954). The Complementary Structure of Deoxyribonucleic Acid. *Proc. Roy. Soc. A 223:* 80–96.

Crick, F. H. C. (1967). *Of Molecules and Men.* Seattle and London: University of Washington Press.

Demerec, M. (1935). Role of Genes in Evolution. *Am. Nat. 69:* 125–138.

Dobzhansky, T. (1970). *Genetics of the Evolutionary Process.* New York: Columbia University Press.

Dubos, R. J. (1976). *The Professor, The Institute, and DNA.* A biography of Oswald T. Avery. New York: The Rockefeller University Press.

Eckhart, W. (1975). The 1975 Nobel Prize for Physiology or Medicine. *Science 190:* 650.

Florkin, M. (1972). *A History of Biochemistry.* Volume *30* of *Comprehensive Biochemistry.* New York: Elsevier.

Freundlich, M. M. (1963). Origins of the Electron Microscope. *Science 142:* 185–188.

Fruton, J. S. (1951). The place of Biochemistry in the University. *Yale J. Biol. and Med. 23:* 305–310.

Fruton, J. S. (1972). *Molecules and Life: Historical Essays on the Interplay of Chemistry and Biology.* New York: Wiley-Interscience.

Garrod, A. E. (1902). The Incidence of Alkaptonuria: A Study in Chemical Individuality. *Lancet 2:* 1616–1620.

Garrod, A. E. (1909). *Inborn Errors of Metabolism.* Reprinted 1963. London: Oxford University Press.

Goldschmidt, R. (1916). Genetic factors and enzyme reaction. *Science 43:* 98–100.

Goldschmidt, R. (1938). The Theory of the Gene. *Scientific Monthly 46:* 268–273.

Griffith, F. (1928). The Significance of Pneumonococcal Types. *J. Hygiene 27:* 113–159.

Haldane, J. B. S. (1942). *New Paths in Genetics.* New York: Harper.

Haldane, J. B. S. (1945). A Physicist Looks at Genetics. *Nature 155:* 375–376.

Hershey, A. D., and Chase, M. (1952). Independent Functions of Viral Protein and Nucleic Acid in Growth of Bacteriophage. *J. Gen. Physiol. 36:* 39–56.

Hess, E. L. (1970). Origins of Molecular Biology. *Science 168:* 644–669.

Hotchkiss, R. D. (1970). Review of *The Coming of the Golden Age* (by G. S. Stent). *Science 169:* 664–666.

Hotchkiss, R. D. (1965). Oswald T. Avery 1877–1955. *Genetics 51:* 1–10.

Ingram, V. M. (1957). Gene mutations in Human Haemoglobin: The Chemical Difference Between Normal and Sickle Cell Haemoglobin. *Nature 180:* 326–328.

Hollaender, A. (1961). Review of Studies on Quantitative Radiation Biology (by K. G. Zimmer). *Science 134:* 1233.

Kendrew, J. C. (1967). Review of *Phage and the Origins of Molecular Biology*. *Sci. Amer. 216:* 141–144.

Kendrew, J. S. (1968). *The Thread of Life: An Introduction to Molecular Biology*. Cambridge, Mass.: Harvard University Press.

Klug, A. (1968). Rosalind Franklin and the Discovery of the Structure of DNA. *Nature 219:* 808–810, 843–844.

Kornberg, A. (1960). Biological Synthesis of Deoxyribonucleic Acid. *Science 131:* 1503–1508.

Leicester, H. M. (1974). *Development of Biochemical Concepts from Ancient to Modern Times*. Cambridge, Mass.: Harvard University Press.

Levene, P. A., and Bass, L. W. (1931). *Nucleic Acids*. New York: Chemical Catalogue.

Levene, P. A. (1957). *Advances in Carbohydrate Chemistry 12:* 1.

Liebig, J. (1847). *Researches on the Chemistry of Food*. Translated by W. Gregory. London: Taylor and Walton.

McClung, C. E. (1902). The Accessory Chromosome—Sex Determinant? *Biol. Bull. 3:* 43–84.

McKusick, V. A. (1960). Walter S. Sutton and the Physical Basis of Mendelism. *Bull. Hist. Med. 32:* 487–497.

Medawar, P. (1973). *The Hope of Progress*. New York: Anchor.

Morgan, T. H. (1910). Chromosomes and Heredity. *Amer. Nat. 44:* 449.

Morgan, T. H. (1910). Sex-linked Inheritance in Drosophila. *Science 32:* 120–122.

Morgan, T. H. (1912). An Attempt to Analyze the Constitution of the Chromosomes on the Basis of Sex-linked Inheritance in Drosophila. *J. Exper. Zool. 11:* 365–412.

Morgan, T. H. (1912). The Explanation of a New Sex Ratio in Drosophila. *Science 36:* 718–719.

Morgan, T. H. (1928). *The Theory of the Gene*. 2nd Ed. New Haven: Yale University Press.

Morgan, T. H., Sturtevant, A. H., Muller, H. J., and Bridges, C. B. (1915). *The Mechanism of Mendelian Heredity*. Reprinted with a new introduction by Garland E. Allen (1972). New York: Johnson Reprint.

Muller, H. J. (1922). Variation Due to Change in the Individual Gene. *Am. Nat. 56:* 32–50.

Muller, H. J. (1923). Mutation. *Eugenics, Genetics and the Family 1:* 106–112.

Muller, H. J. (1927). Artificial Transmutation of the Gene. *Science 66:* 84–87.

Needham, J., ed. (1970). *The Chemistry of Life.* Cambridge: Cambridge University Press.

Olby, R. (1974). *The Path to the Double Helix.* Seattle: University of Washington Press.

Olby, R. (1970). Francis Crick, DNA, and the Central Dogma. *Daedalus 99:* 938–987.

Pauling, L., Itano, H. A., Singer, S. J., and Wells, I. C. (1949). Sickle Cell Anemia, a Molecular Disease. *Science 110:* 543–548.

Rhoades, M. M. (1954). Chromosomes, Mutations, and Cytoplasm in Maize. *Science 120:* 115–120.

Rolleston, H. D. (1936). *The Endocrine Organs in Health and Disease, with an Historical Review.* London: Oxford University Press.

Sayres, A. (1975). *Rosalind Franklin and DNA.* New York: Norton.

Schroedinger, E. (1944). *What is Life?* Cambridge: Cambridge University Press.

Stent, G. S. (1968). That Was the Molecular Biology that Was. *Science 160:* 390–395.

Stent, G. S. (1970). DNA. *Daedalus 99:* 909–937.

Sturtevant, A. H. (1959). Thomas Hunt Morgan, 1866–1945. *Biograph. Memoirs Natl. Acad. Sci. 33:* 283–325.

Sturtevant, A. H. (1913). The Linear Arrangement of Six Sex-linked Factors in Drosophila as Shown by their Mode of Association. *J. Exper. Zool. 14:* 43–59.

Sutton, W. S. (1902). On the Morphology of the Chromosome Group in *Brachystola magna. Biol. Bull. 4:* 24–39.

Van Slyke, D. D., and Jacobs, W. A. (1944). Phoebus Aaron Theodor Levene. *Biog. Mem. Natl. Acad. Sci. 23:* 75–126.

Vries, Hugo de (1910). *Intracellular Pangenesis.* Translated by C. S. Gager. Chicago: Open Court.

Watson, J. D., and Crick, F. H. C. (1953). A Structure for Deoxyribose Nucleic Acid. *Nature 171:* 737–738.

Watson, J. D., and Crick, F. H. C. (1953). Genetical Implications for the Structure of Deoxyribonucleic Acid. *Nature 171:* 964–967.

Weaver, W. (1970). Molecular Biology: Origin of the term. *Science 170:* 581–582.

Wilson, E. B. (1925). *The Cell in Development and Heredity.* 3rd Ed. New York: MacMillan.

Wilson, E. B. (1923). The Physical Basis of Life. *Science 57:* 277–286.

Wyatt, H. V. (1972). When Does Information Become Knowledge? *Nature 235:* 86–89.

INDEX

Abbe, Ernst, 172
Abu, Hanifa al-Dinawari, 83
Academi del Cimento, 144, 162
Academia Secretorum Naturae, 144
Academy dei Lincei, 144
Academy movement (see also
    Scientific societies), 138-151
Academy, Plato's, 29, 31, 35, 71, 81
Achromatic lenses, 170, 171, 172
Acquired characteristics, inheritance
    of, 408, 419, 425, 431
Agassiz, Louis, 205
Age of the earth, 360, 361, 367, 370,
    374, 375, 385
Agnodice, 51
Agricola, Georg Bauer, 369
Agriculture, origins, 3-4
Albertus Magnus, 79, 80
Alcaptonuria, 462-463
Alchemy, 106-112, 238, 239, 289,
    314
    Arab, 82-83
Alcmaeon, 16-17, 180
Alexander the Great, 30, 31, 47-48
Alexandria, 47-55
Alexandrian library, 48, 49, 54
Alexandrian museum, 35, 47-49, 54
Alexandrian science, decline of,
    54-55

Alhazen (or Ibn al Haitham), 156
Allbut, Clifford, 61
Al-Nafis, Ibn, minor or pulmonary
    circulation, 76
Alpha helix, 464
Alphabet, 5
American Academy of Arts and
    Sciences, 150
American Philosophical Society, 150
Amici, Giovanni Battista, 171, 172
Anatomy, 2, 16, 131-132, 288
    Alexandrian, 50-54
    Celsus, 62
    comparative, 96
    Galen, 65-70
    human, 72
    origins, 2
    primitive man and, 2
    Vesalius and, 98-106
Anatomy and art, renaissance, 94-96
Anaxagoras, 13, 21-23, 24, 28
Anaximander, 11-12, 13, 354, 355
Anaximenes, 13-14
Anderson, Thomas F., 454
Aniline dyes, 233, 264, 278-279
Animal chemistry, 324-326
Animal spirits, 292, 294, 295
Anthrax, 257-261, 265-267
Antibodies, 273, 279